Oxford Socio-Legal Studies

Regulating Workplace Safety:
System and Sanctions

REGULATING WORKPLACE SAFETY: SYSTEM AND SANCTIONS

NEIL GUNNINGHAM
Professor Director
Australia Centre for Environmental Law
Faculty of Law
The Australia National University

and

RICHARD JOHNSTONE
Associate Professor
Centre for Employment and Labour Relations Law
Faculty of Law
The University of Melbourne

OXFORD
UNIVERSITY PRESS

*This book has been printed digitally and produced in a standard specification
in order to ensure its continuing availability*

OXFORD
UNIVERSITY PRESS

Great Clarendon Street, Oxford OX2 6DP

Oxford University Press is a department of the University of Oxford.
It furthers the University's objective of excellence in research, scholarship,
and education by publishing worldwide in

Oxford New York

Auckland Bangkok Buenos Aires Cape Town Chennai
Dar es Salaam Delhi Hong Kong Istanbul Karachi Kolkata
Kuala Lumpur Madrid Melbourne Mexico City Mumbai Nairobi
São Paulo Shanghai Taipei Tokyo Toronto

Oxford is a registered trade mark of Oxford University Press
in the UK and in certain other countries

Published in the United States
by Oxford University Press Inc., New York

© N. Gunningham & R. Johnstone 1999

The moral rights of the authors have been asserted
Database right Oxford University Press (maker)

Reprinted 2003

ISBN 0-19-826824-6

General Editor's Introduction

This is a book which turns the results of theoretical inquiry and empirical research to the interests of public policy. It presents us with new ways of thinking about the problem of how best to regulate occupational health and safety. Its organising question is how law can more effectively penetrate organisations so that they become more responsive to regulatory efforts.

The authors start from the position that the current system of implementation and enforcement has outlived its usefulness. The problem with present practice, the authors argue, is that it has been incapable of adjusting to recent, rather rapid, changes in the economy and the character of employment. In particular, current regulation has simply been unable to deal with many workplace problems, has failed to stay abreast of changes in workplace and employment practices, has not got to grips with new workplace hazards, and has often imposed requirements upon employers that are difficult to comply with. To make regulation more sensitive yet more effective the authors conclude that we need to contemplate parallel systems of regulation—a 'two-track' policy—in which traditional regulatory methods are allied with a systems-based approach which seeks to surpass existing standards. The authors do not seek to abandon or to play down the role of the criminal sanction in all of this but argue for its deployment in a more strategic and systematic way than is evident in current practice. They go on to explore what the implications of their dual system would be for inspection, enforcement, and sentencing in a careful and thorough mapping of an integrated system of punishment for regulatory violations.

This book is no parochial survey of local practice, but is based on a searching analysis of the international literature, and on a series of interviews conducted in Australia, Denmark, Sweden, the UK and the USA. It is a wide-ranging, subtle, and sophisticated appraisal that takes proper account of different regulatory problems and different regulatory responses by firms. The work is directed firmly towards policy, but its approach is one that respects, and draws heavily from, socio-legal theory. What the authors have to say will be of particular interest to occupational health and safety policy-makers, as well as to those scholars interested generally in regulation. *Oxford Socio-Legal Studies* has already published a number of important books which offer fresh thinking about regulatory systems, among them Ayres and Braithwaite on *Responsive Regulation*, Haines on *Corporate Regulation*, and Gunningham and Grabosky on *Smart Regulation*. This book is a fitting companion to them.

Keith Hawkins

Preface

In part, this book derives from a project commissioned by the WorkCover Authority, New South Wales, Australia, on the subject of *Enforcement Measures for Occupational Health and Safety in New South Wales.* In completing that project, the authors had an opportunity to interview a wide range of stakeholders in New South Wales and to explore a range of policy questions relating broadly to the redesign of occupational health and safety regulation in that state.

Subsequently, we were able to extend the project internationally, to conduct interviews in Denmark, Sweden, the United States of America, and the United Kingdom, and to draw on the broader international literature in this area. Consistent with its origins, the result is a book which is principally normative in its approach. In particular, it identifies weaknesses in current legal and policy arrangements, and it argues the case for adopting a number of regulatory reforms. In so doing, it builds on experience internationally, recent and important developments in regulatory theory, and on models and approaches constructed during the course of this research. The suggested reforms relate principally to the administration, design, and enforcement of occupational health and safety legislation.

The end product, we hope, will be of interest to policy makers in occupational health and safety, and to social scientists and lawyers interested in regulation more generally. It should be of equal interest to a wide range of specialists and practitioners: personnel and safety officers, union activists, regulatory agencies, and employer and employee associations. Neil Gunningham was responsible for the writing of Chapters 2 to 5 and Richard Johnstone for Chapters 6 and 7. Chapters 1 and 8 and the Appendix were joint enterprises.

In writing the book we have incurred a series of debts. Above all, we are grateful to Patricia Burritt, who provided not only research assistance of the highest order, but also undertook, with energy, efficiency, and good humour, a range of onerous administrative tasks which ensured that the book was kept on schedule, despite a range of other and often conflicting commitments.

Darren Sinclair read the draft from cover to cover and provided us with detailed, insightful, and critical comments which substantially improved the final version. We are also grateful for the comments on various parts of the book provided by Kaj Frick, Michael Quinlan, and Peter Rozen (who was a contributor to our initial project).

Finally, we owe a substantial debt to Jim Cox and Denise Adams, then of the New South Wales WorkCover Authority, for providing us with considerable support, encouragement, and autonomy during the initial phase of the project, and for their efforts to ensure that the resulting Report had a direct policy impact. Without funding from the WorkCover Authority itself, the project would never have got off the ground. We acknowledge this support, while emphasizing that the views expressed in the book are those of the authors alone, and should not be attributed to the WorkCover Authority.

Richard Johnstone also wishes to thank Santina Perrone for research assistance and many valuable discussions about the enforcement of Occupational Health and Safety legislation. Santina's work on the project was funded by a Joint Research Project Award from the University of Melbourne's Collaborative Research Program. Eliza Bergin provided further literature searches and the Centre for Employment and Labour Relations Law and the Faculty of Law at the University of Melbourne provided collegial and financial support to the project. The ideas in the book were shaped by many discussions with colleagues, and in particular benefited from input by Kit Carson, Fiona Haines, Sarah Biddulph, Mary Keyes, Virginia Whalen, Sia Lagos, and Gary Chaplin.

Some aspects of the book have appeared in recent publications: N. Gunningham, 'Towards Innovative Occupational Health and Safety Regulation' (1998) 40 (2) *The Journal of Industrial Relations* 204; 'Towards Effective and Efficient Enforcement of Occupational Health and Safety Regulations: Two Paths to Enlightenment' (1998) 19 (4) *Comparative Labor Law & Policy Journal*; 'From Compliance to Best Practice in OHS: The roles of specification, performance and systems-based standards' (1996) 9 (3) *Australian Journal of Labour Law*; and 'Integrating Management Systems and Occupational Health and Safety Regulation', forthcoming (1999) *Journal of Law and Society*; and R. Johnstone, 'Occupational Health and Safety Prosecutions in Australia: Rethinking State Enforcement of Occupational Health and Safety Statutes', in R. Mitchell and J. Wu (eds.), *Facing the Challenge in the Asia Pacific Region: Contemporary Themes and Issues in Labour Law* (Centre for Employment and Labour Relations Law, The University of Melbourne, Occasional

Monograph No. 5, 1997), pp. 193–214; 'Prosecutions in the Light of the Industry Commission's Report' (1996) October *Law Institute Journal* 54–5; and 'Does the Crime Match the Punishment?' (1997) June *Complete Safety* 21–3.

N.G.
R.J.

Table of Contents

Table of Abbreviations

AMC	Australian Manufacturing Council
BSO	British Standards Organisation
CCP	Cooperative Compliance Program
CEO	Chief Executive Officer
COAG	Council of Australian Governments
EC	European Community
EMAS	Eco Management and Audit Scheme
EMS	Environmental Management System
EU	European Union
FAA	Federal Aviation Agency
GEMI	Global Environmental Management Initiative
HSC	Health and Safety Commission
HSE	Health and Safety Executive
HSW Act	Health and Safety at Work Act 1974 (UK)
IC	Internal Control
ILO	International Labour Organisation
ISO	International Standards Organisation
MIT	Massachusetts Institute of Technology
MOLAC	Ministers of Labour Advisory Committee
NOHSC	National Occupational Health and Safety Commission
OHS	Occupational Health and Safety
OHS&R	Occupational Health, Safety and Rehabilitation
OSHA	Occupational Safety and Health Administration
OSH Act	Occupational Safety and Health Act 1970 (USA)
PIM	Printing Industry of Minnesota Inc
QA	Quality Assurance
SC	Systems Control
SMEs	Small and Medium-Sized Enterprises
SMS	Safety Management System
TQEM	Total Quality Environmental Management
TQM	Total Quality Management
UK	United Kingdom
USA	United States of America
VPP	Voluntary Protection Program
WES	Working Environment Service

Table of Statutes and Conventions

INTERNATIONAL LABOUR ORGANIZATION
CONVENTIONS

REGULATIONS

UNITED KINGDOM

Table of Cases

1 Introduction

Occupational health and safety (OHS) has re-emerged as a major issue on the social policy reform agenda internationally. In 1994, the British Government completed a major *Review of Health and Safety Regulation*, twenty years after the introduction of the Health and Safety at Work Act 1974. In the United States of America (USA), the Clinton Administration released *The New OHSA: Reinventing Worker Health and Safety*, in 1995, as part of its broader 'Reinventing Government' initiative. In Australia, the Industry Commission published its compendious reports: *Workers Compensation in Australia*, and *Work, Health and Safety*, in 1994 and 1995 respectively. In Norway and Sweden, a new systems-based approach to regulation, known as 'internal control (IC)' was introduced in the early and mid-1990s.[1] Within the European Union (EU), a series of OHS Directives[2] has also triggered increasing regulatory action at national level.

While these various reports and legislative changes have quite diverse origins, it is possible to identify within them a number of concerns which transcend national boundaries and national debates. These concerns now permeate the OHS debate, though regrettably, they have so far had only a very limited impact on policy making itself. In essence, at the same time as health and safety legislation is reaching the limits of its efficiency to address traditional problems (the low hanging fruit has all been picked[3]) it is also becoming increasingly ill-suited to accommodate to trends in the economy and in the nature of employment. As we will see, these twin pressures render new regulatory initiatives imperative.

The need for further reform

Turning first to the nature of the OHS legislation itself, the major OHS statutes in many advanced nations are now some 15 to 25 years old.[4] In parts of Europe and Australia, the British Robens Report of 1972 (which we analyse more fully below) had a profound effect on policy making, and in the mid- to late 1970s resulted in widespread legislative change, from the traditional, 'command-and-control' model, imposing detailed obligations on firms enforced by a state inspectorate, to a more 'self-regulatory' regime, using less direct means to achieve broad social goals.

The Robens-influenced legislation replaced detailed technical standards with broad general duties imposed on employers and others, supplemented by more detailed standards in regulations and codes of practice. It provided enforcement agencies with the option of pursuing administrative sanctions in addition to the traditional method of enforcement through criminal prosecution, and introduced, to various degrees amongst the different jurisdictions, self-regulatory mechanisms focusing principally on the development of OHS policies, and the introduction of employee OHS representatives and OHS committees.

At the time it was introduced, this legislation was certainly a major advance on the antiquated system that it substantially replaced, even though its logic was not properly developed and its fundamental assumptions were flawed. As we will see, these shortcomings relate not just to methods of OHS standard setting, but also to conceptions of the causes of work-related illness and injury, workplace relations, the part workers play in developing and implementing OHS solutions, and to the limited role given to the criminal law in enforcing OHS standards.[5] Similarly, in the USA the Occupational Safety and Health Act of 1970 is increasingly showing its age and the limitations of the assumptions upon which it was based.[6] Ironically, neither under the Robens nor the USA models has the limited role of the criminal law in the OHS arena been seriously re-evaluated by policy makers, even though there has been a concerted movement towards greater 'law and order' in more traditional areas of crime in Australia, Europe and the USA.

And notwithstanding some very significant improvements in OHS performance (particularly resulting from post-Robens legislation) many traditional health and safety problems have still not been solved. On the contrary, according to the *Second European Survey on Working Conditions*,[7] a quarter of the workforce still complains of high-level noise and air pollution, a third of extremes of temperature and of heavy loads and nearly a half of painful postures induced by faulty ergonomical design of workstations. Some 40 per cent of workers in the EU still work in discomfort and over half have no control over comfort (lighting, temperature, etc.).[8]

However, the second problem—the failure of OHS legislation to accommodate to changes in the economy and in the nature of the workforce—is arguably even more serious. Essentially, the legislation of the 1970s and early 1980s was designed to address the OHS issues of that era, and it is increasingly ill-suited to deal with important changes that have taken place since that time. For example, the paradigm case, and model for OHS regulation, was implicitly manufacturing (and to a more limited extent, construction), with the structure and administration of OHS legislation primarily directed at physical hazards relevant to (but not exclusive to) this sector, and at workers in medium to large workplaces. This was understandable since manufacturing held such a dominant role in the economy and the largest proportion of workers were employed in large enterprises. However, over the last decade, the picture has changed dramatically.

Manufacturing itself is contracting as a proportion of the economy, and has dwindled in significance, while the service sector is rapidly expanding. As a result there is also a shift from traditional accidents (slipping, falling, machinery accidents, etc.) towards those involving psychosocial problems, like mental health and musculoskeletal disorders, which are due to a combination of factors such as workstation design, work organization, job content, and time patterns.[9]

There has also been a very substantial increase in the proportion of small enterprises.[10] Such firms are characterized by: simple management structures (often just one general manager who carries out all management duties); a high chance of failure; and high rates of serious and fatal injuries.[11] The prevention of occupational injury and disease in these firms is likely to require very different regulatory strategies from those which have traditionally been applied to large enterprises.

Even within large firms, much has changed. New management philosophies have challenged basic assumptions about the way firms should conduct their business. Derived in large part from successful Japanese manufacturers, such philosophies include total quality management (TQM), quality assurance (QA), just-in-time, lean production, and agile production.[12] Some of these approaches have the potential to improve OHS performance. However, many others may have the opposite effect. For example, the trend towards fragmentation of large firms, and for division into autonomous business units,[13] results not only in less central concern and control over OHS but also in a further emphasis on short-term profit (sometimes the only business target against which the firms are judged) which, in itself, is likely to have adverse OHS implications.

A number of new hazards have also arisen, which are by no means limited to, or principally concerned with, manufacturing. Stress, occupational overuse syndrome, violence in the workplace, and the psychosocial issues mentioned earlier are of increasing concern, particularly in the fast growing service sector. Different technologies are also having important

OHS implications (for example, over two-thirds of workers use comput-
ers). Moreover, a growing proportion of workers in the EU are now
reporting increasing pressure and pace of work and tighter deadlines,[14]
while empowerment at work remains limited (a third of EU workers
report they have no influence on the organization or pace of work, while
42 per cent could not choose when to take a break[15]). Other important
changes include increased automation, changes to the physical environment (for example, air conditioning), and changes to chemicals and other
substances present in the workplace.[16]

Also of concern is the substantial increase in outsourcing, in 'non-
standard' jobs and 'enforced flexitime', and in casual, part-time, and
self-employment.[17] As Quinlan argues, the increasing prevalence of
outsourcing, self-employment, and the use of contingent workers, 'can
have a significant effect on OHS because of the competitive pressures on
subcontractors (resulting in corner cutting, work intensification and excessive hours), disorganization (more attenuated control systems in the
workplace, under-resourced operators, having strangers on site etc.), and
undermining regulatory controls'.[18] Predictably, OHS problems themselves remain unevenly spread: between sectors; between genders; between countries; and between status.

We also note the difficulties experienced by OHS regulators in enforcing OHS statutes, and the problems facing organizations in complying
with statutory obligations. Organizational ignorance of OHS law and
failure to understand its requirements remains a significant impediment
to the success of OHS regulation.[19] This is particularly acute for small
businesses, faced with a plethora of regulatory requirements including
taxation, superannuation, workers' compensation, consumer and environmental protection as well as OHS. Individuals operating small businesses often express bewilderment at the demands that these regulatory
requirements impose upon their time. Even large organizations have
difficulty in keeping up with regulatory requirements.[20]

Where duty holders have knowledge of the existence of standards,
many have difficulty seeing the relevance of abstract standards to the
situations actually confronting them in their workplaces. Others may be
uncertain about exactly what measures they need to take to comply with
statutory standards, and whether the measures they have taken have met
the regulatory requirements. Even where business organizations are
aware of their OHS obligations, they may attempt to avoid compliance
if they perceive that it is not affordable, or if the required measures are
seen as being inefficient. Managers of businesses are under pressure to

manage their organizations as efficiently as possible, and will resist OHS compliance measures which are perceived to be overly costly.[21] For smaller businesses holding a weak competitive position in the market for their goods or services, the constraints within which they operate often restrict their ability to devote resources to OHS or to place OHS at the core of their business activities. Often they are forced to choose between OHS and short-term profit and survival.[22]

Compliance with OHS regulation is more likely where organizations can see the benefits to themselves of adherence to statutory standards, and are motivated to improve their OHS performance.[23] Incentives to comply with, or even exceed, regulatory provisions, can come in many forms, as we demonstrate in Chapter 3. Large organizations sometimes experience difficulty in ensuring that even where the organization has grasped its responsibilities, an appropriate organizational culture is introduced, maintained, and implemented through the actions of the individuals and groups acting on behalf of the organization. Often there is workforce resistance to safety measures.[24] Where the organization has outsourced its operations, or has a transient workforce, the organization's reduced control over these operatives may exacerbate difficulties in implementing OHS management systems.

A final impediment to compliance lies in the role of the OHS enforcement agency itself. Business organizations' commitment to comply with their regulatory obligations may be undermined if they perceive procedural injustices in the agency's approach to enforcement of the requirements. Such procedural injustices will include inconsistent implementation of statutory provisions, inconsistent enforcement decisions, and lack of impartiality. Further, compliance is less likely where the enforcement agency fails to implement the best mix of deterrent, incapacitative, and persuasive enforcement measures, or to recognize that business organizations vary considerably in their structure and motivation for compliance.[25]

Finally, while the problems of health and safety at work are growing in complexity and challenge, the two institutions most capable of combating these problems, organized labour and state regulators, are both confronting increased pressures and difficulties in fulfilling this role. Trade unionism, often the most powerful countervailing force to the emphasis of management on short-term profit, is in decline, a trend itself connected to shifts in the size of firms, to downsizing, to mass redundancies, and to the use of contingent labour, to the increasing decentralization of industrial relations, and to relative growth of different sectors.

The resources of regulators are also declining. The trend towards smaller government and fiscal restraint, driven by the tenets of neo-liberalism, has spread to the large majority of Western Democracies. OHS regulators have not been immune from this trend, and indeed in some cases, as in the USA under the Reagan administration, they have suffered disproportionately as a consequence of it.

The role of the state

At the same time that the problem of OHS regulation is becoming increasingly complex, the capacity of the regulatory state to deal with those issues, at least using existing regulatory models, is itself increasingly being challenged. Ever since the early British Factories Acts, enacted in the first half of the nineteenth century, and copied by North American, European and Australian jurisdictions, governments have accepted that, at least where worker health and safety was concerned, the state has an 'activist'[26] role allowing it to pursue and impose particular views of the good, or safe, society, by enacting statutory standards and establishing and maintaining enforcement agencies to enforce those standards. As Damaska notes,[27] unlike the state's 'reactive' disposition, in which the state sees its task as simply supporting existing social practice, protecting order, and being a neutral arbiter between conflicting and private interests,[28] the activist state sees society, or aspects of it, as defective and in need of improvement, and the role of the state to infiltrate spheres of social life to implement programmes of material and moral betterment. 'In contrast to law in the reactive state, whose stress is on defining forms in which freely chosen goals may be pursued, activist law is directive, sometimes even hectoring: it tells citizens what to do and how to behave.'[29]

The particular form of activist law adopted in the regulation of OHS in Europe, North America and Australia has been the 'command-and-control' or instrumental model of regulation. That model relies on what Teubner and others would call 'material law', a form of law having broad goals and using specific direct means to achieve those goals. It establishes specific prescriptions through detailed rules to control the judgment, operating standards, and behaviour of regulated business organizations, and imposes penalties via criminal prosecution upon those who fail to comply with those rules.[30] Increasingly, it is realized that this model has failed to have the desired impact upon the level of workplace injury and disease. Critics of command-and-control regulation assert that it:

. . . puts government in the position of making the business and operating decisions that the regulated industry should make. It displaces the judgments of
people who know their businesses with those of regulators who may not. The
critics assert that while government is better able to determine what society's
goals should be and build the necessary incentives and accountability mechanisms to get companies to achieve them, the private sector is in a better position
to determine what mix of technologies, process changes, or management practices will achieve the regulatory goals.[31]

Problems with this form of regulation have been accentuated by the
ambiguous aims of state enforcement.[32] At first glance, the adoption of
an activist command-and-control model of OHS regulation, with its
theoretically adversarial, formalistic, and legalistically burdensome orientation, appears antithetical to the minimalist statism required by the
market economy. In practice, while the state appeared to recognize the
structural disadvantage of workers, and regulated to protect workers'
health and safety, an outwardly adversarial criminal process in fact was
transformed (except in the USA[33]) into a consensus-based process of
negotiated compliance between state inspectors and management, with
minimal involvement from worker representatives.

Specifically, since the introduction of the traditional OHS model of
regulation, enforcement practices and strategies have varied across
countries, but the overall pattern has been that most European and
Australian OHS regulatory agencies have developed informal techniques (including advice, education, persuasion, and negotiation)
to secure compliance with the detailed standards.[34] As Glasbeek has
argued, this approach to the enforcement of OHS standards reflected a
conception of the market as an autonomous, self-regulating sphere, and
enforcement rested on a threshold of voluntary compliance within the
market place, rather than on the threat of externally imposed legal
sanctions.[35] The criminal law lost any significant role in enforcing OHS
standards, and in most countries has been used sparingly, usually as a last
resort when more informal methods have failed, or in response to serious
injuries or fatalities at work.[36] Put differently, while eager to mitigate the
incidence of workplace injury and illness by specifying standards of
behaviour, the state has been reluctant to enforce these standards with
the full force of the criminal law against respectable employers engaged
in productive economic activity.[37] In contrast the fairly clear and un-
equivocally activist approach to conventional areas of crime has highlighted the disparities in the state's treatment of conventional and OHS
crime.

In seeking to find new solutions to improve the regulation of workplace health and safety, the environment, and other related areas, policy makers have increasingly turned away from the command-and-control model. Recently, commentators have argued that there is a limit to the extent to which it is possible to add more and more specific prescriptions (the traditional approach to OHS standard setting) without this resulting in counter-productive regulatory overload. Moreover, as Teubner and others have argued,[38] there is a crisis in the capacity of traditional forms of direct regulation to grapple with, and resolve, social problems of increasing difficulty and sophistication, such as the regulation of high-technology, high-hazard facilities.[39] Traditional regulation (material law and the command-and-control model) is increasingly unresponsive to the demands of the enterprise, unable to generate sufficient knowledge to function efficiently, and unable to control the adverse OHS consequences of commercial organizations. Material law, according to Teubner, has largely failed because the complexity of modern society is incapable of being matched by a legal system of comparable complexity capable of harnessing direct goal-seeking law to accommodate social goals. In sum: 'the complexity of society outgrows the possibilities of the legal system to shape the complexity into a form fitting to the goal-seeking direct use of law'.[40] Baldwin and others have also criticized the Robens model in terms of serious flaws in the regulatory techniques used.[40a]

In line with these general criticisms of material law, from the end of the 1960s the command-and-control model of OHS regulation came under critical scrutiny, particularly in the report of the influential British Robens Committee.[41] The Robens Report was strongly critical of the command-and-control model,[42] and argued that:

There are severe practical limits on the extent to which progressively better standards of safety and health at work can be brought about through negative regulation by external agencies. We need a more effectively self-regulating system. . . . The objectives of future policy must therefore include not only increasing the effectiveness of the state's contribution to safety and health at work but also, and more importantly, creating the conditions for more effective self-regulation.[43]

The Robens model consensus-based assumptions were quite explicit— the Committee argued that there is a:

. . . greater natural identity of interest between 'the two sides' in relation to safety and health problems than in most other matters. There is no legitimate scope for 'bargaining' on safety and health issues, but much scope for constructive discussion, joint inspection, and participation in working out solutions.[44]

Many commentators have criticized this assumption, largely on the grounds that it is inconsistent with industrial history since the early nineteenth century, and with more detailed studies of the causes of workplace illness and injury which lay the blame with process failures and production pressures.[45]

The OHS legislation emanating unreflectively from the Robens Report is susceptible to criticism both on the above grounds and because it is based on flawed notions of workplace power relations. The most questionable assumptions were the acceptance of conventional wisdom that 'accidents' were caused by 'apathy', that employers and employees shared a common interest in OHS, and that there was little scope for the criminal law in the enforcement of OHS standards.[46] Largely, for these reasons, the activist model of OHS regulation adopted by most Western nations, to overcome the weaknesses of the traditional command-and-control model, itself proved ineffective both in motivating employers and other duty holders to improve OHS management, and in sanctioning those who failed to comply with minimum OHS standards. Nevertheless, the resulting 'Robens model', reflecting the workplace democratization movements in parts of Europe, North America, and Australia, as well as the philosophy of 'rolling back the state',[47] has been influential in Europe, Canada, and Australia.[48]

As Wilthagen has argued,[49] this new style legislation involved 'the regulation of self-regulation', and displayed what Teubner describes as 'reflexive rationality': 'the state withdrew from taking full and primary responsibility for the working environment, increasing the responsibility of self-reliance of employers and employees'. Law still has a role in regulating social processes, but rather than taking responsibility for substantive outcomes (as it did in the command-and-control model), under a reflexive model the legal system, one of a number of operationally closed, self-referential subsystems, seeks to use indirect means to achieve broad social goals, and, in relation to OHS, aims to encourage organizations to establish processes of internal self-regulation to monitor, control, and replace economic activities injurious to health and safety and to partially resolve the conflict between business priorities, legal obligations, and OHS management concerns. Modern OHS regulatory models, while moving away from the command-and-control model, still, however, contain elements of 'substantive rationality', in that OHS standards still specify substantive outcomes to be achieved by employers, largely in the form of general duties in the principal OHS statutes, and a combination of specification, process-based, and performance standards in regulations, codes of practice, and other subordinate or supplementary standard setting instruments.[50]

A further theoretical framework within which to analyse recent and future developments in OHS regulation is provided by Ayres and Braithwaite's work on 'responsive regulation'.[51] They attempt to transcend the impasse between those calling for stronger state regulation of business and those favouring deregulation 'by seeking to understand the intricacies of interplays between state regulation and private orderings'.[52] Basing their work within a republican theory in which the state, markets, community, and associations (for example trade unions, civil liberties groups, and similar organizations) each exercize countervailing power over each other, and where citizens participate directly in regulation,[53] Ayres and Braithwaite advocate forms of regulation that are responsive to industry structure, the differing objectives and motivations of industrial actors, and to 'how effectively industry is making private regulation work'.[54]

Most distinctively, responsiveness implies not only a new view of what triggers regulatory intervention, but leads us to innovative notions of what the response should be. Public regulation can promote private market governance through enlightened delegations of regulatory functions. . . . Central to our notion of responsiveness is the idea that escalating forms of government intervention will reinforce and help constitute less intrusive and delegated forms of market regulation. . . . Responsive regulation is not a clearly defined program or a set of prescriptions concerning the best way to regulate. On the contrary, the best strategy is shown to depend on context, regulatory culture, and history.[55]

New challenges have also been raised by recent work, particularly that of Fiona Haines,[56] on the development of 'organizational virtue'. Virtue in the context of OHS, according to Haines, is to be examined at the level of the organization, by examining the 'culture' ('a normative device which shapes the way individuals reflect and act within their organization') of the organization, and its 'overt behaviour' in 'operationalizing' virtue.[57] How can OHS regulators recognize, encourage, and 'nurture' firms with organizational cultures that accept that good OHS management is integral to their business, and which are prepared to allocate resources towards improving OHS? How can regulators identify, and modify, external constraints on organizations (such as cut-throat competition) which force organizations to place their own short-term survival ahead of good OHS management?

The aims of this book

The shifts in workplace problems, in the nature of the workforce itself, and in the capacity of existing institutions and forms of regulation to accommodate to these trends, raise many different and complex challenges for OHS policy. These are likely to have profound implications for OHS regulation, enforcement, and management.[58]

How will policy makers meet these challenges? So far, the response has been very diverse. Some agencies continue much as before.[59] Others have embarked on a brave new world where 'internal control' by enterprises themselves, is to be layered on top of conventional inspection.[60] Yet others, more constrained in resources, have sought a hybrid approach, whereby some sectors are audited on the basis of their OHS management systems, while others remain the subject of traditional, and largely reactive, inspection.[61] A few, in Osborne and Graebler's terminology, seek smarter regulation rather than less regulation.[62]

Whether, and to what extent, these various initiatives will be successful, is largely an empirical question, and sound empirical evidence is regrettably, extremely scarce. We simply do not know how well many of these experiments are working. However, what very limited empirical evidence is available (which we review in Chapters 2 and 3) suggests we still have a long way to go before any agency can claim anything like clear success, with any particular approach.

In this fluid and rather unsatisfactory situation, this book seeks to advance the debate concerning the most appropriate forms of OHS regulation for the twenty-first century. In embarking on this exercize in regulatory and social policy design, we both survey and draw on the evidence of what works and does not work internationally, and upon our own qualitative empirical work. From this, we seek to construct models of the forms of regulation, enforcement, and penalties most suited to address some of the problems identified above in advanced industrialized nations in Western Europe, North America, and Australia.

We emphasize 'some' of the problems, because no single book could credibly claim to deal with all the complex and myriad changes to the workforce and to the workplace that have taken place in recent years, and this one is no exception. Rather, our particular focus is on the potential of different types of regulation to address many (but certainly not all) of these rapid and profound changes in the workplace; upon the capacity of existing approaches to be modified, improved, and better targeted where more radical alternatives are unlikely to be successful; and upon innovative means of using third parties as surrogate regulators. We also aim to remedy the near absence of debate over the role of prosecutions and criminal penalties for contraventions of OHS statutes, which has failed to keep pace with the major developments in approaches to OHS standard setting in recent decades.[63] In particular we re-examine the purpose of criminal sanctions in the OHS regulation debate, re-evaluate the place of prosecution in OHS enforcement strategies, and look at possible approaches to combining prosecutions of corporations with prosecutions of individual managers and directors with particular responsibility for corporate non-compliance with OHS standards. Finally, we suggest a broader range of criminal sanctions which can be imposed on OHS offenders, and outline ways of guiding courts in the imposition of penalties against individuals and corporations convicted of OHS offences.

One theme which permeates the entire book is our attempt to answer the question: how can law influence the internal self-regulation of organizations in order to make them more responsive to OHS concerns? It is in responding to this question that we see the very considerable potential of OHS management systems to engage with, and offer solutions to, a number of the major problems confronting OHS policy today. Acknowledging the increasing limitations of direct regulation identified above, we see the potential for law instead to stimulate modes of self-organization within firms in such a way as to encourage internal self-critical reflection about their OHS performance. That is, we see the potential of systems-based regulation to function as a form of 'reflexive law'. However, while we find the concept of 'reflexivity' a valuable sensitizing concept, our quest is not to advance the theoretical literature on reflexive law[64] but rather to offer a more pragmatic and policy-oriented contribution to regulatory redesign in the area of OHS.

One issue that has been ignored in the debate about 'reflexive' and management-based models of OHS regulation, is that these developments have profound implications for the way in which the penal provisions in OHS statutes will have to develop to ensure that OHS enforcement does indeed provide appropriate signals to induce employers to adopt safety management systems (SMSs). In the later chapters of the book we argue that the trend of the last 160 or so years towards downplaying the role of criminal enforcement of the OHS statutes should be reversed, and that the role of criminal prosecution in modern OHS regulatory systems needs to be rethought. We explore the issues involved in developing a new model of OHS prosecution, so that prosecutions play an appropriate role in a 'reflexive' model of OHS regulation, and in ensuring that employers focus on systems of work and on continually improving their approaches to OHS management. In this endeavour we draw upon some of the key ideas of 'responsive regulation' developed by Ayres and Braithwaite.

In particular, the critical issues which this book seeks to address (which include a number highlighted at a major workshop on OHS policy conducted by the European Foundation for the Improvement of Living and Working Conditions[65]) include the following:

- is management 'more likely to consistently meet (and hopefully exceed) its statutory requirements if it has a coherent set of strategies to address OHS rather than an array of *ad hoc* or specific policies targeted at specific hazards'?[66]
- are tripartite/bipartite mechanisms for worker, union, and employer involvement in OHS management (a characteristic of European but not American systems) fundamentally important to successful systems-based innovations?
- in a 'reflexive' model of OHS regulation, what are the overall purposes of OHS regulation, of OHS enforcement, and of OHS prosecutions in particular?
- how does criminal prosecution, of both contraventions of the obligations in the OHS statutes, and for manslaughter and similar crimes, fit into an overall OHS enforcement agency approach to secure compliance with the standards outlined in OHS statutes and to provide incentives for employers to go beyond compliance with minimum standards?
- how should OHS enforcement agencies approach the question of prosecuting individual managers and directors to whom can be attributed particular responsibility for corporate contravention of OHS statutes?
- which new forms of corporate sanctions, beyond the traditional sanction of the fine, can OHS policy makers include in OHS statutes in order to fulfil the purposes of criminal regulation of OHS?
- how can one ensure that regulatory strategies promote more effective

management of OHS on the part of employers in a diverse and rapidly changing environment, and avoid the risks of mere 'paper' systems?
- how should regulation be designed in those circumstances where SMSs and similar approaches are unlikely to be successful;
- in particular, what strategies are likely to be most appropriate for small and medium-sized enterprises (SMEs)?; and
- how can regulatory instruments be redesigned in order to better target the worst employers and others most in need of attention, and more generally, how should resources be best deployed between large, SMEs?

As will be apparent, the book is principally normative in its approach. It identifies weaknesses in current arrangements, and it argues the case for adopting a number of regulatory reforms and innovations drawing from experience internationally, on recent and important developments in regulatory theory, and on the models and approaches we have constructed during the course of this research. Its focus is on the administration, enforcement, and design of OHS legislation. The book is empirically-based. Initially, extensive empirical work was conducted in Australia.[67] Subsequently, the scope of the project was expanded internationally, with the focus on OHS regulation in some of the advanced economies in Western Europe and the USA. Semi-structured interviews were conducted with regulators, employers, trade union officials, and with academics involved in policy making, in Denmark, Sweden, the United Kingdom (UK), and the USA. The book draws extensively upon this material, which is complemented by our review of the various recent reports on OHS regulatory reform, upon which we also build, and upon the regulatory and social policy literature internationally.

Necessarily, given the breadth of the tasks in which we engage, we have had to be selective, and we have drawn our material from, and relate our policy recommendations most closely to, regulation as it has evolved in Australia, Scandinavia, North America, and the UK. These jurisdictions have been our main reference points, and while we hope that readers in parts of continental Europe other than Scandinavia will gain insights from this account, we do not claim close knowledge of their systems of OHS regulation.

As might be expected, different aspects of international experience are emphasized in different places in the book. For example, much of the early debate on OHS regulatory policy was shaped by the British experience, and in particular the Robens Report, and we draw substantially on this experience and on that Report. Developments on SMSs seem most advanced in Europe, particularly Scandinavia, and so we rely heavily upon empirical evidence gleaned from Scandinavia (and to a lesser extent the UK) in examining the role of systems-based approaches. The debate over corporate sanctions generally has been most vigorous in the USA, and we similarly derive insights from USA experience in relation to such issues as sentencing guidelines and corporate compliance. Throughout, we have drawn substantially upon our own knowledge and substantial empirical experience of the various Australian systems, particularly where this is likely to resonate with developments which are taking place in a number of other jurisdictions internationally.

The structure of the book

We begin, in Chapter 2, by examining perhaps the biggest shift of the last decade in OHS regulation internationally, from a prescriptive or performance-based approach to regulation to one on the use of process-based regulation and, more particularly, upon the use of SMSs (a form of reflective regulation). We examine the design of OHS standards and explore the relevant roles of, and relationship between, specification, performance, process and systems-based standards. Such an examination is a necessary first step to deciding what types of standard, and what types of regulation and enforcement, are likely to provide optimal OHS outcomes. Our conclusion is that there is a case for a 'two-track' system of regulation: one involving traditional regulation, and another involving a systems-based approach intended to encourage enterprises to go 'beyond compliance' with existing regulatory standards.

In Chapters 3, 4, and 5, we explore the implications of this 'two-track' approach for inspection and enforcement and, drawing on the empirical evidence, we make various recommendations for achieving efficient and effective regulatory outcomes. In Chapter 3, we recommend a variety of incentives be introduced to encourage firms to adopt a systems-based approach to safety and we suggest a new regulatory approach to systems-based firms, which seeks to harness the capacities of firms for self-regulation, but maintains an important oversight role for the regulatory authority. Mindful that not all firms are suited to a systems-based approach, in Chapter 4 we provide a strategy for dealing with such firms, including many SMEs. We also outline the role of an OHS enforcement pyramid, to guide OHS enforcement agencies in developing an escalating enforcement response to non-compliance with OHS standards. The pyramid assumes firms will seek to comply with OHS standards, but if they do not, it sets out increasingly vigorous sanctions for non-compliance. In Chapter 5, we explore, in some detail, how to design a regulatory structure appropriate for those who opt for a management systems-based approach under Track Two, and how regulators should respond to this development.

In Chapters 6 and 7 we develop a model which aims to reassert a strong role for criminal sanctions in OHS regulation. We argue that while the primary purpose of criminal sanctions at the top of the enforcement pyramid should be general deterrence, OHS regulators should ensure flexibility, in the purposes of prosecution (with particular emphasis on organizational reform), prosecution strategies, and the range of sanctions conferred on the courts. We suggest that prosecution and sentencing guidelines should be public and transparent, so as to give business organizations incentives to develop OHS compliance programmes and SMSs.

Finally, in Chapter 8, we summarize our main arguments and draw together our broader policy prescriptions into a single and coherent package. We outline some of the matters we have not examined in this book, and canvass a few of the more complex and intractable issues which will need to be faced by OHS regulators in the next decade. We conclude by outlining further areas which OHS researchers and policy makers will need to examine if OHS regulation is to play a significant part in influencing organizational behaviour to improve workers' health.

As will be apparent, the book is broad-based and policy-oriented and does not purport to address the detailed content of OHS regulation in any jurisdiction. However, in order to provide a necessary context for readers without any knowledge of the existing legal provisions, the Appendix reviews the legal, institutional, and industrial relations environment in selected jurisdictions. The countries considered are the USA, the UK, Sweden, Denmark, and Australia.

NOTES

1   See below p. 53.
2   Of these, the most important is the EEC Directive on the Introduction of Measures to Encourage Improvements in the Safety and Health of Workers at Work, Directive 89/391, discussed in Chs. 2 and 3 below.
3   Consistent with this, a number of national authorities found that, during

the 1980s, in spite of efforts to improve working conditions, the statistics on work accident and disease did not improve. See, for example, O. Saksvik, *The Norwegian Internal Control Regulation*, SINTEF IFIM, Institute for Social Research in Industry, Trondheim, Workshop on Integrated Control/Systems Control, Dublin, 29–30 Aug. 1996. However, it must be acknowledged that OHS statistics are notoriously hard to interpret. For example, while, in the UK, deaths from accidents at work have reached their lowest level since recording began, this is generally attributed to changes in British industry (the shift from the high risk industries such as manufacturing and mining towards service industries). See D. Walters (ed.), *The Identification and Assessment of Occupational Health and Safety in Europe*, Volume 1: The National Situations (European Foundation for the Improvement of Living Conditions, Dublin, 1996), at p. 196.

4   For an overview see D. Walters (ed.), *The Identification and Assessment of Occupational Health and Safety in Europe*, Volume 1: The National Situations (European Foundation for the Improvement of Living Conditions, Dublin, 1996). See also R. Baldwin and T. Daintith, *Harmonisation and Hazard: Regulating Health and Safety in the European Workplace* (Graham and Trotman, London, 1992); Health and Safety Executive (UK), *The Regulation of Health and Safety in Five European Countries*, Contract Research Report 84 (HMSO, London, 1996); and L. Vogel *Prevention at the Workplace* (Trade Union Technical Bureau, Brussels, 1998).

5   See below pp. 6–7.

6   See, for example, US General Accounting Office, & A. Gore, *The New OSHA: Reinventing Worker Health and Safety* (Washington DC, 1995).

7   European Foundation for the Improvement of Living and Working Conditions, *Second European Survey on Working Conditions* (Dublin, 1996).

8   *Newsletter of the European Trade Union Technical Bureau for Health and Safety*, Nov. 1996, p. 1.

9   On shifting patterns of employment see J. McClean, J. Parks, & D. L. Kidder, '"Till Death Do Us Part": Changing work relationships in the 1990s' (1994) 1 *Trends in Organizational Behaviour* 111–36.

10   Health and Safety Executive (UK), *Health and Safety in Small Firms* (HMSO, London, 1995), p. 15. For example, in the UK, between 1979 and 1993 the total number of small businesses doubled from 1.8 million to 3.6 million. By the late 1990s 99% of all firms were firms who employ fewer than 50 people. These counted for 44% of total private sector employment, compared to 35% in 1979. Of these, 9 out of 10 employ fewer than 5 people.

11   Accidents involving fatal or serious injuries are 40% more likely to happen in premises with fewer than 50 workers than in premises with more than 100 workers.

12   In essence, these management approaches emphasize: (i) the elimination of waste, for example wasted time repairing faulty products or wasted resources with unnecessarily large stock; (ii) utilizing employees as quality assurers, whereby each individual is responsible for correcting mistakes when they occur; (iii) only producing things as they are needed; and (iv) a 'value stream' philosophy that encompasses the firm's entire

operations, from suppliers to customers, not just discrete products and processes.

13   Sometimes the main company simply sets broad business targets and leaves the subunits with considerable autonomy as to how to achieve them.

14   *Newsletter of the European Trade Union Technical Bureau for Health and Safety*, Nov. 1996, p. 1.

15   Ibid.

16   M. Quinlan, *The Development of Occupational Health and Safety Control Systems in a Changing Environment*, Paper presented to Workshop on Integrated Control/Systems Control, European Foundation for the Improvement of Living and Working Conditions, Dublin, 1996.

17   Ibid.

18   Ibid.

19   H. Genn, 'Business Responses to the Regulation of Health and Safety in England' (1993) 15 *Law & Policy* 219 at 225–7.

20   See J. Braithwaite, *Improving Regulatory Compliance: Strategies and Practical Applications in OECD Countries*, Regulatory Management and Reform Series No. 3 (OECD, Paris, 1993).

21   J. Braithwaite, *Improving Regulatory Compliance: Strategies and Practical Applications in OECD Countries*, Regulatory Management and Reform Series No. 3 (OECD, Paris, 1993); H. Genn, 'Business Responses to the Regulation of Health and Safety in England' (1993) 15 *Law & Policy* 219 at 229; and B. Hutter, *Compliance: Regulation and Enforcement* (Clarendon Press, Oxford, 1997), pp. 183–4.

22   See F. Haines, *Corporate Regulation: Beyond 'Punish or Pursuade'* (Clarendon Press, Oxford, 1997), Chs. 2 and 7; and H. Genn, 'Business Responses to the Regulation of Health and Safety in England' (1993) 15 *Law & Policy* 219 at 224 and 228–9.

23   H. Genn, 'Business Responses to the Regulation of Health and Safety in England' (1993) 15 *Law & Policy* 219, pp. 222–4.

24   H. Genn, 'Business Responses to the Regulation of Health and Safety in England' (1993) 15 *Law & Policy* 219 at 230. Hutter lists low worker morale as an impediment to compliance: B. Hutter, *Compliance: Regulation and Enforcement* (Clarendon Press, Oxford, 1997), p. 184.

25   J. Braithwaite, *Improving Regulatory Compliance: Strategies and Practical Applications in OECD Countries*, Regulatory Management and Reform Series No. 3 (OECD, Paris, 1993); and S. Shapiro & R. S. Rabinowitz, 'Punishment versus Cooperation in Regulatory Enforcement: A Case Study of OSHA' (1997) 49 *Administrative Law Review* 713 at 718–19.

26   See M. Damaska, *The Faces of Justice and State Authority: A Comparative Approach to the Legal Process* (Yale University Press, New Haven and London, 1986), p. 72.

27   Ibid. p. 82.

28   The 'reactive' state contemplates no notion of separate interest apart from social and individual (private) interests. It does not envisage engaging in self-conscious governmental direction. The state is a neutral arbiter when individual or group conflicts arise. Accordingly, there are only social and

individual problems, and the state only springs into protective action when someone complains seeking redress, and when someone else refuses to meet these demands. State law does not set out what individuals should do substantively. Rather it sets down procedures to make arrangements binding and enforceable (M. Damaska, *The Faces of Justice and State Authority: A Comparative Approach to the Legal Process* (Yale University Press, New Haven and London, 1986), pp. 73–6).

29   M. Damaska, *The Faces of Justice and State Authority: A Comparative Approach to the Legal Process* (Yale University Press, New Haven and London, 1986), p. 82. Of course, the state is not a monolithic entity, and need not be consistently activist or reactive. It may choose, as most Western governments do, to be uninvolved in some aspects of social life (a *laissez-faire* approach), and managerially interventionist in others, such as OHS.

30   See D. J. Fiorino, 'Towards a New System of Environmental Regulation: The case for an industry sector approach' (1996) 26 *Environmental Law* 457 at 463; E. Bardach & R. Kagan, *Going by the Book: The problem of regulatory unreasonableness* (Temple University Press, Philadelphia, 1982), p. 66.

31   D. J. Fiorino, 'Towards a New System of Environmental Regulation: The case for an industry sector approach' (1996) 26 *Environmental Law* 457 at 464.

32   See N. Gunningham, *Safeguarding the Worker* (Law Book Company, 1984), Ch. 4; and R. Johnstone, *Occupational Health and Safety Law and Policy* (LBC Information Services, 1997), Ch. 2, and the sources cited therein.

33   See E. Bardach & R. Kagan, *Going by the Book: The problem of regulatory unreasonableness* (Temple University Press, Philadelphia, 1982).

34   See B. M. Hutter, *Compliance: Regulation and Environment* (Clarendon Press, Oxford, 1997), pp. 12–19; and T. Wilthagen, 'Reflexive Rationality in the Regulation of Occupational Health and Safety', in R. Rogowshi and T. Wilthagen, *Reflexive Labour Law* (Kluwer, Deventer, 1994), pp. 350–1.

35   H. Glasbeek, *The Maiming and Killing of Workers: The one-sided nature of risk taking in capitalism*, Jurisprudence Centre Working Papers, Department of Law, Carleton University, Ottawa, 1986.

36   See B. M. Hutter, *Compliance: Regulation and Environment* (Clarendon Press, Oxford, 1997), pp. 12–19; T. Wilthagen, 'Reflexive Rationality in the Regulation of Occupational Health and Safety', in R. Rogowshi and T. Wilthagen, *Reflexive Labour Law* (Kluwer, Deventer, 1994), pp. 350–1 and especially p. 362; and R. Johnstone, *The Court and the Factory: The Legal Construction of Occupational Health and Safety Offences in Victoria*, unpublished Ph.D. thesis, the University of Melbourne, 1994.

37   Furthermore, it has operated within a context where the state has taken largely reactive approaches to the regulation of economic activity in general. These ambiguities and inconsistencies in regulatory motivation and orientation have limited the effectiveness of government approaches to regulating workplace health and safety.

38   G. Teubner, 'Substantive and Reflexive Elements in Modern Law' (1983) 17 *Law & Society Review* 239; and G. Teubner, L. Farmer, & D. Murphy (eds.), *Environmental Law and Ecological Responsibility: The concept and practice of ecological self-regulation* (Wiley, UK, 1994).

39   See C. Perrow, *Normal Accidents: living with high risk technologies* (Basic Books, New York, 1984).

40   C. Koch & K. Nielsen, *Danish Working Environmental Regulation: How reflexive—How political?* A *Scandinavian case*, Working Paper, June 1996, Technical University of Denmark, Lyngby, Denmark.

40(a) R. Baldwin *Rules and Government* (Oxford University Press, Oxford, 1995) Chs. 5 and 6. As Baldwin puts it: 'The Robens legacy demonstrates quite graphically how distorations and mistaken assumptions can be built into a regulatory structure and how difficult it is, even decades later, to reform legislation so as to make it more effective and legitimate' (p. 301).

41   *Report of the Committee on Safety and Health at Work 1970–72* (HMSO, London, 1972).

42   See Ch. 2 below.

43   Robens Committee (Committee on Safety and Health at Work), *Report of the Committee on Health and Safety at Work 1970–72* (HMSO, London, 1972), p. 12.

44   Ibid. p. 21, para. 66.

45   See, for example, T. Nichols & P. Armstrong, *Safety or Profit: Industrial accidents and the conventional wisdom* (Falling Water Press, Bristol, 1973), reproduced in T. Nichols (ed.), *The Sociology of Industrial Injury* (Mansell, London, 1997), Ch. 3; N. Gunningham, *Safeguarding the Worker* (Law Book Co., Sydney, 1984), Ch. 11, and M. Quinlan & P. Bohle, *Managing Occupational Health and Safety in Australia* (Macmillan, Melbourne, 1991), p. 223.

46   See, for example, N. Gunningham, *Safeguarding the Worker* (Law Book Co., Sydney, 1984), Ch. 11.

47   T. Wilthagen, 'Reflexive Rationality in the Regulation of Occupational Health and Safety', in R. Rogowshi and T. Wilthagen, *Reflexive Labour Law* (Kluwer, Deventer, 1994), pp. 350–1, and especially pp. 353-4.

48   Central features of this approach, some or all of which have been incorporated in legislation in various parts of Canada, Europe, and Australia, are: less detailed regulatory standards; a focus on 'general duties'; tripartite administrative agencies; a greater use of administrative sanctions (improvement and prohibition notices) rather than prosecution; and 'a statutory duty on every employer to consult with his employees or their representatives at the workplace on measures for promoting safety and health at work, and to provide arrangements for the participation of employees in the development of such measures' (Robens Committee (Committee on Safety and Health at Work), *Report of the Committee on Health and Safety at Work 1970–72* (HMSO, London, 1972), para. 70).

49   T. Wilthagen, 'Reflexive Rationality in the Regulation of Occupational Health and Safety', in R. Rogowshi and T. Wilthagen, *Reflexive Labour Law* (Kluwer, Deventer, 1994) pp. 350–4.

50   For further discussion of this, see Ch. 2.

51   I. Ayres & J. Braithwaite, *Responsive Regulation: Transcending the Deregulation Debate* (Oxford University Press, Oxford, 1992). See also J. Braithwaite, 'Responsive Business Regulatory Institutions', in C. A. J. Coady & C. J. C. Sampford (eds.), *Business Ethics and the Law* (The Federation Press, Sydney, 1993), pp. 83–92.

52   I. Ayres & J. Braithwaite, *Responsive Regulation: Transcending the Deregulation Debate* (Oxford University Press, Oxford, 1992), at p. 3.

53   Ibid. p. 17.

54   Ibid. p. 4.

55   I. Ayres & J. Braithwaite pp. 4–5.

56   See F. Haines, *Corporate Regulation: Beyond 'Punish or Pursuade'* (Clarendon Press, Oxford, 1997), especially Chs. 2, 7–10. See also J. Braithwaite, 'Responsive Business Regulatory Institutions', in C. A. J. Coady & C. J. C. Sampford (eds.), *Business Ethics and the Law* (The Federation Press, Sydney, 1993), pp. 83–92; C. Stone, *Where the Law Ends: The Social Control of Corporate Behaviour* (Harper and Row, New York, 1975); and C. Stone, 'Corporate Regulation: The Place of Social Responsibility', in B. Fisse & P. French (eds.), *Corrigible Corporations and Unruly Law* (Trinity University Press, San Antonio, 1987), pp. 13–38.

57   F. Haines, *Corporate Regulation: Beyond 'Punish or Pursuade'* (Clarendon Press, Oxford, 1997), at p. 63. See also pp. 61–5, 70–8.

58   M. Quinlan, *The Development of Occupational Health and Safety Control Systems in a Changing Environment*, Paper presented to Workshop. on Integrated Control/Systems Control, European Foundation for the Improvement of Living and Working Conditions, Dublin, 1996.

59   Many Australian jurisdictions fall within this category. In the USA, the Clinton administration in 1994 signalled a shift in regulatory philosophy from a confrontational comment and control approach to one based on cooperative partnership. See further the Appendix to this book.

60   See the Swedish approach, described at p. 53 below.

61   This is the UK response, described at pp. 158–62 below.

62   See the Maine 200 and other targeting initiatives referred to at pp. 81–2 below.

63   A notable exception is the recent report from the Industry Commission, Australia, entitled *Work, Health and Safety* (1995), particularly pp. 99–126.

64   Indeed, within systems theory, such semi-autonomous subsystems are treated as particularly difficult to change from outside.

65   These have been ably summarized in M. Quinlan, *Reshaping Regulation and OHS Management Systems: Lessons from Europe*, (forthcoming).

66   M. Quinlan, *The Development of Occupational Health and Safety Control Systems in a Changing Environment*, Paper presented to Workshop on Integrated Control/Systems Control, European Foundation for the Improvement of Living and Working Conditions, Dublin, 1996.

67   Most of this research took place in New South Wales, where the authors undertook a consultancy for the New South Wales WorkCover Authority on the redesign of that jurisdiction's OHS regulation.

2 From Compliance to Best Practice in OHS: The Roles of Specification, Performance, and Systems-Based Standards

Introduction

In designing OHS standards, it is vitally important to determine what *types* of standards to adopt. For example, what kinds of measures are most likely to achieve best policy outcomes? What techniques are most likely to influence organizational behaviour, to be flexible, produce safety and health benefits at an acceptable cost, provide practical guidance to employers, and be easy to enforce? Decisions about these issues will have major implications not only for regulators, duty holders, and potential victims of work-related injury and disease, but also for the overall effectiveness of the regulatory regime. As Andrew Hopkins has put it: 'in many circumstances the real problem facing policymakers is not to select the best strategy for achieving compliance but to decide what it is the regulated are to be asked to comply with'.[1]

This Chapter examines the various types of standards that might be invoked to protect OHS. In particular, it is suggested that three main options are available: specification, performance, and process/systems-based standards.[2] These options are not mutually exclusive. Nevertheless, they are conceptually quite distinct. The Chapter evaluates the strengths and weaknesses of each approach, and the extent to which different approaches are appropriate in different contexts or for different types of enterprise. It also considers whether it is possible to impose minimum acceptable standards while at the same time encouraging continuous improvement beyond those standards.

The first part of the Chapter addresses the first two approaches, which are well established, and the relationship between them—for example, must individual approaches be used in isolation or are complementary combinations possible. In the second part, it is argued that there is considerable potential to develop a 'third phase' of regulation, which, building on the development of 'process-based' standards,[3] encourages enterprises to adopt a systems-based approach which addresses OHS across an entire enterprise, and facilitates best practice and continuous improvement. In subsequent Chapters, we go on to explore, in some detail, the means by which this might be done, and the implications for regulatory policy and administrative practice. Finally, we make recommendations as to the type(s) of regulation that should be adopted in the future, taking account of the fact that different types of enterprises may respond best to different types of regulation.

How should standards be classified?

Before examining the relative merits of these three types of standards, it is necessary to say something about the distinction between them (in particular between specification and performance standards) and to acknowledge that other forms of classification are possible, and indeed valuable.

To begin, it is important to define our terms. A specification standard is a standard which tells the employer precisely what measures to take and which requires little interpretation on the employer's part. Such a standard is defined in terms of the specific types of safeguarding methods that must be used in a specific situation and in terms of its emphasis on the design and construction of these safeguards.[4] In contrast, a performance standard is one which specifies the outcome of the OHS improvement but which leaves the concrete measures to achieve this end open for the employer to adapt to varying local circumstances.[5] As such, performance standards are outcome-based and the means of achieving that outcome are not prescribed.

Specification and performance standards are 'ideal types', hypothetical constructs, unlikely to be found in their pure form. That is, 'pure' specification and performance standards can best be regarded as polar extremes on a continuum. Any real world standard can be located on the continuum somewhere between the two poles (though most will clearly fall much closer to one pole then the other, justifying the categorization of standards as either specification- or performance-based).

For example, a standard requiring guard rails to be 'at least 100 cm in height, with at least one lower bar in the middle and of at least 100 kp strength' would be readily recognizable as a specification standard. Yet it is still silent about many details: what materials must be used; how the railings should be constructed; or other technical specifications. In the absence of legislative guidance, courts are likely to interpret these matters against a performance (outcome) standard: what is adequate to guard against accident by falls? Thus even a seemingly clear-cut specification standard contains some performance-related elements.[6]

Phase one: Specification standards

Specification standards have their origins in the very earliest forms of OHS legislation. For example, detailed provisions concerning the guarding of dangerous machinery, and, in particular, requirements that mill-gearing, and moving and dangerous parts 'shall be securely fenced', can be traced back to the Factories Amendment Act 1844 (UK) and are still recognizable today.[7] Other requirements, containing specific prescriptions and proscriptions in respect of dangerous plant, substances, and processes, are contained in a variety of industry-specific or process-specific legislation or regulations.[8] This approach is still used, although, for reasons discussed below, it is far less popular than was previously the case.[9]

As specification standards have the virtue of identifying precisely what is required of the employer, they enable employees and inspectors to readily ascertain whether the employer has breached those standards. They have particular attractions to SMEs, which may lack the technological sophistication or resources to apply broader-based (and necessarily less precise) performance standards to the particular circumstances of their own operation. Unions, too, sometimes prefer specification standards, because they have a greater chance of identifying breaches and of bringing pressure to bear for improvements, and because they distrust the lack of detail and certainty inherent in broader-based alternatives.

These standards also offer administrative simplicity and ease of enforcement. For example, technology to monitor emissions of harmful substances is often cumbersome, expensive, and not very accurate. As a result, regulated firms may 'resist enforcement through legal challenges related to evidentiary and technical issues. Also, agency monitoring and enforcement resources are often quite limited. With specification standards, enforcement personnel need only check fuel supply invoices or determine whether control equipment is operating.'[10] Empirical evidence confirms the reluctance of inspectors to relinquish the detail of specification standards in favour of broader-based (and perhaps more ambiguous) performance standards.[11]

Specification standards are particularly important where there is a high degree of risk and there are specific controls which are applicable to all circumstances where the risk occurs, which are essential to control the risk. Such circumstances commonly arise in relation to the design, selection, maintenance, and use of plant and materials, and in relation to the control of dangerous goods. Indeed, even one major government report, which is generally hostile to this approach, acknowledges that 'in some circumstances, mandatory technical requirements are appropriate. For example, it may be more efficient to have key safety elements included in the design of plant and equipment than to engineer them in subsequently.'[12]

There are, however, serious disadvantages to the use of specification standards across the board. Inevitably, they must be extremely detailed to cover all kinds of machines and plant layouts, and, even then, may be incapable of preventing many work injuries.[13] Such an approach tends to result in a mass of intricate detailed law, difficult to comprehend or keep up to date. Moreover, because such standards are prescriptive they do not allow duty holders to seek least cost and/or innovative solutions and accordingly are unlikely to be cost-effective in the majority of circumstances. Finally, they have 'a static, machine-type and substance-type focus, whereas many hazards arise not from these static features of a workplace, but from the organization of work therein'[14] and cannot be dealt with in this manner. Hazards such as occupational stress, manual handling, or occupational overuse syndrome, demonstrably lend themselves to a different approach.

Phase two: Performance standards

The introduction of performance standards was a direct consequence of the 1972 Robens Report,[15] in which it was argued that there were too many Acts and Regulations and that, as a result, employers had difficulty identifying what their legal obligations were. The more general problem with the proliferation of specific regulations was said to be that it: 'encourages rather too much reliance on state regulation, and rather too little on personal responsibility and voluntary, self-generating effort'.[16] Robens's solution, intended to dispel apathy and to offer employers and others signposts as to how to achieve the 'more effectively self-regulating system' which he believed to be desirable, was the introduction of a series of general duties of care, complemented with codes of practice and, to a lesser extent, by regulations of a different nature from those previously applied.

An illustration of the application of general duties, in particular their very broad scope, is provided by section 2 of the UK Health and Safety at Work Act 1974 (HSW Act). Under this section, employers are required to ensure, so far as is reasonably practicable, the health, safety, and welfare at work of all their employees. This is further elaborated in subsection (2) to include:

(a) the provision and maintenance of plant and systems of work that are, so far as is reasonably practicable, safe and without risks to health;

(b) arrangements for ensuring, so far as is reasonably practicable, safety and absence of risks to health in connection with the use, handling, storage and transport of articles and substances;

(c) the provision of such information, instruction, training and supervision as is necessary to ensure, so far as is reasonably practicable, the health and safety of his employees;

(d) so far as is reasonably practicable as regards any place of work under the employer's control, the maintenance of it in a condition that is safe and without risks to health and the provision and maintenance of means of access and egress from it that are safe and without such risks; and

(e) the provision and maintenance of a working environment for his employees that is, so far as is reasonably practicable, safe, without risks to health, and adequate as regards facilities and arrangements for their welfare at work.

General duties (such as the European Community Framework Direction on OHS[17]) have the capacity to address new hazards as they emerge, to enable new information to be taken account of, and to allow new technologies to be adopted at an early stage. They are also concerned 'with influencing attitudes and with creating a framework for better safety and health organization and action by industry itself'.[18] As such, they promise to substantially overcome many of the serious deficiencies of specification standards.

Conventionally, general duties have been treated as a form of performance standard.[19] However, some have argued that they provide insufficient guidance to duty holders as to the outcomes required of them, and for this reason should not be so classified. For example, Brooks has suggested that because the general duties are not simple, straightforward, and simply verified,[20] they lack the characteristics of genuine performance standards (consider a noise standard prescribing a limit of 85 dB(A) over a time-weighted eight-hour day, which does provide a clearly identified outcome). Although there is substance in this point, general duties are closely connected to the codes of practice and regulations that flesh out the principles they articulate, and they must be considered as an integral part of a performance-based package.

Codes of practice are an important component of this package, providing employers with non-mandatory guidance as to how to meet the principles and performance-based requirements set out in the general duties and in regulations. Codes were intended by Robens to fill in much of the detail which was lacking in general duties, but to do so in a more flexible and participatory fashion than had occurred in the past.

Under most of the post-Robens legislation, codes have a quasi-legal status,[21] in that while failure to comply with a code does not in itself involve a breach of the Act, it nevertheless has evidentiary value.[22] That is, the onus is on the duty holder to prove, if challenged, that the action taken was at least as good as that set out in the code. This solution has the considerable attraction of providing more detailed guidance as to *one* acceptable way to comply with the general duties, while preserving duty holders options to devise their own means of satisfying those duties.[23] As one senior regulator has noted:

This flexibility is particularly important when there is more than one satisfactory way to achieve a certain level of health and safety, or when technology changes at a faster rate than the code of practice can be updated. We are showing industry one way of meeting the standard, but freeing them from constraints of making it the only way.[24]

Moreover, many of the codes themselves are both process-[25] and performance-oriented rather than prescriptive, again preserving flexibility in how OHS goals are achieved.[26] This allows firms to select the least costly or least burdensome means of achieving compliance. In most cases, the general duty states the principle (in very broad terms) and the codes identify *one* non-mandatory way of achieving it. This is an important issue for SMEs which, in the absence of specification stand-ards, may still require specific guidance on how to identify and resolve problems.

In some circumstances regulations may be preferred to codes as a means of expanding on general duties (for example, higher risk activities where the general duty alone has proved to be an incomplete way of ensuring OHS). If so, they are intended to be short, easily understood performance standards, prescribing what outcomes are to be achieved (for example, exposure to no more than 2 fibres of asbestos per millilitre, over a time-weighted eight-hour day) but not how to achieve them.

A considerable attraction of performance standards principally in the form of general duties is that because they focus on the outcomes to be achieved rather than on the precise hazards to be controlled or the means of controlling them, they can accommodate to changes in technology and the creation of new hazards (unlike specification standards which commonly fail to keep pace with technological change). Moreover, because the general duties are so broad, they do not result in problems which 'fall between the cracks' of the numerous industry-specific and detailed specification standards.

Performance standards are also likely to be less resource-intensive for the regulatory agency. As one American commentator has pointed out:

> OSHA must know a great deal about an industry when it promulgates a design standard, and this information usually comes from the industry itself. These enormous information requirements virtually ensure that the pace of regulation will be slow. Performance standards require less information about risk reduction technologies and should therefore require fewer agency resources.[27]

On the other hand, because they are sometimes imprecise, performance standards are potentially more difficult to enforce than specification standards and inspectorates have commonly experienced considerable difficulties of adjustment. The difficulties are likely to be exacerbated in those jurisdictions where most inspectors still come from a trade background, and as a result are more comfortable enforcing specification standards where far less professional judgment and discretion is required.

Towards optimal standards design

Specification versus performance standards?

While most commentators have welcomed the shift away from specification standards towards performance standards,[28] this move is not without its critics. Brooks, in particular, has attacked Robens-style legislation as 'inadequate and inappropriate. There are alarming indications also that it is ineffective and possibly counterproductive.'[29] In her view, the performance standards package recommended by Robens, and, in particular, the general duties of care, fails to provide duty holders with the necessary guidance as to *how* to discharge their obligations. If such guidance can only be provided by detailed regulations and codes of practice (many of which are specification standards), then it is argued that the entire performance-based approach is redundant, or worse, counter-productive.

For example, Brooks suggests that in the large majority of cases based on a breach of general duty provisions, a specification standard already covers the hazard involved—and in the minority of cases where a specification standard could not appropriately address the hazard, a code of practice could do so, without need for a general duty requirement. Moreover, the existence of general duties may have a pernicious effect, for their existence may detract from the need to draft more, rather than less, specification standards. The expansion of specification standards, Brooks suggests, is the preferable approach.

However, some at least of Brooks's arguments have proved to be seriously overstated.[30] After over a decade of experience with the general duties and associated codes of practice, there is now considerable evidence that employers, regulators, and courts have generally accommodated to them and found the degree of guidance they offer to be more than adequate to the majority of circumstances.[31] Many (but by no means all) unions also support this approach.[32] Similarly, the inspectorates (after a considerable period of adjustment) have accepted the performance standards approach and are increasingly comfortable working within it. For example, it is now unusual for inspectors to resort to specification standards in addition to general duty requirements in bringing prosecutions.[33] In the UK, about a third of all prosecutions are now for general duty violations.[34]

As indicated earlier, the general duty requirement also has the considerable advantage of filling in the many and inevitable cracks between the various specification and performance (outcome) standards spelt out in codes and regulations. Importantly, and contrary to Brooks's predictions, the experience of the last decade shows that, even though the general duties are indeed very broad, they still provide a valuable framework within which duty holders must operate, which will be particularly important where no more practical guidance is available from other sources.

In any event, the choice is not as stark as one between vague (and in Brooks's view, largely meaningless) general duties on the one hand and specification standards on the other. Rather, as indicated above, general duties are complemented in many instances by performance-based codes of practice. However, there is still room for specification standards in *some* circumstances, particularly to provide guidance to those (usually small) employers who do perceive a need for the sort of highly detailed prescription that specification standards provide. By incorporating these in non-mandatory codes of practice or in technical guidance documents, these needs can be accommodated, but without imposing a prescriptive straitjacket on the many who do not require such an approach. That is, the preferences of some enterprises for certainty through prescribed standards can be reconciled with more flexible performance-based standards for others. Equally important are two further and related issues: first, whether, and to what extent, to curb the proliferation of regulations; and second, how to achieve the right balance between performance and specification standards.

The first issue concerns the general problem of regulatory overload. As long ago as 1972, Robens argued that there was already too much legislation, and that this made it difficult for employers to know what was required of them. Although most commentators agree, some dismiss the argument, pointing out that only a minority of the total number of regulations in force will apply to a particular employer, and suggesting that the 'problem' is largely a myth.[35] On the contrary, the general problem is far more serious than conceded by the contrarians, and it has become substantially worse since Robens wrote in 1972.

The extent of the problem as it currently applies to OHS, is well documented. For example, in the UK, a 1994 review found that:

... over time, a mass of legal requirements, recommendations (in approved codes of practice) and suggestions (in guidance) about health and safety paperwork has grown up ... of course very few firms are covered by more than a few of these requirements and recommendations. However, there is: confusion amongst companies and others about exactly what is required or recommended; considerable inconsistency between the various requirements and recommendations; and doubt whether all of them are still necessary, especially given the risk assessment and control, training and information provisions of modern legislation.[36]

However, there is also a broader context, namely the total burden imposed on employers by all forms of social regulation. In other words: 'it is the cumulative effect of many regulations that weighs business down'.[37] Rising volumes of new laws and rules produce regulatory fatigue, and threaten to reduce economic growth and investment by imposing non-tariff barriers to trade and inhibiting rapid responses to global competitive pressures.[38]

Against this backdrop, there are strong arguments for managing regulation, not in ways that detract from its social goals, but so that it achieves those goals at least cost and so that the sheer weight of regulation does not become counter-productive. An expansion of specification standards is unlikely to further this goal. Such standards are typically highly detailed, complex, and lengthy, and often also highly specific with the result that there have to be many of them to 'cover the field'. Performance and process standards, on the other hand, can normally be designed so that they are much shorter and impose a much lesser regulatory burden.

As regards the second issue, it follows from the above that there are good reasons for shifting the balance substantially in favour of performance standards. This might best be achieved by limiting the use of specification standards in regulations and codes to the small minority of exceptional situations where the same ends cannot be adequately achieved by performance or process standards. As indicated above, these circumstances arise where there is a known high degree of risk, and where specific controls which are applicable to all circumstances where the risk occurs are essential to control the risk.

However, there remain the needs of many small employers, (and indeed of many firms in the construction industry) who lack the skills, knowledge, or sophistication to devise their own least cost solutions to OHS problems. Such firms will continue to require technical information and detailed practical guidance, but this, in contrast to the previously over-prescriptive specification standards approach, could be provided for through technical data sheets and other advisory material. Such material could continue to be issued not only by regulatory agencies, but also by independent standard setting bodies or even by industry itself. However, given the danger that the latter approach may result in lowest common denominator solutions, for reasons identified below, this is not regarded as the preferred strategy.

The particular advantage of using technical data sheets, rather than codes of practice, is that the former are less likely to be regarded (albeit erroneously) as *de facto* regulations, thereby avoiding the problem of regulatory overload. Because these documents would not have any formal legal status,[39] it would also be easier to modify them quickly and reduce the danger of their becoming rapidly outdated. In this way, the accumulated wisdom that is often contained in specification standards would not be lost but would be located in a different form. This is similar to the developing American approach whereby most standards are written in performance language but 'include specification guidelines in an appendix that small employers can follow to ensure compliance'.[40]

Finally, there is an important role for manufacturers and suppliers in disseminating the information contained in codes and technical data sheets. Although manufacturers in many jurisdictions already have a duty which includes the provision of information, this could be more sharply focused so that it becomes clear that it includes issuing any relevant technical data sheet or code of practice in respect of the plant or substance itself.

The role of codes of practice

Much recent standards development, particularly in countries which have adopted the Robens model, has taken place through the vehicle of codes of practice, which have the considerable advantages identified above.[41] However, such arrangements have been seriously criticized on three basic grounds: that they have become far too prescriptive in practice; that they are too general to be of practical value; and that they rely unnecessarily on government initiative.[42] In each case the critics have proposed specific solutions to remedy these problems. However, as we will argue, in each case, the medicine they prescribe is far worse than the disease. Each of these issues will now be addressed in turn.

First, there is the issue of the codes' unduly prescriptive nature. It was clearly never intended that codes of practice should be interpreted this way, quite the contrary. Nevertheless, it must be acknowledged that the codes (for example in countries such as the UK, Canada, or Australia)[43] have provided less flexibility in practice than was intended. This is largely because, in the eyes of business, the reversal of the onus of proof (making the duty holder demonstrate that, if they have not complied with the code, they have devised an alternative which is at least as good) has had the effect of making the codes into *de facto* regulations.

This has led one government report to propose that the present quasi-legal codes be replaced with codes which are entirely voluntary.[44] However, this would be counter-productive for a variety of reasons. Codes with purely voluntary status lack 'clout'. In contrast, the present structure has very considerable advantages in terms of its combination of outcomes orientation, practical guidance, and flexibility described above. It is clear that these advantages have not been fully realized to date but the problems involved are neither intractable nor inherent. Rather, there is evidence, both in Australia and the UK, that a central problem with the current operation of the codes is that their operation and advantages are not well understood by duty holders.[45] The solution to this difficulty is likely to be more education of duty holders so that the legal implications of the codes are better understood, rather than either a greater emphasis on specification standards (as Brooks suggests) or the replacement of the existing codes with industry-based voluntary standards (as the Industry Commission advocates).

Second, it has also been suggested that the codes are so general that they commonly fail to provide industry with the practical guidance it needs in order to know how to comply. Specifically, it is said that most codes of practice focus on how to manage a particular hazard in all workplaces in all circumstances. Yet few hazards lend themselves to one or two control measures that can be effectively adopted in every workplace. The result may be codes that are too broad to be of practical help in many workplaces. Instead, the development of voluntary standards and codes *at individual enterprise and at industry levels* has been advocated on the basis that 'these are the best ways to promote best practice because they can accommodate well to the circumstances of individual workplaces'.[46] Such codes are said to be particularly needed by small to medium-sized enterprises.

Again, this proposal is misguided. In dismissing the value of hazard-specific codes, the critics seriously underestimate the contribution that such standards make, both in their own right, and as essential building blocks in the regulatory framework. There is evidence that 'generic codes of practice for particular hazards are useful, and in some cases essential, starting points for an industry code . . . providing a set of principles and performance requirements which, where relevant, can be adopted by the industry to its needs'.[47] In particular, hazard-specific codes provide important directions about how to handle such hazards. Not only do they assert the general principles of hazard identification, risk assessment, and risk control, but they also provide hazard identification checklists which 'draw the attention of those responsible for managing risks to the types of hazards they need to address, which are different for different types of hazards and may be easily overlooked in a generic risk assessment'.[48] They also commonly address a hierarchy of preferred control options and issues of design, information, and training. They not only focus attention on the types of hazard expected, but also provide a process for resolving them.

Accordingly, the jettisoning of the current general codes would be seriously counter-productive. There may be some value in supplementing the generic regulations/codes with industry-specific codes, where this is practicable. However, commonly it is not,[49] and there may be more to be gained by promoting and explaining the generic approach to individual industries, and in providing industry-specific information, advice, and training, than in industry-specific regulation.

Finally, it has been argued that government developed generic codes should be replaced by codes which not only are voluntary, but which are *industry-developed*. The value of such codes would be to help define what would be regarded by the courts as 'reasonably practicable' (the standard required of duty holders in complying with the general duty requirements) even though a particular employer or other duty holder did not use them in its workplace. They would *de facto* have the force of law, because 'when the courts consider whether the duty of care has been met, they will turn to such codes as representing industry custom and practice'.[50]

Again, there are serious problems with this approach. Unfortunately, industry associations cannot be relied upon to develop codes in the public interest. On the contrary, such associations are notorious for developing lowest common denominator approaches[51] and there would be considerable temptation to do so in these circumstances, resulting in a *de facto* lowering of the general duty standard of care. As one trade union organization has put it: '[our] experience has been that employers have tended to take the opportunity to write down the legislation, rather than as a means to apply [industry codes] with specific guidance for a particular industry'.[52] Similarly, industry-based codes might also result in a bias towards technologies or processes in which the industry has an investment.

Even codes which were co-sponsored by government would run these risks if co-sponsorship meant that government merely assesses the code and advises its sponsor whether it is capable of providing an adequate system of dealing with risk, given its underlying assumptions, but 'the agency is not in a position to confirm or deny the validity or presence of those assumptions'.[53] In the absence of a genuine tripartite approach to the writing of codes we believe such proposals are seriously flawed and likely to result in lower standards of OHS.

Limitations of the first two phases

Notwithstanding the considerable contribution that both specification and performance standards have made to improving OHS, both approaches have a substantial limitation: namely they only require enterprises to achieve *minimum* standards and provide no incentives or encouragement to go beyond those minima. A specification standard, for example, may prescribe a specific way of guarding a machine utilizing a particular method or technology. But once that has been done there is no requirement or even encouragement to go further to devise methods that might achieve far higher levels of OHS. Similarly, performance standards prescribe particular outcomes but do not imply that further improvements are necessary, once that outcome has been achieved. For example, a performance standard may prescribe that a maximum noise exposure of 85 dB(A) (over a time-weighed eight-hour day) is the outcome required. Yet while this standard will on average achieve substantial reductions in work-related hearing loss, we know that nevertheless a significant number of workers will still suffer such loss, even under this standard—i.e. exposure limits do not guarantee that all workers will be protected. Moreover:

. . . the [current] legislative approach focuses chiefly on the hardware. It is relatively easy for the regulatory authority to verify, for example, whether a machine is adequately guarded—and to convince a magistrate (should this be necessary) that it has not been. The legislative approach does not address the workplace OHS culture, except perhaps in a negative way by encouraging minimum compliance and avoidance of inspection/audit by the regulatory authority.[54]

To summarize, phase one and phase two standards do not encourage enterprises to improve OHS over and beyond the legal limits prescribed. They do not encourage continuous improvement[55] or industry best practice. Nor do they directly encourage enterprises to develop a safety culture or to 'build in' safety considerations at every stage of the production process.

Certainly, for some enterprises, particularly those with little expertise, sophistication, or commitment to OHS (for example, many small firms, those with high labour mobility), it may not be realistic, at least in the short term, to expect compliance with anything more than the legal minima, and even bringing them up to this standard may be a considerable achievement. Performance and specification standards both continue to make a substantial contribution in this regard.

However, there are many other enterprises which, potentially at least, could achieve far more than those minima. An important role of law is to encourage them to do so. Yet not only do phase one and phase two standards fail to achieve this goal but there is some evidence that they may even discourage efforts to go beyond compliance with existing minimum standards. For example, in 1995, Shell argued that:

The prescriptive and mandatory emphasis of codes of practice stifles enterprise and efficiency on the part of employers' compliance activities . . . Additionally, it promotes a compliance mentality—that compliance with regulations and codes of practice are the extent of OHS management. . . . There is little inclination to extend programs beyond the minimum to avoid prosecution.[56]

For some enterprises, at least, there is considerable potential in developing a third phase of standard design—one which provides not only considerable flexibility and enables enterprises to devise their own least-cost solutions, but which gives them direct incentives to go 'beyond compliance' with the minimum legal standards prescribed under phases one and two. Such an approach may also prove effective in improving the safety performance of some, but not all, categories of below-compliance performers. This third phase is now considered.

Phase three: Systems-based standards

The precursor: Process-based standards

By the mid-1990s a new phase of regulation could be identified, based on the development of OHS management systems (see below). The precursor to this development was the evolution of process standards in the late 1980s: standards which address *procedures* for achieving a desired result. These standards specify the processes to be followed in managing nominated hazards and have been most used in respect of hazards that do not lend themselves to measurement, such as manual handling or safe working practices, or to address risk assessment more generally.

In Europe, the most striking example of process standards is the European Community Health and Safety Framework and daughter Directives, which require every employer to carry out an assessment of *all* risks to health and safety, both of their employees and anyone else who might be affected by the work.[57] The assessment must be 'suitable and sufficient', implying that it 'will recognise all foreseeable hazards, assess the risks arising from these measures and identify both the measures needed to control the risks and any other appropriate action necessary to comply with the employer's duties'. Necessarily, the Framework Directive has resulted in a variety of process-based legislative measures at

---

Box 1

**Demands on management in the executive
order on workplace assessment**

(1) a workplace assessment shall be conducted for all work
(2) all elements of relevance for health and safety at work shall be included
(3) if the assessment reveals risks, preventive measures must be initiated
(4) certain preventive measures shall be preferred to others in accordance with the preventive principles
(5) the assessment shall be included in the planning and organizing of work
(6) the safety organization and/or the employees shall participate in the assessment, but the employer has the responsibility
(7) expert opinion shall be procured if there is doubt how to prevent risks
(8) the safety organization shall be involved in deciding which experts to rely on
(9) the workplace assessment shall be accessible to all employees
(10) the assessment must be audited when changes of relevance for safety and health are made.

Source: The Danish Executive Order on Workplace Assessment

---

national level, designed to implement it. The Danish Executive Order on Workplace Assessment, set out in Box 1, is a pertinent example.

Other nations outside of the EU have also begun to adopt process-based regulations. For example, in Australia, a series of national process-based standards have been introduced relating to: Manual Handling, Certification of Operators of Dangerous Equipment, Plant, Noise, Hazardous Substances, Major Hazardous Facilities, and Storage and Handling of Dangerous Goods. The most important characteristic of the regulations or standards resulting from this initiative is their consistent approach to managing hazards by incorporating the three fundamental steps of hazard identification, risk assessment, and risk control. Thus employers are obliged to review operational procedures to assess risks at the workplace and, in consultation with their workers, to regularly evaluate and improve those procedures. These steps may be regarded as both process- and principle-based.

The Control of Substances Hazardous to Health Regulations 1988 (COSHH) (UK) take a similar approach, requiring employers to systematically assess the risks associated with their operations, identify and implement appropriate control strategies, and monitor the operation of those measures. Most recently, the EU's Control of Major Accident Hazards (COMAH) Directive (Seveso II) 1996 requires the provision by the facility (and assessment by the government regulator) of comprehensive installation safety reports.

Although the 'identify, assess, control' approach is the most obvious example of the process-based approach it is by no means its only manifestation. Other examples include the safety case regime adopted by various nations for offshore oil platforms,[58] and the requirements for suppliers of hazardous substances to provide Material Safety Data Sheets to users, for health monitoring for highly hazardous substances, and for the management of major hazardous facilities. On a broad interpretation, even the provisions of much earlier European legislation, requiring worker participation in OHS decision making, and the provision of training, could also be regarded as examples of process-based regulation.

Process-based approaches address Robens's criticism of old style regulation as failing to take account of organizational factors. However, they represent only one aspect of a full-blown organizational and systems-based approach because they are largely confined to specific hazards or processes, and/or to the process of risk assessment and control. Even when these requirements are incorporated generally, they still concern only one aspect of a fully developed and comprehensive systems-based approach. That is, although process-based standards have something in common with a broader systems-based approach, and can be integrated within it, it is only the latter which provides a coherent strategy for addressing OHS in an entire enterprise, or across an entire facility.

What is a systems-based approach?

A systems approach involves managing OHS, product quality, or any other problem, in terms of systems of work rather than concentrating on individual deficiencies. That is, it involves the assessment and control of risks and the creation of an inbuilt system of maintenance and review.[59] Its focus is on the organizational structure, responsibilities, practices, procedures, processes, and resources for implementing and maintaining OHS management. A management system thus 'spans the entire organisation by relating the organisation to its environment, setting the goals, developing comprehensive, strategic, and operational plans, designing the structure, and establishing control processes'.[60] Of particular importance will be the setting of objects and targets, the establishment of a management programme, procedures for achieving the targets, and measurement techniques to ensure that they are reached. In effect, this approach is a direct application of Robens's exhortation that regulation should be:

... predominantly concerned not with the detailed prescriptions for innumerable day to day circumstances but with influencing attitudes and with creating a framework for better safety and health organisation by industry itself.[61]

If successful, a SMS will achieve far more through co-operation, through changing the norms of the business community, and through developing a safety culture, than any of its predecessors have achieved through other means. However, because such a development has such recent origins, neither the approach itself nor its implications have yet to be subject to detailed consideration. For these reasons, we examine in some depth the potential role of a management systems-based approach and its relationship with OHS legislation and regulatory practice.

Two related forms of management system dominate the field: quality assurance (QA) and total quality management (TQM). As indicated in Figure 1, the former is concerned with maintaining certain minimum standards, on setting up the system and managing it to a certain standard, while the latter focuses on planning and continuous improvement. In a sense, TQM is about going forward and doing better, while QA is about the avoidance of going backwards. More recent approaches to quality systems recognize a complementarity between the two approaches. For example, the ISO 14000 series, which applies to the related issue of environmental protection, builds in elements of both approaches.[62]

In its broad sense QA refers to any action directed toward providing consumers with products of appropriate quality. It can be defined as:

... the total effort involved in planning, organising, directing, and controlling quality in a production system with the objective of providing the consumer with products of appropriate quality. It is simply making sure the quality is what it should be.[63]

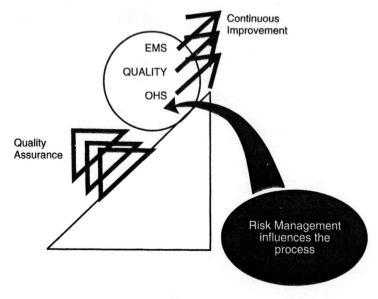

Figure 1

QA involves two related systems. The management system is concerned with the planning, design, organization, and implementation of QA programmes, while the technical system involves: the assurance of quality and reliability in engineering design; the sciences of testing and measurement; the planning and design of manufacturing processes; and the control of incoming materials, intermediate production, and finished goods.[64] Quality audits are central to QA. Such audits provide evidence that products conform to specifications or that operations conform to procedures.

As applied to health and safety in the workplace, QA is concerned with the production process which includes the physical facilities and information and control systems that are required to convert resources into products or services. Quality efforts and assurance here is focused on providing evidence that procedures have been introduced, into the management system of the enterprise, to assure management that the production process meets with all of the necessary criteria for safe operation within the workplace. One important practical step towards providing QA in OHS is the preparation of a risk management policy.[65] Once risks are determined in an organization, the procedures and work practices to minimize these risks in all areas of the workplace can be developed and written into the risk policy.

Of greater contemporary significance is TQM, which emerged as a critique of previous forms of QA rather than as an entirely new development, and has subsequently become extremely influential.[66] Traditionally, TQM aspires to provide management with a framework 'on which to build "quality" into every conceivable aspect of organizational work'.[67] It is a business discipline and philosophy of management which institutionalizes planned and continuous business improvement.[68]

More specifically, a TQM systems-based approach details not only what objectives will be sought but also the activities to be carried out, and how achievement is to be measured and maintained. In this manner, the organizational structure and the enterprise culture will gradually be transformed so that it supports the quality objectives sought. This approach is consistent with corporate compliance strategies which emphasize management commitment, education and awareness, and implementation and control (see Box 2 below).[69]

The original focus of TQM as an enterprise was on the manufacture of products and customer satisfaction. However, current formulations of the TQM philosophy have shown that this system can be usefully applied to other objectives in an enterprise, including OHS.[70] In such circumstances, we way may define this approach as a SMS. The SMS is in essence a specific application of the generic TQM model.

For example, it has been demonstrated that 'policy, planning, implementation, and assessment' can be built into a SMS so that OHS is always considered when equipment is purchased and installed, when people are trained in its use, and when it is operated; at the time that work is organized and allocated; and as chemicals are purchased, distributed, and used.[71] Again, it has been noted above that a fundamental aspect of TQM is the promotion of continuous improvement in the performance of processes and people in terms of cost and efficiency in the workplace.[72] The adoption of this management philosophy into the area of OHS will ensure that procedures and practices on OHS are regularly assessed, reviewed, and improved as and where necessary by the enterprise.[73]

Workplace participation and involvement can also be incorporated within a SMS so that management and employees are required to become responsible and accountable for their part in the planning and

Box 2

OHS Management System Model

AN OHS MANAGEMENT SYSTEM
'involves the setting of objectives and targets, the establishment of a management program and procedures for achieving these, and measurement techniques to ensure that they are reached'.

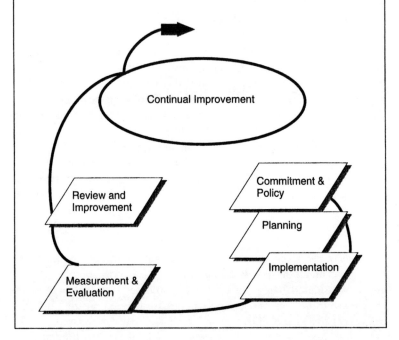

functioning of the overall system of workplace health and safety which is developed and adopted by the enterprise.[74] The principles of a TQM system can further be used to encourage the presence of a workforce which is well-educated and trained in OHS and is personally involved in the process of implementing OHS objectives and procedures within the enterprise. For example, the outcome of this objective should, in an emergency situation, bring about 'a prompt and effective response [which] results from people knowing what to do. The "first respondents" in an emergency depend on training, a reliable communications system, and well maintained equipment. If any of these elements do not work as intended, the "system" will not work. This system includes not only people, but training, drills, and emergency equipment.'[75]

Finally, and crucially, it must be emphasized that the introduction of a SMS is likely to provide substantial advantages to an enterprise not just in terms of OHS, but also in terms of profit and a range of other benefits such as higher employee morale, greater productivity, and a sense of pride in the corporation.[76]

A number of 'off the peg' management systems now have relevance to OHS, although no international standard yet addresses OHS specifically. In 1997 the International Standards Organisation (ISO) considered developing a SMS standard, but suspended this initiative in the face of opposition. Those opposing the establishment of such a standard argued that it was premature, that current national legislation was more suitable for dealing with health and safety issues, and it would be costly bringing added paperwork. However, the British Standards Organisation (BSO) has developed a guideline (as distinct from a standard): BS 8800. Work is also under way toward a standard dealing directly with OHS management systems for Australia and New Zealand.[77] The Committee addressing the development of this standard has agreed that 'the standard should promote the effective management of OHS within an organisation in a way that is integral with the organisation's operations'. The draft will be compatible with existing management systems within the organization, including quality management, and will follow the continuous improvement model adopted by the ISO in the Environmental Management System standard. It will also be compatible with a QA approach.

Strengths and weaknesses of SMSs

There is evidence that such SMS an approach can have substantial results in terms of reducing workplace injury and disease. For example, there is evidence that 'the best OHS outcomes are delivered by employers who have enterprise SMSs [which are] based on the principles of total quality management',[78] and that 'best practice' enterprises manage risk as part of a TQM approach to the running of their operation.[79] The Norwegians have had very considerable success applying what was essentially a systems-based approach (albeit a top-down one) to the safety of North Sea Oil installations.[80] In Germany, a 1998 report on the progress of eighteen companies which had introduced integrated safety management four years previously showed that injury rates had dropped significantly, with the greatest improvement found in companies which used performance appraisal, had strong committed leadership, and used some form of incentive plan for all workers.[81]

The potential value and importance of SMSs is also highlighted by John Braithwaite's research on coal mine safety. His study of the internal compliance systems of the companies with the best coal mine safety records in the USA found that: 'what was consistently evident across all five safety leaders in the study was clearly defined accountability for safety performance, rigorous monitoring of safety performance, and systems for communicating to managers and workers that their safety performance was not up to standard'.[82] Conversely, common organizational defects which were most frequently a cause of disasters were: lack of a plan to deal with a hazard, or poor planning; a generalized pattern of inattention, or sloppiness, in relation to safety matters; poor communication, or reporting systems; inadequate training; and inadequate definition of responsibilities.[83]

Other research in related areas also acknowledges the importance of a systems-based approach where this is practicable.[84] For example, organizational theorist Charles Perrow's seminal work on major technological disasters suggests that 80–90 per cent of the failures relate to the management or organizational system and only 10–20 per cent are based on operator error or equipment failure.[85]

In addition to the benefits to workers in securing and maintaining minimum standards (QA), or in 'locking in' a corporate commitment to continuous improvement (TQM), there are also considerable benefits to both employers and regulators in the adoption of SMSs. For employers, the most obvious advantage is the financial benefit which commonly, but not invariably,[86] results from a substantial reduction in work related injury and disease (particularly since a SMS is likely to achieve this in a flexible and cost-effective manner). For this reason, many leading companies have concluded that improved OHS, through a systems-based approach, is an important factor in achieving business objectives.[87] However, there are also likely to be other benefits, including the commercial advantages of improved public image, and better industrial relations.[88]

A further attraction for employers in adopting a SMS is that it provides an opportunity for self-regulation within an enterprise (providing regulators adopt a flexible approach as described below), and for the enterprise itself to devise ways of reducing or preventing injury and disease at work. Rather than simply following prescriptive government regulations, management is encouraged to take the initiative and respon- sibility for deciding how to satisfy regulatory requirements. And because a management system involves identifying responsibility at each level of the organization, including the top, it directly makes responsible, and engages, top management. It is argued that this approach contributes towards the notion of SMSs becoming 'a habit and an inherent part of company culture and thereby becoming central rather than peripheral to management decision-making'.[89] As one British regulator put it: 'you've got to get OHS into the bloodstream of the organization'. A SMS is the principal vehicle capable of achieving this objective.

Finally, from a regulator's perspective, a SMS increases the likelihood of improved OHS performance even in the absence of frequent inspec- tion. This can ease the tension between regulator and regulated, particu- larly where, as in the USA, an adversary approach to regulation exists. For example, as Bardach and Kagan note: 'regulatory toughness in its legalistic manifestation creates resentment and resistance, undermines attitudes and information sharing practices that could otherwise be coop- erative and constructive, and diverts energies on both sides into pointless and dispiriting legal routines and conflicts'.[90] In contrast, because a SMS passes responsibility back to the regulated, the regulators can adopt an oversight role which only changes to a punitive one if the regulated demonstrably fail to live up to their own commitments.

However, while SMSs can deliver very substantial benefits in terms of improved OHS, enhanced profitability, and more efficient use of regula- tory resources, these benefits can only be obtained if the system is properly implemented. Superficial or tokenistic attempts to introduce a SMS may well be totally ineffective and even counter-productive.[91] Consequently, the dangers of implementation failure must not be underestimated.[92]

A lack of understanding or, more likely, a lack of commitment (in terms of effort or finance) to SMSs amongst management will seriously reduce the chances of success.[93] Indeed, many large and experienced corporations strongly support the view that management commitment is the 'most important single feature' of a successful SMS.[94] Moreover, SMSs present a challenge to conventional management techniques be- cause, as indicated earlier, they cannot be simply grafted into existing organizational structures and systems.[95] Problems arise where organiza- tions attempt to introduce SMSs partially and at a low cost, without effective analysis, planning, or appropriate implementation. A lack of commitment and understanding of the SMS process amongst middle and senior management in an enterprise will also weaken the success of this approach. Both Spector and Beer and a commissioned report of the British Health and Safety Executive (HSE) report that most failures of such systems are a result of management's half-hearted commitment and a lack of understanding of the dynamics of organizational change.[96]

A commitment to the 'long haul', rather than an expectation of short-term benefits, is also important to the success of SMSs. A 'full-blown' SMS will take several years to introduce, develop, and institutionalize into any organization.[97] Therefore, to be successful, a SMS 'requires a strategic and long term approach to improving health and safety and a great deal of energy from individuals'.[98] For these reasons, organizations with a preoccupation with short-term profit may have particular difficulty introducing a SMS successfully.

It must also be recognized that even the successful application of a SMS does not necessarily imply a commitment to achieving the highest standards of OHS. For example, there is a conflict in the field of quality management 'between, on the one hand, an established straightforward association of quality with superior or exceptionally high standards of goods and services, and on the other hand, the quality gurus' conception of quality as meeting reliable and consistent standards which may or may not be identified as exceptionally high'.[99] In fact, Deming supported 'the development of "uniform and dependable" work practices that are congruent with delivering products or services at low cost with a quality suited to the market'.[100] Thus it would be quite feasible for an enterprise to commit itself to a SMS which aspired to levels of safety performance which add little, if anything, to existing regulatory requirements.

A further problem lies in the capacity to successfully translate TQM-based approaches from product quality to OHS. In its generic form, TQM is driven by customer demands (i.e. the ability of products and services to fulfil customers' expectations). And since companies which meet customer expectations are likely to be more successful than those who do not, the use of TQM can readily be integrated with the enterprise's core objectives. But in the case of OHS it would be remarkable if the customer were to consider OHS in manufacture as an important consideration for the products they purchase, leaving a precarious link between OHS, customer demand, and quality management. Thus it must be recognized that there are different sources of motivation for developing quality OHS systems and quality product systems. In the case of the former, it is likely to be the connection between improved OHS performance and core objectives such as profit (which is far from being a complete fit[101]), government regulation, and worker and trade union pressure that provide the impetus for such systems, suggesting that voluntarism alone is likely to be less successful in achieving improvements in OHS than in product quality.

At a political and ideological level, TQM has been also been criticized as an attempt by senior management to indoctrinate employees into certain managerial ideologies that are devised to serve corporate interests rather than those of employees. As Kamp and Le Blansch put it:

> One of the most prominent traits of the management systems is the rational planning model. Here goal-setting is crucial, and in this process knowledge and expertise play an important role. Knowledge in this connection is primarily technical, and data, which can be quantified, is preferred. After goal-setting follows a rational planning process and actors are assumed to act accordingly. As discussed within the organisation theory for decades (see e.g. Mintzberg and Waters 1985, Cohen et al 1972, Pfeffer 1981, Mintzberg 1994), this model poses a simplistic understanding of the actors behaviour. The social construction of goals are not discussed, it is taken for granted that these will be established by 'the legitimate authority' of the company. As the organisation is conceived as prevalently harmonic, authority is genuinely legitimate and not contested, and power and conflicting interests in the organisation are neglected.
>
> Secondly, the bureaucratic (or Tayloristic) traits are unmistakable. Central features are: regulating behaviour through central goal-setting and a hierarchical set of formalised procedures, together with strong emphasis on control. So the problem is diagnosed as managements lack of control, and the solution is centralisation of knowledge and rule-based control. In this conception the management is conceived as a monolith, not as different groups which may have different and maybe conflicting preferences and rationales.[102]

Corporate management is seen to be given the opportunity to adapt TQM to its own ends, reinforcing the existing hierarchy, gaining increased control over innovation, without any change to the collaborative ethos, or serious threat to internal labour markets.[103] Webb points out that the employee participation which is also a fundamental aspect of any TQM system does not necessarily provide a means for employees to question corporate objectives or witness the input from all levels of employees into the decision-making process.[104] Critics of TQM have issued warnings that the system may in fact bind employees into a situation where they must monitor their own progress and undergo surveillance and sophisticated reporting requirements set up by management which do not necessarily result in improvements in OHS in the workplace.

There may be particular problems in terms of promoting employee commitment and participation, and in revising existing organizational attitudes and belief systems.[105] For example, Dawson argues that the multicultural workforce in a country such as Australia does not lend itself easily to this attempt to manage organizational culture 'towards some questionable common features'.[106] In Japan, TQM was established in the context of a homogeneous society within which cultural uniformity was easily accepted. In contrast, in the context of a heterogeneous workforce, there is the added dimension of a changing ethnic composition. The diverse range of interest groups in a multicultural community will reflect different allegiances, expectations, and values which may be at odds with the unitary notion of culture in TQM. In essence, where enterprises are not characterized by high trust and loyalty to common goals, compliance may be problematic.

A further challenge to the concept of an enterprise-wide approach to a TQM system for OHS is found in reference to the small enterprise. Whiting and Horrigan point out that, to obtain the greatest advantage, TQM systems should be adaptable to any industry, and to both large and small enterprises.[107] However, many attempts to introduce a TQM system into OHS in a small enterprise have proved to be problematic, and only very recently have OHS management systems been designed which are not too complex and time-consuming for the small business.[108] Moreover, the employers of small firms are unlikely to have the financial capacity or resources to employ an OHS professional or afford high consulting fees for advice and guidance on the implementation of a successful TQM system for OHS within the enterprise and, even if an appropriate system was installed, it may be difficult to provide resources to sustain it. As the very large majority of workplaces in most countries are in fact small (and the proportion of such workplaces is increasing as a result of structural changes in the economy), the challenge to develop SMSs appropriate to the needs and capabilities of these firms is a crucial one.[109]

There is also a more general issue: the potentially very large gap between the way the system designers think (which assumes rational and motivated employees who undertake diligently the things the system requires them to do), and actual behaviour. For example, employees who are not directly involved in decision making, who are apathetic or disinterested, may fail to perform many of their intended functions, or may do so in a manner that falls far short of the system designer's expectations. For example, it may well be not only that systems are most appropriate to technically complicated and complex areas, but also adherence to the sorts of technical specifications and procedures that lie at the heart of SMSs will be much higher when dealing with long-term employees. The latter's commitment to the organization and its internal culture may be reasonably high, whereas when those who must carry out the procedures are contractors and subcontractors, they will largely lack any such commitment. As Skaar has pointed out, the Norwegian internal control approach has turned out to build:

> . . . on ways of thinking taken from the worlds of engineers, where internal controls are seen as a technical/administrative tool of steering aiming at effective goal achieval. Within this way of thinking the attention is focussed at the tool in itself, more than on the users and the areas where the tool is used.[110]

A final problem is that an enterprise may think it has a system in place, but, in the circumstances, that system is unlikely to be effective. For example (viewed in terms of the theory of firm behaviour), it is significant that firms typically separate safety, occupational health, and environment from issues of production output and efficiency.[111] This can be explained in terms of the distinction between the primary and secondary objectives of an organization. As a Massachusetts Institute of Technology (MIT) Report put it:

> An organisation's primary objectives include its internally stated goals (which may differ from its articulated goals to the outside world) and the prevention or management of perceived threats to its survival or growth. Examples of an organisation's primary objectives might include profit maximisation, sales growth or product quality. Secondary objectives are things the organisation does because it has to, not because it wants to, in order to pursue and accomplish primary objectives. Examples of secondary objectives might include the satisfactory provision of staff procurement and training, information systems (eg, telephone and computer), internal environment (eg heating and lighting) and legal compliance.[112]

Usually, the organization's primary objectives are associated with its core production activities while peripheral activities relate to its secondary objectives. In the past, OHS and regulatory issues have typically fallen into the latter category (profit and sales goals taking priority over OHS).[113] Even when (as part of a systems-based approach) responsibility for safety, health, and environment is incorporated into strategic decision making and to the highest echelons of the organization[114] (as is increasingly occurring), there may still be serious difficulties depending on how the firm chooses to address problems which threaten secondary objectives. In the terminology of the (MIT) Report:

> . . . a firm may attempt an 'orthogonal solution', meaning one which improves performance without contributing to other secondary objectives or modifying the firm's core technology. Or it may respond to the problem with an 'aligned solution', which improves performance relating to the problem while also enhancing the way in which the organisation approaches its core objectives ['either a change in the organisation's core production that remedies the threat to the secondary objective or a redefinition of the organisation's primary objectives to encompass the former secondary objective'[115]]. Orthogonal solutions are often limited in effect and may fail to deal with the problem in a convincing way.[116]

One important policy issue is therefore whether it is possible to encourage firms to make OHS a primary rather than a secondary objective, or, where it remains a secondary objective, to adopt aligned rather than orthogonal solutions. This problem is highlighted by Danish evidence of what is referred to as the 'sidecar' role of OHS, which Jensen, Stranddorf, and Moller describe as follows:

> Even though the law emphasises the responsibility for work environment is with management and in the line organisation, the actual situation in most firms leaves the responsibility to the safety organisation, and gives it a marginal importance in relation to the decision-making processes of the firm. This situation has been described as the 'sidecar-functioning' of work environment, implying: (1) that it has not been integrated as an important aspect of the decision-making processes of the firm; and (2) that the safety organisation primarily considers minor problems and simple solutions.[117]

Finally, the particular kind of system will be crucial to the results achieved. Those which merely focus on quantitative, measurable variables are likely to focus on injuries and downplay disease. Those which emphasize technical expertise may exclude employees as active participants. Similarly, those which adopt a 'top-down' approach will not reap the benefits of a fully engaged, committed, and participating workforce. Nor, will a behaviourist approach which results in victim-blaming and failure to address the systemic nature of many OHS problems (a common criticism of the Du Pont approach).[118]

Particular criticisms have been made of:

... narrowly defined safety management systems approaches, such as those based on ISO 9000 or 14004 [which] do not seek to change relations in the workplace and risk merely replacing the myth of the careless worker with the myth of the safety management system. This has little effect on OHS and the experience of the workforce ... A narrowly conceived safety management system approach also risks reinforcing the individualisation of risk, because it creates the illusion of control—that the only way an injury or disease could occur is if an individual does not follow the current procedures.[119]

While the problems identified above cannot be lightly dismissed, there is nevertheless considerable evidence, cited earlier, that they can be overcome by those enterprises which take SMSs seriously, in particular, by having senior management commitment and concern for direct worker participation in decision making,[120] investing in the necessary initial resources, and accepting that deep-seated change does not happen overnight. It involves 'the development of a system to suit the culture, environment and risks of the enterprise. The challenge is to promote the realization that improving OHS management requires tailor-made solutions and is not merely a question of risk management—effective OHS management includes the integration of OHS into all management procedures and processes.'[121] Under this scenario, even the problem of transforming secondary objectives into primary ones can be successfully addressed. Indeed, one of the most graphic illustrations of such successful transformation comes from the closely related area of quality systems, where many American companies have successfully moved from reactive quality control to proactive QA through the introduction of QA systems and what the MIT team would describe as a shift from an orthogonal to an aligned solution.[122] The challenge for policy design is to take advantage of the very considerable benefits of OHS management systems, while avoiding the pitfalls of implementation failure confronting the unwary. Where this occurs, the system generates and reinforces a culture which emphasizes joint responsibility for safety, open communication, and top management commitment integrated into the organization's core concerns.

The relationship between SMSs and performance and specification standards

Given the considerable attractions of SMSs, it is important to consider how their application might be integrated or reconciled with the more conventional performance and specification standards. It might be argued that in order to encourage the use of SMSs, governments should create a regulatory framework which encourages rather than inhibits commercial drive and innovation. This might imply that governments should avoid being prescriptive and instead confine themselves to setting targets or boundaries. In other words, they should leave it to business itself to determine how best to reach those OHS targets, relying on the attraction of potential benefits such as increased profits, competitive advantage, improved industrial relations, and enhanced public image to stimulate companies to achieve those targets voluntarily.

In this context, is there any reason why those who agree to adopt a systems-based approach should also be required to comply with specification and performance standards? According to one government report, the fact that enterprises contemplating adopting a SMS must also comply with existing specification and performance standards, is a serious impediment to the adoption of a systems approach. Specifically, it is argued that the volume, complexity, and prescriptive detail involved in complying with all three types of standard is substantial. The *additional* effort that is required in adopting a SMS is so large as to seriously detract from an enterprise's capacity to adopt such an approach. As the Report puts it: 'existing regimes do not sufficiently encourage development of quality management approaches to safety. Many provisions discourage such systems. Some are incompatible with them.'[123]

There are a number of serious flaws with this approach. First, no convincing evidence has been offered why, *in principle*, complying with existing standards would impose an onerous additional burden. As indicated earlier, the continuing trend is towards outcome-based standards which are not prescriptive and which do not inhibit industry in how it achieves those ends. Specification standards are likely to have a much reduced role, largely being confined to serious hazards in circumstances where they are cost-effective.

Second, there is little evidence, *in practice*, that the need to comply with existing specification and performance standards, has inhibited the adoption of SMSs. For example, Worksafe Australia's 1992–4 studies of OHS best practice enterprises do not imply any incompatibility between existing regulation and the application of management systems, and in at least one case achievement of best practice was stimulated by the need to meet existing regulatory requirements.[124] Conceivably, the Worksafe study is not representative in this regard, because it is enterprises that fall well below best practice which are most likely to confront serious problems of regulatory overload. However, there is no evidence in support of this proposition,[125] and the experience of Norway and Sweden, which operate a systems-based approach for *all* enterprises, suggests to the contrary.[126]

Finally, it would be wrong to assume that, provided a SMS demonstrably works, it will *necessarily* achieve standards which surpass individual performance and specification standards. This would ascribe to systems-based approaches both process *and* outcome-based characteristics. On the contrary, while SMSs are centrally concerned with process, this does *not* guarantee outcomes (although the two are closely related in most cases). For example, environmental management systems (EMSs) have come in for criticism on precisely these grounds. ISO 14001, in particular, has been described as a driver's licence rather than as an indication of how well you drive, and is allegedly capable of being manipulated to conceal far more than it reveals.[127]

What is needed is both encouragement and incentives for SMSs (because where these work well they encourage a safety culture and practices which go beyond compliance) and a continuing insistence on outcomes (i.e. the continuing application of the general duties and outcome-based codes of conduct). Indeed, for these reasons the Swedish inspectorate, despite requiring all enterprises to adopt a SMS approach (known as Internal Control (IC)[128]), continues to inspect enterprises in terms of traditional performance standards as well as in terms of their management systems.

As indicated earlier, while we support the culling of many specification standards (with more detailed guidance going into technical advice documentation rather than codes) there are nevertheless a minority of situations where specification standards continue to be both necessary and cost-efficient and even those who adopt a systems-based approach should remain subject to them. In most situations, however, a combination of performance and systems-based standards will be the most appropriate and complementary combination. The current approach adopted in Norway and Sweden under their system of IC[129] (described more fully below) and under the Australian National Standards, described above, both of which incorporate both process and outcome standards,[130] (or in the Scandinavian case, both systems and outcome standards), has much to recommend it.

NOTES

1 A. Hopkins, 'Compliance With What?: The fundamental regulatory question' (1994) *34 British Journal of Criminology* 431.

2 It is not suggested that this classification is exhaustive. Other types of standards can also be identified, such as documentation standards (which arguably are a subcategory of process standards), and technical standards (which are a subcategory of specification standards). The three types discussed in the text are by far the most important for present purposes.

3 These are standards which address *procedures* for achieving a desired result. These standards specify the *processes* to be followed in managing nominated hazards. See further pp. 36–38 below.

4 P. W. McAvoy, *OSHA Safety Regulation*, Report of the Presidential Task Force (American Enterprise Institute, Washington DC, 1977).

5 Specification and performance standards are further examined at pp. 24–32 below and a description of the extent to which each is used in selected jurisdictions is contained in the Appendix.

6 Kaj Frick (personal communication, June 1996) has gone further, arguing that standards could be further disaggregated and reclassified in terms of: (i) design/prescribed measures (e.g. height of guard rails); (ii) exposures (e.g. a maximum of 20 p.p.m. of a chemical in air, maximum weight to be lifted); and (iii) effects (e.g. injury by falling)—while recognizing that actual standards can, and often do, combine all three of these elements in different ways.

Frick suggests that the difference between categories (i) and (ii) above is only one of degree of specification (as, in the example of the guard rail above, or with manual handling requirements, many or few elements may be specified in detail, the remainder may be exposure-based). More important is the difference between (ii) and (iii), i.e. between technical specifications of exposure (e.g. no more than 2 p.p.m. of a chemical in the work environment) and its effects on humans, usually employees (e.g. general duty of care standards to 'ensure health, safety, and welfare . . .'). As Frick puts it: 'there is within performance standards an absolute division between those which refer to technical exposure and those which refer to OHS effects on employees'. In practice, most standards are couched in terms of exposures rather than effects (i.e. type (ii) rather than type (iii)). This is largely because, as Hopkins puts it : 'it is not possible to impose straight-forward performance standards requiring employers not to harm their workers, since employers are not necessarily in a position to comply totally with such a legal requirement'. As a result, those standards that are effects-based are generally vague and value-laden.

Frick's proposed reclassification is particularly useful in suggesting how exposure and effects standards may differentially accommodate to changes in technology and standards, encourage innovation, and be resource-intensive. However, for present purposes, the classification adopted in this Chapter, between specification, performance, and system-based standards, has greater capacity to identify the progression from, and tension between, the different types of standards, and the reasons for advancing beyond the first two types to the newer style of process- and system-based standards. It is, moreover, consistent with the terminology of the debate as it has evolved in North America, most of Europe, and Australia.

7 See, for example, s. 27 of the Factories, Shops and Industries Act 1962 (New South Wales).

8 For example, in Belgium, legislation continues to emphasize detailed regulations rather than the broader framework legislation. The Italian and German legislative systems also focus on detailed requirements and have officially accommodating broader framework provisions. See D. Walters (ed.), *The Identification and Assessment of Occupational Health and Safety in Europe*, Volume 1: The National Situations (European Foundation for the Improvement of Living Conditions, Dublin, 1996), at p. 132.

9 See generally D. Walters (ed.), *The Identification and Assessment of Occupational Health and Safety in Europe*, Volume 1: The National Situations (European Foundation for the Improvement of Living Conditions, Dublin, 1996).

10 R. B. Stewart, 'Regulation, Innovation, and Administrative Law: A conceptual framework' (1981) 69(5) *California Law Review* 1256.

11 La Trobe/Melbourne Occupational Health and Safety Project (W. G. Carson, W. B. Creighton, C. Henenberg, & R. Johnstone), *Victorian Occupational Health and Safety: An assessment of law in transition* (Department of Legal Studies, La Trobe University, Victoria, 1989).

12 Industry Commission, Australia, *Work, Health & Safety: Inquiry into Occupational Health & Safety*, Volume I Report and Volume II Appendices, Report No. 47 (AGPS, Canberra, 1 Sept. 1995), in Vol. I p. 75.

13 P. W. McAvoy, *OSHA Safety Regulation*, Report of the Presidential Task Force (American Enterprise Institute, Washington DC, 1977).

14 A. Brooks, 'Rethinking Occupational Health and Safety Legislation' (1988) Sept. *The Journal of Industrial Relations* 348 at 353.

15 Robens Committee (Committee on Safety and Health at Work), *Report of the Committee on Health and Safety at Work 1970–1972* (HMSO, London, 1972).

16 Ibid. p. 7.

17 The European Union's 'framework' Directive for the Introduction of Measures to Encourage Improvements in Safety and Health of Workers furthers the policy of the single European Act of encouraging improvements in health and safety of workers through 'the harmonisation of conditions in this area, while maintaining the improvements made'. It lays down a series of principles that employers in each of the member countries should apply in developing protective and preventive measures. These include priority to the avoidance rather than the control of risk, and the importance of combating risk at source rather than through ameliorative measures. Five 'daughter' Directives cover workplace conditions, safe use of work equipment, manual handling, personal protective equipment, and display screen equipment.

Each Directive creates a legal relationship between the EU and the member state that 'is binding as to the result achieved upon each member state to which it is addressed but shall leave to the national authorities the choice of form and methods' (Treaty of Rome, Article 189). It was originally believed that Directives did not provide any directly enforceable rights for individuals. However, a series of landmark decisions of the European Court of Justice have established that private individuals can enforce Directives in certain circumstances, which now arguably extend to employees in the private sector (*Franovich* v. *Italian Republic* [1992] IRLR 84).

18 Robens Committee (Committee on Safety and Health at Work), *Report of the*

Committee on Health and Safety at Work 1970-1972 (HMSO, London, 1972), p. 7.

19 See, for example, E. A. Emmett, *National Uniformity/Regulatory Reform: A more effective approach to occupational health and safety* Occasional Paper No. 4, Worksafe, Sydney, May 1993.

20 A. Brooks, *Occupational Health and Safety Law in Australia*, 4th edn. (CCH, Sydney, 1993), p. 935. Note, however, that the general duties are modelled on the common law duty of care and, as such, the standard is a very familiar one to lawyers.

21 Robens intended the codes to be industry-based and purely voluntary, whereas, under current legislative approaches, they are state-based and quasi-regulatory.

22 There is some variation between jurisdictions as to the legal effect of the codes. See A. Brooks, *Occupational Health and Safety Law in Australia*, 4th edn. (CCH, Sydney, 1993); and D. Walters (ed.), *The Identification and Assessment of Occupational Health and Safety in Europe*, Volume 1: The National Situations (European Foundation for the Improvement of Living Conditions, Dublin, 1996).

23 See generally R. Johnstone, *Occupational Health and Safety Law and Policy: Text and materials* (LBC Information Services, Sydney, 1997), Ch. 7.

24 E. A. Emmett, 'New Directions for Occupational Health Safety in Australia' (1992) 8(4) *Journal of Occupational Health Safety in Australia and New Zealand* 293 at 296.

25 See further pp. 32–34 below.

26 However, Brooks (A. Brooks, *Occupational Health and Safety Law in Australia*, 4th edn. (CCH, Sydney, 1993), p. 937) correctly points out that there is considerable variation in the presentation style of codes of practice which may give rise to confusion.

27 T. O. McGarity and S. A. Shapiro, *Workers at Risk: The failed promise of occupational safety and health administration* (Praeger, Westport, 1993).

28 See, for example, Industry Commission, Australia, *Work, Health & Safety: Inquiry into Occupational Health & Safety*, Volume I Report and Volume II Appendices (AGPS, Canberra, 1 Sept. 1995), Vol. II, pp. 357–9.

29 A. Brooks, 'Rethinking Occupational Health and Safety Legislation' (1988) Sept. *The Journal of Industrial Relations* 348 at 352.

30 It must be remembered that these views were first expressed in 1986 (see A. Merritt, *Guidebook to Australian Occupational Health and Safety Laws*, 2nd edn. (CCH Australia Ltd., Sydney, 1986)). However, see also A. Brooks, 'Rethinking Occupational Health and Safety Legislation' (1988) Sept. *The Journal of Industrial Relations* 348 at 352; and A. Brooks, *Occupational Health and Safety Law in Australia*, 4th edn. (CCH, Sydney, 1993).

31 For example, the Australian Industry Commission, reporting in 1995, found that most employers were satisfied with, and in the case of larger employers, preferred, a performance-based approach, including the general duties.

32 Industry Commission, Australia, *Work, Health & Safety: Inquiry into Occupational Health & Safety*, Volume I Report and Volume II Appendices (AGPS, Canberra, 1 Sept. 1995), Vol. I pp. 72 and 90.

33 Personal communication with Mr J Cox, WorkCover Authority, NSW, Dec. 1995. See also Industry Commission, *Work, Health & Safety: Inquiry into Occupational Health & Safety* (AGPS, Canberra, 1 Sept. 1995), p. 110; and A. Hopkins, 'Patterns of Prosecution', in *Occupational Health and Safety Prosecutions in Australia: Overview and Issues*, R. Johnstone (ed.) (Centre for Employment and Labour Relations Law, The University of Melbourne, Melbourne, 1994), p. 2.

34 *Employment Gazette*, 1992, 73.

35 See, for example, A Brooks, 'Rethinking Occupational Health and Safety Legislation' (1988) Sept. *The Journal of Industrial Relations* 348 at 352.

36 Health & Safety Executive (UK), *Review of Health and Safety Legislation* (London, HMSO, 1994), p. 8; and see generally D. Walters (ed.), *The Identification and Assessment of Occupational Health and Safety in Europe*, Volume 1: The National Situations (European Foundation for the Improvement of Living Conditions, Dublin, 1996), for an overview of the European situation.

37 Secretary of State for Employment (UK), *Building Business . . . Not Barriers*, a White Paper presented to Parliament (HMSO, London, 1986), p. 4.

38 S. Jacobs, 'The Future of Regulatory Reform', Paper presented at *From Red Tape to Results: International perspectives on regulatory reform*, Conference organized by the New South Wales Government, Sydney, June 1995.

39 Although they might still be used as evidence of industry practice by a defendant seeking to establish that they did what was 'reasonably practical'. Nevertheless the *perception* of technical data sheets is likely to be that they are less constraining than codes or regulation.

40 L. Davey, 'Specification versus performance language in safety standards' (1990) Fall *Job Safety and Health Quarterly*.

41 See generally the Appendix below. There is no precise equivalent in the USA to codes of practice.

42 See, for example, Industry Commission, Australia, *Work, Health & Safety: Inquiry into Occupational Health & Safety*, Volume I Report and Volume II Appendices (AGPS, Canberra, 1 Sept. 1995).

43 See the Appendix below.

44 Industry Commission, Australia, *Work, Health & Safety: Inquiry into Occupational Health & Safety*, Volume I Report and Volume II Appendices (AGPS, Canberra, 1 Sept. 1995).

45 See Health & Safety Commission (UK), *Review of the Health and Safety Regulation*, Main Report (Volume 1) and Summary of findings and of the Commission's response (Volume 2) (HMSO, UK 1994).

46 See examples cited in Industry Commission, Australia, *Work, Health & Safety: Inquiry into Occupational Health & Safety*, Volume I Report and Volume II Appendices (AGPS, Canberra, 1 Sept. 1995), Vol. II p. 366 and Vol. I p. 48.

47 Department of Industrial Relations, *Occupational Health & Safety* (Portfolio Response) Submission Number DR338 to the Industry Commission, 31 May 1995 p. 4. See generally the evidence of the effectiveness of the Manual Handling Regulations cited by the South Australian Government, *Occupational Health and Safety*, Submission Number DR275 to the Industry Commission.

48 South Australian Government, *Occupational Health and Safety*, Submission Number DR275 to the Industry Commission.

49 If such codes are to be developed by government then resources are limited and priority is likely to be given to generic standards. If they are to be developed by industry, then it must be said that industry progress in this direction, to date, has been extremely slow.

50 Industry Commission, Australia, *Work, Health & Safety: Inquiry into Occupational Health & Safety* Volume I Report and Volume II Appendices (AGPS, Canberra, 1 Sept. 1995), Vol. I, p. 50.

51 On the reasons for this and for a case study on self-regulation in one industry, see N. Gunningham, 'Environment, Self-Regulation, and the Chemical Industry: Assessing Responsible Care' 1995 17(1) *Law and Policy* 57.

52 Australian Council of Trade Unions (ACTU), *Occupational Health & Safety*, Submission Number DR336 to the Industry Commission, 1 June 1995, p. 11, in Industry Commission, *Work, Health & Safety: Inquiry into Occupational Health & Safety*, Volume I Report and Volume II Appendices (AGPS, Canberra, 1 Sept. 1995), Vol. I, p. 94.

53 Industry Commission, Australia, *Work, Health & Safety: Inquiry into Occupational Health & Safety* (AGPS, Canberra, 1 Sept. 1995), Vol. I, p. 94.

54 Occupational Health and Safety Authority of Victoria, *Occupational Health & Safety*, Submission Number 176 to the Industry Commission, 25 Oct. 1994, p. 6.

55 Continuous improvement may be defined as an approach to improving organizational performance beyond the level of minimum compliance with standards, leading to superior OHS performance through integrating OHS principles into workplace practice.

56 Shell Company of Australia Limited, *Occupational Health & Safety* Submission Number 67 to the Industry Commission, 1994, pp. 9–10; and in Industry Commission, Australia, *Work, Health & Safety: Inquiry into Occupational Health & Safety*, Volume II Appendices, Report No. 47 (AGPS, Canberra, 1 Sept. 1995), p. 365.

57 EEC Directive on the Introduction of Measures to Encourage Improvements in the Health and Safety of Workers at Work, Directive 89/391/EEC.

58 Cullen, Lord (Chairman), *Piper Alpha Inquiry* (HMSO, London, 1990).

59 B. A. Bottomley, 'Systems Approach to Prevention', Paper presented at *Future Safe Conference*, Sydney, May 1994, p. 2.

60 F. E. Kast & J. E. Rosenzweig, *Organisation and Management: A systems approach*, 2nd edn. (McGraw-Hill, Tokyo, 1974), p. 113.

61 Robens Committee (Committee on Safety and Health at Work), *Report of the Committee on Health and Safety at Work 1970–1972* (HMSO, London, 1972), p. 7.

62 T. Tibor & I. Feldman, *Implementing ISO 14000: A practical, comprehensive guide to the ISO 14000 environmental management standards* (McGraw-Hill, New York, 1997); and R. A. Reiley, 'The New Paradigm: ISO 14000 and its regulatory reform' (1997) 22 *The Journal of Corporation Law* 535–69. See below p. 43.

63 J. R. Evans & W. M. Lindsay, *The Management and Control of Quality* (West Publishing Company, US, 1989), p. 12.

64 Ibid pp. 21–2.

65 National Safety Council of Australia (1996).

66 B. Robinson, 'Victorian Legislation', Paper presented at the *Environmental Management Systems, Programs and Due Diligence Conference*, Sydney 27 and 28 Mar. 1995, p. 3. See also A. Wilkinson & H. Willmott (eds.), *Making Quality Critical: New perspectives on organizational change* (Routledge, London, 1995), p. 8.

67 The notion of 'quality' has been attributed to all kinds of management techniques and initiatives and the diverse and fluid meanings ascribed to quality initiatives and programmes in organizations has led to there being no single theoretical formulation of this philosophy of management. The word 'quality' conveys the suggestion of subtle and vague factors that are not easy to quantify. In terms of TQM, 'quality' is defined as conformance to specifications. Specifications describe what is expected or required of an industry involved in a process to provide a product or service to a customer. Quality can be defined in terms of customer satisfaction, universal standards, professional standards, standards that fit particular purposes, or as the participants themselves define it (A. Wilkinson & H. Willmott (eds.), *Making Quality Critical: New perspectives on organizational change* (Routledge, London, 1995), p. 2).

68 S. Hill, 'From Quality Circles to Total Quality Management', in A. Wilkinson & H. Willmott (eds.), *Making Quality Critical: New perspectives on organizational change*, (Routledge, London, 1995), p. 36.

69 See, for example, J. A. Sigler & J. E. Murphy (eds.), *Corporate Lawbreaking & Interactive Compliance: Resolving the regulation-deregulation dichotomy* (Quorum Books, US, 1991), pp. 80–1; J. Juran, *Juran on Planning for Quality* (Free Press, New York, 1988); K. Shikawa, *What Is Total Quality Control? The Japanese Way* (Prentice-Hall, Englewood Cliffs, NJ, 1985); T. Fisher, 'A "Quality" Approach to Occupational Health, Safety and Rehabilitation' (1991) 7(1) *Journal of Occupational Health Safety—Australia and New Zealand*, 23 at 24; Global Environmental Management Initiative (GEMI) *Total Quality Environmental Management* (Washington DC, 1992) p. 4; G. Blake, *TQM and Strategic Environmental Management in Executive Enterprises* (Publications Co. Inc., 1992) pp. 1–4 at p. 3; and generally W. E. Deming, *Out of the Crisis* (Cambridge University Press, Cambridge Mass., 1986); and W. E. Deming, *Quality, Productivity and Competitive Position* (MIT Press, Cambridge, Mass., 1982). See also M. Walton, *The Deming Management Method* (Dodd Mead & Co., New York, 1986). When incorporated into the broader approach now known as Total Quality Environmental Management (TQEM) (which applies TQM principles to environment) the further components are: identify your customers; continuous improvement; and 'do the job right first time'. See further Global Environmental Management Initiative (GEMI), *Total Quality Environmental Management* (Washington DC, 1992) and in particular:

• the structure of the organization using it and the performance appraisal and reward systems in place;

- the ability and the will of senior managers to translate the philosophy of TQM into specific forms relevant to the business;
- the political-economic operating environment of the business as constructed by managers; and
- the dynamics of trust and distrust between the workforce and senior management over the introduction of TQM.

(J. Webb, 'Quality Management and the Management of Quality', in A. Wilkinson & H. Willmott (eds.) *Making Quality Critical: New perspectives on organizational change* (Routledge London, 1995), pp. 105–26 at p. 122).

See also R. Cole (ed.), *The Death and Life of the American Quality Movement* (Oxford University Press, New York, 1995), p. 2; P. Merrylees, 'Identifying and Implementing the Key Elements in an Effective Compliance System', Paper presented at the *Environmental Management Systems, Programs and Due Diligence Conference*, Sydney, 27 and 28 Mar. 1995, p. 2; R. Chang, *TQM Fever*, an interview presented by Business Report on ABC National Radio, July 1995. Shell Australia highlights the importance of this particular commitment in terms of successful OHS performance in its approach which insists that management takes an active role in this planning process. This participation highlights the importance the business places on OHS (Shell Company of Australia Limited, *Occupational Health & Safety Submission Number 67* to the Industry Commission, 1994; Industry Commission, Australia, *Work, Health & Safety: Inquiry into Occupational Health & Safety*, Volume I Report and Volume II Appendices, Report No. 47 (AGPS, Canberra, 1 Sept. 1995), Vol. I p. 84); J. George, 'How Can OH&S Contribute To Overall Efficiency At The Enterprise Level?', Paper presented at the *Strategic OH&S Management Conference*, Sydney, 13 and 14 Sept. 1993, p. 5; A. Wilkinson & H. Willmott (eds.), *Making Quality Critical: New perspectives on organizational change* (Routledge, London, 1995), p. 20; and R. Stace, 'TQM and the Role of Internal Audit' (1994) 64(6) *Australian Accountant* 26.

70 J. F. Whiting & H. R. Horrigan, 'Validating OHS Performance', Paper presented at the *Proactive OH&S Management Conference*, Sydney, 9 & 10 Mar. 1994, p. 3.

71 M. Burdeu, 'Policy Into Practice: integrating health and safety management into management systems', Paper presented at the *OH&S In The Public Sector Conference*, Sydney, 31 Aug. and 1 Sept., 1995, p. 3.

72 P. Harper, *Turning Rhetoric into Practice: integrating TQM into plant operations* (Executive Enterprises Publications Co. Inc. (edn.)), pp. 95–106 at p. 97.

73 Consistent with the above, Jensen suggests that generally OHS measures are considered a cost to the firm and not adopted (P. L. Jensen, J. Stranddorf, & N. Moller, *Developing Safety Management in Practice*, Department of Working Environment, Technical University of Denmark, Lyngby, Denmark, undated, p. 5). However, 'if proposed measures are included in the normal procedures for budgeting and work environment issues are integrated in the relevant planning procedures in the firm then economic considerations do no in general stop work environment measures' (p. 11).

74 Industry Commission (1994), p. 29. See comments, such as those made by the Construction Forestry Mining and Energy Union, which emphasized that 'the sharing of safety goals by employer and employee was the critical component in establishing as effective safety culture' in an enterprise (CFMEU Submission (1995), p. 3 to the Industry Commission, 1994; and Industry Commission, Australia, *Work, Health & Safety: Inquiry into Occupational Health & Safety*, Appendices, Report No. 47 (AGPS, Canberra, 1 Sept. 1995), p. 187).

75 Global Environmental Management Initiative (GEMI), *Total Quality Environmental Management* (Washington DC, 1992), p. 4.

76 See, for example, J. A. Sigler & J. E. Murphy (eds.), *Corporate Lawbreaking & Interactive Compliance: Resolving the regulation-deregulation dichotomy* (Quorum Books, US, 1991).

77 Committee SF/1; and see Environment Institute of Australia, (1995) *Newsletter*, No. 31, Dec., 6; and Committee SF/F—Occupational Health and Safety Management, Draft Australian and New Zealand Standard, July 1997. See also S. P. Levine & D. T. Dyjack, 'Critical Features of an Auditable Management System for an ISO 9000-Compatible Occupational Health and Safety Standard' (1997) 58(4) *American Industrial Hygiene Association Journal* 291–8; A. Scot, 'Europe Weighs Its Standards Options—ISO Versus EMAS' (1997) 159(13) *Chemical Week* 33–4; and Standards Association of Australia, Joint Technical Committee SF/1, Occupational Health and Safety Management, *Occupational Health and Safety Management Systems: General guidelines on principles, systems and supporting techniques* (Standards Australia, Homebush, NSW, 1997).

78 Industry Commission, Australia, *Work, Health & Safety: Inquiry into Occupational Health & Safety*, Volume I Report and Volume II Appendices, Report No. 47 (AGPS, Canberra, 1 Sept. 1995), p. 8. The quality management approach adopted by Shell Australia in its Health, Safety and Environment Management System is particularly cited in Industry Commission, *Work, Health & Safety: Inquiry into Occupational Health &* Safety, Volume I Report and Volume II Appendix, Report No. 47 (AGPS, Canberra, 1 Sept. 1995), p. 84.

79 Industry Commission, Australia, *Work, Health & Safety: Inquiry into Occupational Health & Safety*, Volume I Report and Volume II Appendix, Report No. 47 (AGPS Canberra, 1 Sept. 1995), pp. 192–4. See also W. Burgess, *'Best Practice' and OH&S, Victorian Institute of Occupational Health and Safety* (Ballarat, Victoria, 1997); and C. Gallagher, 'Occupational Health and Safety Management' in *Belts to Bytes*, Conference Proceedings, Work Cover, Adelaide, Australia, 1994.

80 See further O. Saksvik & K. Nytro, 'Implementation of Internal Control of Health, Environment and Safety in Norwegian Enterprises', Paper presented to *Seventh European Congress on Work and Organisational Psychology*, Gyor, Hungary, Apr. 1995; and O. Saksvik & K. Nytro, *Implementation of Internal Control of Health, Environment and Safety: An Evaluation and a Model for Implementation* (SINTEF IFIM, Institute for Social Research in Industry, Trondheim, Norway, 1995). But cf. Woolfson.

81 G. Zimolong, 'What Successful Companies Do', Paper presented at *Safety In Action Conference*, Melbourne, Australia, Feb. 1998.
82 J. Braithwaite, *Crime, Shame and Reintegration* (Cambridge University Press, Cambridge, 1989).
83 J. Braithwaite, *To Punish or Persuade: Enforcement of coal mine safety* (State University of New York Press, Albany, United States, 1985), p. 19.
84 Ibid.
85 See C. Perrow, *Normal Accidents: Living with high risk technologies* (Basic Books, New York, 1984).
86 This crucial qualification is explored in more detail at pp. 72–9. See also N. Gunningham, *Safeguarding the Worker* (Law Book Co., Sydney, 1984), Ch. 11; and A. Hopkins, 'Patterns of Prosecution', in R. Johnstone (ed.), *Occupational Health and Safety Prosecutions in Australia: Overview and Issues* (Centre for Employment and Labour Relations Law, University of Melbourne, 1994), Chs. 5 and 10.
87 F. Hilmer et al., *Working Relations: A fresh start for Australian enterprises* (Business Council of Australia, 1992).
88 R. Kagan & L. Axelrod (eds.), *Regulatory Encounters: Multinational corporations and American adversarial legislation* (in press); R. Brown, 'Theory and Practice of Regulatory Enforcement: Occupational Health and Safety Regulation in British Columbia' (1994) 16(1) *Law and Policy* 63; and S. S. Shapiro & R. S. Rabinowitz, 'Punishment Versus Cooperation in Regulatory Enforcement: A case study of OSHA' (1997) 49(4) *Administrative Law Review* 731–62.
89 T. C. R. van Someren, J. van der Kolk, K ten Have, & P. T. Calkoen (KPMG Environment/IVA), *Company Environmental Management Systems, Interim Evaluation 1992*, Summary on behalf of Ministerie van Volkshuisvesting Ruimtelijk Ordening en Milieubeuheer, The Hague/Tilburg, Jan. 1993, p. 17.
90 E. Bardach & R. Kagan, *Going By The Book: The problem of regulatory unreasonableness* (Temple University Press, UK, 1982), p. 119.
91 N. Burke, 'Gaining Organisational Commitments to OH&S by Integrating Safety Onto Your Business Plans', Paper presented at *Proactive OH&S Management Conference*, Sydney 9 and 10 Mar., 1994, p. 3. Any enterprise planning the introduction of an effective TQM system for the control of OHS will need:

• top-down integration with the normal business planning cycle;
• compatability with the corporate OHS direction and priorities;
• meaningful participation by all levels of the organization in plan development;
• review and correction processes at appropriate organizational levels;
• planned activities that result in meaningful and measurable tasks and targets at the individual level, forming part of the annual staff performance appraisal; and
• regular review of plans to ensure adequate completion.

92 R. Chang, *TQM Fever*, an interview presented by Business Report on ABC National Radio, July 1995.
93 Industry Commission, Australia, *Work, Health & Safety: Inquiry into Occupa-*

tional Health & Safety, Volume II Appendices, Report No. 47 (AGPS, Canberra, 1 Sept. 1995) p. 186; BHP (1994), p. 17 in Industry Commission, Australia, (1995), Vol. II p. 18; and Shell Company of Australia Ltd., *Occupational Health & Safety* Submission Number 67 to the Industry Commission, 1994, p. 8.

94 See, for example, Health Industry OHS WorkCover Advisory Committee Submission (1994), p. 28 in Industry Commission, Australia, (1995), Vol. II; and the Department of Transport (WA) (1994), p. 4 in Industry Commission, Australia, (1995) Vol. II, also called for the 'positive and active commitment by management at all levels as the single most important element of a successful occupational health program'.

95 J. Osborne & M. Zairi, *Total Quality Management and the Management of Occupational Health and Safety*, Health and Safety Executive, Research Report 153 (London, 1997).

96 B. Spector & M. Beer, 'Beyond TQM Programmes' (1994) 7 *Journal of Organisational Change* 63–70; and J. Osborne & M. Zairi, *Total Quality Management and the Management of Occupational Health and Safety*, Health and Safety Executive, Research Report 153 (London, 1997).

97 S. Hill, 'From Quality Circles to Total Quality Management', in A. Wilkinson & H. Willmott (eds.) (1995), pp. 3–53 at p. 48.

98 M. Burdeu, 'Policy Into Practice: Integrating health and safety management into management systems', Paper presented at the *OH&S in the Public Sector Conference*, Sydney, 31 Aug. and 1 Sept. 1995, p. 4.

99 A. Wilkinson & H. Willmott (eds.), *Making Quality Critical: New perspectives on organizational change* (Routledge, London, 1995), p. 3.

100 W. E. Deming, *Out of the Crisis* (Cambridge University Press, Cambridge, 1986).

101 As we have indicated, in some circumstances there will be a strong connection between improved OHS and improved profitability but this is likely to be the case in some industries more than others (chemical, manufacturing versus construction) and more for some risks than others (accidents versus disease). See further N. Gunningham, *Safeguarding the Worker* (Law Book Company, Sydney, 1984), Ch. 11.

102 A. Kamp & K. Le Blansch, 'Integrating Management of OHS and the Environment: Participation, Prevention and Control', Paper delivered to *Policies for Occupational Health and Safety Management Systems and Workplace Change Conference*, Amsterdam, 21–4 Sept., 1998.

103 A. Tuckman, 'The Yellow Brick Road: Total Quality Management and the Restructuring of Organizational Culture' (1994)15 (3) *Organizational Studies* 727 at 741.

104 J. Webb, 'Quality Management and the Management of Quality', in A. Wilkinson & H. Willmott (eds.), *Making Quality Critical: New perspectives on organizational change* (Routledge, London, 1995), p. 123.

105 P. Dawson, 'Managing Quality in the Multi-Cultural Workforce', in W. Wilkinson & H. Willmott (eds.), *Making Quality Critical: New perspectives on organizational change* (Routledge, London, 1995), p. 190.

106 Ibid.

107 J. F. Whiting & K. R. Horrigan, 'Validating OHS Performance', Paper

presented at the *Proactive OH&S Management Conference*, Sydney, 9 and 10 Mar. 1994, p. 4.

108 It should be noted that the new environmental management standard ISO 14000 is best suited to 'mature enterprises'. The standard can be costly to install into the management of an enterprise and in this respect may be less attractive to the smaller firm. The Kean Report (Committeee of Inquiry into Australia's Standards and Conformance Infrastructure), *Linking Australia Globally: Overview* (AGPS, Canberra, 1995), notes that 'unthinking insistence on it can discriminate unfairly against new firms and small business'.

109 Australian Council of Trade Unions (ACTU), *Occupational Health & Safety*, Submission Number DR336 to the Industry Commission, 1 June 1995, p. 10.

110 S. Skaar, P. Lindoe, T. Claussen, E. Jersin, & R. Tinmannsvik, *Internkontroll—orkenvandring eller veien til det forjettede land? [Internal Control— wandering in the desert or the road to the promised land?]* (SINTEF IFIM, Trondheim, 1994), p. 43.

111 For example, in the chemical industry, at least in the past, safety and environmental issues have 'usually [been] given to safety professionals expert in secondary prevention but not particularly expert in process design or choice of material imputs' (N. A. Ashford et al., 'The Encouragement of Technological Change for Preventing Chemical Accidents: Moving firms from secondary prevention and mitigation to primary prevention'. A Report to the US Environmental Protection Agency, Centre for Technology, Policy and Industrial Development at MIT, Cambridge, Mass., July 1993, v-1).

112 N. A. Ashford et al., 'The Encouragement of Technological Change for Preventing Chemical Accidents: Moving firms from secondary prevention and mitigation to primary prevention'. A Report to the US Environmental Protection Agency, Centre for Technology, Policy and Industrial Development at MIT, Cambridge, Mass., July 1993, v-8. See also P. Cebon, *The Missing Link: Organisational behaviour as a key element in energy/environment regulation and university energy management*, unpublished thesis, MA, Massachusetts Institute of Technology.

113 For example, evidence from The Netherlands suggests that many companies with an average to severe impact on the environment say they already have an OHS management programme. However, on closer examination it was found that in most cases there was simply a collection of practically-oriented measures consisting of disconnected projects and plans and that the companies 'lack any coherent and systematic approach to the programming of environmental management'. The study concluded 'the integration of environmental management into general corporate management proves to have been successful in only a few cases' (T. C. R. van Someren, J. van der Kolk, K. ten Have, & P. T. Calkoen (KPMG Environment/IVA), *Company Environmental Management Systems, Interim Evaluation 1992*, Summary on behalf of Ministerie van Volkshuisvesting Ruimtelijk Ordening en Milieubeuheer, The Hague/Tilburg, Jan. 1993, p. 14). Other evidence cited of lack of integration is that although companies are aware of what

permits require, most of them have not translated this awareness into company procedures and responsibilities. Note, however, that large companies with serious environmental impacts have a much better record than others (p. 19) and the chemical industry is far ahead of the rest of the field in developing environmental management.

114 On the changing relationships between strategic, production, and R & D, see P. Groenewegen & P. Vergagt, 'Environmental issues as threats and opportunities for technological innovation' (1991) 3(1) *Technological Analysis and Strategic Management* 43–55 at 43.

115 N. A. Ashford et al., 'The Encouragement of Technological Change for Preventing Chemical Accidents: Moving firms from secondary prevention and mitigation to primary prevention'. A Report to the US Environmental Protection Agency, Centre for Technology, Policy and Industrial Development at MIT, Cambridge, Mass., July 1993, v-9.

116 For example, if chemical safety is only a secondary objective, the organization might be expected to appoint a safety engineer who reports to, but is not part of, senior management. In this situation, the safety engineer's responsibilities will typically be limited to secondary accident prevention and mitigation devices and procedures. Indeed, the safety engineer will often only be designated after a plant has been fully designed. At this point, the safety engineer has few options but to accept the chemical hazards that the given technology presents and to recommend add-on technologies to control those hazards. By comparison, where accident prevention is a primary objective, chemical safety becomes one of the organization's 'products' and responsibility and concern about accident prevention pervades the organization from senior management throughout the workforce. In this case, the safety engineer will be designated at the outset of plant and process design and will participate in the making of design decisions as well as all other core activities affecting chemical safety (N. A. Ashford et al., 'The Encouragement of Technological Change for Preventing Chemical Accidents: Moving firms from secondary prevention and mitigation to primary prevention'. A Report to the US Environmental Protection Agency, Centre for Technology, Policy and Industrial Development at MIT, Cambridge, Mass., July 1993, v-11).

117 P. L. Jensen, J. Stranddorf, & N. Moller, *Developing Safety Management in Practice*, Department of Working Environment, Technical University of Denmark, Lyngby, Denmark, undated, p. 5.

118 R. Workytch, & C. Van Sandt, 'Occupational Health and Safety Management in the United States and Japan: The Du Pont Model and the Toyota Model', Paper delivered to *Policies for Occupational Health and Safety Management Systems and Workplace Change Conference*, Amsterdam, 21- 4. Sept. 1998.

119 A. Shaw, 'What Works? The strategies which help to integrate OHS management within business development and the role of the outsider', Paper delivered to *Policies for Occupational Health and Safety Management Systems and Workplace Change Conference*, Amsterdam, 21–4 Sept. 1998, p. 9.

120 H. Shannon, et al., 'Overview of the Relationship Between Organisational and Workplace Factors and Injury Rates' (1997) 26(3) *Safety Science* 201-17.

121 A. Shaw, 'What Works? The strategies which help to integrate OHS management within business development and the role of the outsider', Paper delivered to *Policies for Occupational Health and Safety Management Systems and Workplace Change Conference*, Amsterdam, 21–4 Sept. 1998, p. 9.

122 The powerful potential of EMSs (and by implication OHS management systems too) to achieve equally dramatic results is amply demonstrated by the impressive achievements of a number of companies who took part in demonstration projects (President's Council on Environmental Quality (PCEQ) (Quality Environmental Sub-committee), *Total Quality Management: A framework for pollution prevention* (Washington DC, 1993)). The results, as one commentator has noted, are nothing less than 'stunning success stories of rapid and dramatic reductions in pollution that are uncommon, if not unprecedented, in the history of US environmental law'. The same author concludes that the President's Council Report 'signifies a basic transformation in the history of environmental regulation in the United States'. See E. D. Elliott, 'Environmental TQM: Anatomy of a pollution control program that works!' (1994) 92 *Michigan Law Review* 1840 at 1843.

123 Industry Commission, Australia, *Work, Health & Safety: Inquiry into Occupational Health & Safety*, Volume I Report and Volume II Appendices, Report No. 47 (AGPS, Canberra, 1 Sept. 1995), Vol. I, p. 44.

124 See also Department of Industrial Relations Submission (1995) to Industry Commission.

125 Industry Commission, Australia, *Work, Health & Safety: Inquiry into Occupational Health & Safety* (AGPS, Canberra, 1 Sept. 1995), Vol. I, p. 85.

126 O. Saksvik & K. Nytro, '*Implementation of Internal Control of Health, Environment and Safety: An Evaluation and a Model for Implementation*' (SINTEF IFIM Institute for Social Research in Industry, Trondheim, Norway, 1995); O. Saksvik, *The Norwegian Internal Control Regulation*, SINTEF IFIM, Institute for Social Research in Industry, Trondheim, Workshop on Integrated Control/Systems Control, Dublin, 29–30 Aug. 1996; K. Flagstad, *The Functioning of the Internal Control Reform*, Doctoral Thesis, NTH, Trondheim, 1995; P. Lindoe, *Internal Control* (Stavanger College, Rogaland Research, Stavanger, 1995); T. Larsson, *Systems Control Development in Sweden*, SAMU, Institute for Social Science, Uppsala, Workshop on Integrated Control/Systems Control, Dublin, 29–30 Aug. 1996; and C. Frostberg, *Internal Control of the Working Environment*, Analysis and Planning Division, National Board of Occupational Health and Safety, Sweden, December 1994, p. 5.

127 B. Johnston, 'Plying Industry's Green Standard' (1993) 21(2) *New Scientist* 10.

128 See further below p. 74–7.

129 In Norway, regulation concerning the internal control of health, environment, and safety, came into force in 1992, requiring systematic actions (at the enterprise level) to ensure and document that the activities of health and safety are performed in accordance with the requirements specified. In Sweden, see C. Frostberg, *Internal Control of the Working Environment*, Analysis and Planning Division, National Board of Occupational Health and Safety, Sweden, December 1994, p. 5; T. Larsson, *Systems Control Development in Sweden*, SAMU, Institute for Social Science, Uppsala, Workshop on Inte-

grated Control/Systems Control, Dublin, 29–30 Aug. 1996; and National Board of Occupational Health and Safety, *Internal Control of the Working Environment in Small Enterprises: A Monitoring Project of the Occupational Health and Safety Administration, 1995–1997* (Stockholm, March 1997).

130 The main exception is the standard and code on manual handling.

3 Towards a Systems-Based Approach: Voluntarism, Legislation, or Incentives?

Accepting that a systems-based approach to health and safety improvement is desirable, then how might it best be achieved? In broad terms, policy makers wishing to encourage such an approach are faced with three options: leave it to the market (i.e. rely on the enlightened self-interest of enterprises in voluntarily implementing management systems); require by law that all or some enterprises implement SMSs (or at least make it optional for them to do so); or provide incentives (including subsidies) to enterprises to implement SMSs but with no element of compulsion. Each of these options will now be examined.

The limits of voluntarism

Given the considerable advantages that the introduction of a systems-based approach can have for improved OHS, for public and employee relations, and, as a result of all these, for profits, can enterprises be relied upon to introduce such systems voluntarily? It can be argued that industry does indeed have a self-interest in improving OHS, and that, for this reason, one should rely solely on market forces to achieve optimal outcomes.

This view would likely be endorsed by those leading companies who, over the last few years, have sought to move 'beyond compliance'. Such firms have adopted a proactive stance on OHS, involving a voluntary commitment to levels of OHS performance that substantially exceed existing legislative requirements. The use of OHS management systems has been crucial in achieving such results.[1]

As with the parallel debate on environment,[2] it has been argued that going 'beyond compliance' is both good for business and good for OHS—that there is a happy coincidence between private profit and public interest. If so, there is little need for government intervention and enlightened self-interest can be relied upon, without more, to achieve improved OHS performance.

Regrettably, the benefits of this approach are seriously overstated. There are substantial limits to the achievement of 'win-win' outcomes. Specifically, there are circumstances in which the economic benefits of investing in OHS are tenuous, and where the costs to business of implementing OHS protection measures will not be offset by any resulting savings from improved economic performance. That is, there are situations where OHS and private profit do not coincide. The reasons why this is so are well documented and the details need not be repeated here.[3]

However, it is important to stress that the single largest impediment to improved OHS performance is probably the emphasis of corporations on short-term profitability. When a large group of Chief Executive Officers (CEOs) in the USA were asked what prevents their companies doing a better job on health, safety, and environmental issues, over 50 per cent cited the pressure to achieve short-term profits as the main reason.[4] Because corporations are judged by markets, investors, and others principally on short-term financial performance, they have difficulty justifying investment in OHS improvements with primarily long-term pay-offs. Individual managers face a similar dilemma. They too will be judged by senior management essentially on short-term performance, and if they cannot demonstrate tangible economic success in the here and now, there may be no long term to look forward to.[5]

That is, even in those circumstances where a firm does have a self interest in improving OHS, it may have difficulty in doing so voluntarily, if benefits cannot be demonstrated in the short term. As indicated above, the benefits of introducing a SMS are almost always medium- to long-term, making it more difficult for enterprises to justify the introduction of such measures.

Moreover, even when it might appear economically rational for a firm to invest in OHS (even in the short term) it will not necessarily do so. There is considerable evidence that enterprises behave with far less rationality than mainstream economic theory would have us believe. In practice, some enterprises may simply not realize the benefits to be gained from improved OHS performance, others may prove to be incompetent or irrational.[6] Sometimes, it may simply be that a firm can only make so many changes or have so many priorities at a particular time, and that OHS, for various reasons, is not one of them (i.e. their behaviour is more consistent with 'bounded rationality' rather than full rationality).[7] Often, because accidents happen in ones and twos and the costs to the company are not reflected directly in corporate accounts, health and safety issues are 'invisible' for many enterprises.[8] As Tombs puts it: 'it remains the case that many companies either fail to understand the costs of accidents or do not act on such understandings. Corporations are not rational entities neither always capable of engaging in accurate forms of calculation, nor always able to act upon such calculation.'[9]

The empirical evidence is entirely consistent with this analysis. Despite the fact that the safety performance of the majority of enterprises could be substantially improved without loss of profit by adopting a systems-based approach, few of them, left to their own devices, have chosen to adopt such an approach.[10] Nor have they taken other proactive measures to improve their OHS performance. This is reflected in substantial evidence from international studies of generally poor managerial and supervisory skills in conjunction with lack of training and lack of senior management commitment to OHS.[11]

For example, there has been a disappointing response on the part of employers in the UK to the HSW Act 1974 (UK) requirement that employers provide written safety policies. This provision is intended to encourage a more systematic approach to OHS on the part of management. Yet the HSE reported that safety policies often contained insufficient information on hazards or on procedures for hazard control, inadequate specifications of managerial responsibility, and a lack of reference to safety training. This conclusion has been supported by subsequent studies, which have also noted the limited practical impact of many safety documents on OHS awareness and practice and the frequent failure of employers to keep their operation under review.[12]

Similarly, a major study in The Netherlands of the related area of environmental protection noted that it was only with the introduction of legislation that many companies began to take a serious interest in a systems-based approach. As the study put it: 'the main factors influencing the introduction of an EMS are government requirements concerning (technical) environmental measures laid down in licences and the possibility of companies being obliged by law to introduce such a system'.[13] A subsequent study by Aalders and Wilthagen comes to a similarly pessimistic conclusion about the limits of voluntarism and of industry's willingness to take up OHS management systems in The Netherlands.[14]

In the USA, there is also evidence of the failure of voluntarism. For example, insufficient progress after some years of the government regulator seeking to develop a co-operative, consensus-building programme[15] led one research group in 1993 to conclude that a government regulator 'even in the most cooperative of atmospheres with industry, should not expect the [chemical processing industry] to change its safety culture from secondary to primary prevention without considerable prodding or encouragement'.[16]

In conclusion, it seems unlikely that voluntary approaches alone will be enough to change industry behaviour substantially. It is therefore necessary to consider other avenues through which the principles and philosophy of the TQM system might be introduced successfully into the OHS management of an organization. Should it be the responsibility of government to educate and encourage personnel involved in the management of the workplace to create responsible systems for OHS? Is it appropriate to offer some forms of incentive, accreditation, and approval where the introduction of a systems-based approach to OHS in the workplace has proved successful? Is there a need to introduce some form of persuasion, by coercion by means of law, or by some combination of a number of approaches?

From a regulatory agency's perspective, a priority is to ensure that company plans are effective, genuine, and implemented, and to ensure an enterprise has genuinely engaged in this approach. In the absence of such assurances, there is a danger that firms will merely put together a cosmetic 'paper system' to keep the regulators at bay. How can an agency encourage and reward companies adopting a systems approach, and how can inspectorates take advantage of such an approach and at the same time remain independent? Can a systems approach work across the board or is it much better suited to some types of enterprise than others? If so, what are the implications for legislation and enforcement practice?

The role of legislation

As indicated above, there is sometimes a conflict between safety and profit. In these circumstances, voluntarism cannot be expected to deliver significant OHS improvements. Even where firms might gain a competitive advantage from improved OHS performance, not all of them will realize this, or necessarily behave in an economically rational manner.[17] Accordingly, provided the benefits exceed the costs,[18] there remains a legitimate role for governmental intervention.[19] A crucial question then becomes whether this intervention should be by means of regulation or whether an incentives-based approach could achieve the desired outcomes at less cost or in a less interventionist manner that is more acceptable to business. In this first instance, we will examine how OHS legislation could be modified so as to compel enterprises to develop a SMS.

Designing a mandatory systems-based approach

An example of legislation which *mandates* the adoption of a SMS, is the regime for off-shore oil safety adopted in the aftermath of the Piper Alpha oil rig disaster (which occurred off the British coast in 1988, and claimed 165 lives). The principal recommendation of the Cullen Inquiry, which was appointed to investigate the disaster, was that a facility operator should be required by regulation to submit a 'safety case' for each of its installations demonstrating that it had evaluated risks and set up a system of management and control to deal with them.

A leading part of each safety case was to be a SMS 'setting out the safety objectives of the operator, the system by which these objects are to be achieved, the performance standards which are to be met and the means by which the adherence to these standards are to be monitored'.[20] Cullen emphasized the need for safe systems arising out of a safety culture supported by senior management and implemented by properly trained workers under the leadership of effective and disciplined management.

In order to implement these recommendations, Cullen recommended that:

(i) The main regulations . . . should take the form of requiring that stated objectives are to be met (referred to as 'goal setting' regulations) rather than prescribing that detailed measures are to be taken. (ii) In relation to goal setting regulations, guidance notes should give non-mandatory advice on one or more methods of achieving such objectives without prescribing any particular method as a minimum or as the measure to be taken in default of an acceptable alternative. (iii) However, there will be a continuing need for some regulations which prescribe detailed measures.[21]

This general approach is largely embodied in the Safety Code in the Control of Industrial Major Accidents Hazards Regulations 1984 (UK), and in the European Community Health and Safety Framework and daughter Directives.[22] A broadly similar approach was also introduced in Scandinavia in light of positive experience in the offshore oil sector.

In Norway, regulations were introduced in 1991 and made operational in 1992, making it mandatory for every enterprise, regardless of size or number of employees, to establish systems of 'internal control' to ensure and document that the activities of health and safety are performed in accordance with requirements specified in the legislation such as the Working Environment Act. The actions must be described in administrative procedures that document that the activities of health and safety control are performed in accordance with requirements specified in laws and regulations. This represents a partial shift from on-site inspections, where the authorities issued detailed regulations for technical and organizational factors, to a systematic approach whereby the enterprise must *also* demonstrate that health and safety programmes, procedures, and routines are in place and functioning well. This implies clearly set out organization, documentation of plans and results, and the verification of health and safety routines through internal and external audits.[23] See further Box 3 below.

Similarly, in Sweden, 1992 provisions require all employers to plan, direct, and follow up their activities in such a way that the requirements of the work environment legislation are satisfied, with IC activities applying to the whole of the working environment.[24] Under these provisions, employers must: have a working environment policy; allocate tasks, powers, and resources so that working environment requirements

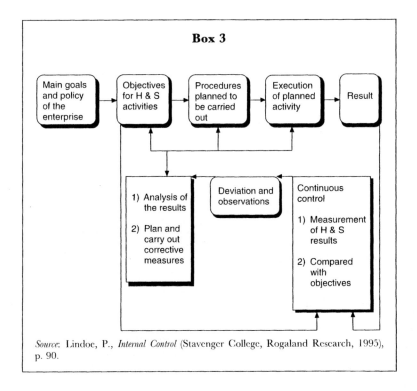

Box 3

Source: Lindoe, P., *Internal Control* (Stavenger College, Rogaland Research, 1995), p. 90.

are met; give employees an opportunity to participate in IC; assess health risks and risks of occupational accidents; take action to deal with risks or, where measures cannot be taken immediately, draw up a schedule for their introduction; ensure directors, managers, and employees have the knowledge to meet working environment requirements; and have routine procedures for and follow up internal controls. In 1994, the National Board of OHS declared that systems control was no longer one among many inspection techniques but was now 'the modern paradigm for state supervision of the working environment'.[25]

Assessments of the effectiveness of the mandatory 'safety case' approach for off-shore oil installations have been mixed. There is some evidence that this approach has been accompanied by at best, no improvement, and at worst by a deterioration in safety performance offshore.[25a]

Would a mandatory approach be effective?

In favour of legislation, it can be said that many enterprises *do* respond positively to the dictates of government, in circumstances where they are most reticent to make changes voluntarily. For example, the major Dutch study referred to earlier showed that, for many companies, stringent licensing and enforcement by the government is the principal motive for introducing EMSs.[26] Given the considerable benefits of management systems, and in particular their demonstrated capacity both to influence the culture of an enterprise and to achieve continuous improvement in OHS, the prospect of legislating to achieve these goals should at least be given serious consideration.

Nevertheless, the practical obstacles to successfully implementing a legislative approach are considerable. In terms of effectiveness, the most substantial problem is that firms who are unwilling to develop a management system voluntarily may respond to compulsion by complying with the letter of the law rather than its spirit. That is, they may simply adopt 'paper systems' which appear to meet the legal requirement, but which in practice are little more than empty shells. Frick encapsulates the problem neatly, by distinguishing between IC (quality management of OHS as part of the economic and organizational culture of the organization itself) and Systems Control (SC) (regulation of IC by the political, bureaucratic, and legal culture). He points out that: 'Our sole instrument is SC. IC, is an indirect effect, a management reaction determined by much more than SC. Only if we understand and address these complexities, can SC become more than an inspection routine which makes employers issue impressive folders.'[27] We explore the problem of ensuring *genuine* compliance in more detail below.

This problem is not unique to mandated systems, but may also apply to incentive-based approaches. However, it will be at it most severe where enterprises are *required* to adopt a system, because, in these circumstances, one is dealing with conscripts rather than volunteers, and many of the former will lack commitment to the legislative requirement. As one experienced observer put it: 'any time a program is compulsory, the participants will have an incentive to minimise the impact. This will invite a form of gamesmanship in which the corporation seeks to comply with the legal requirements by doing nothing more than the minimum. Such approaches tend to stifle innovation and limit the effectiveness and flexibility of the programs.'[28] They can result in what Blewett and Shaw describe as 'a trend away from encouraging people to think in problem-solving terms and a shift towards complicated and paper heavy OHS systems and away from simple, consultative or team-based models . . . Ultimately this is planning for mediocrity and it will inevitably achieve second best.'[29]

Second, there are doubts as to the extent to which something as individualistic as a management system lends itself to an across the board, uniform, legislative approach. To quote the Dutch experience in the related area of environmental protection: 'An environmental management system is an integral part of a company's overall management practices and not an additional component. This means that EMSs must be integrated into the structure and culture of individual companies and must therefore be tailor-made. The general nature of such systems cannot therefore be prescribed by law and there is no point making their introduction mandatory.'[30] Moreover, as the same report put it: 'application of detailed statutory regulations would inevitably mean imposing a framework on reality and would lead to an attempt to distort the reality of company practice to fit a predetermined framework—an attempt which must, virtually of necessity, be doomed to failure'.[31] (This point is in our view exaggerated. Much depends on the provisions of the legislation in question.)

Third, it is not clear that government possesses the capacity to enforce mandatory provisions, particularly given the wide variation in the content and design of management systems in different companies. Some agencies, at least, have expressed major reservations about the implications of implementing a legislated systems-based approach in terms of 'the need to refocus scarce resources away from inspection and enforcement activities to approval of a potentially large range of disparate, enterprise specific SMSs'.[32] Such an approach is more likely to be practicable in countries where regulatory resources are relatively large (for example, most of the Scandinavian countries) than in the many other countries where such resources are seriously strained.

Finally, for practical reasons systems approaches could more readily be imposed on certain sectors (for example, chemicals and major hazardous facilities) or on certain types of organizations (for example, large and sophisticated enterprises). Indeed, it is arguable that it would be premature to insist that the SMEs adopt a systems approach—at least unless it is substantially modified.[33]

Unfortunately, only very limited empirical evidence is available as to the effectiveness of SMSs. In Norway, one year after their mandatory regulations were introduced, 66 per cent of Norwegian enterprises had not even prepared for the introduction of IC and 58 per cent had not heard of the IC regulations. SMEs are facing further problems.[34] Saksvik and Nytro suggest that many small enterprises lack a general understanding of the basics of 'the strategic management systems approach' and for this reason, amongst others, are often very ill-equipped to adopt such an approach.[35] Moreover, as Lindoe notes: 'supervisory agencies do not have sufficient capacity to follow up small companies and the process of introducing IC had come to a halt'.[36] However, the results for those who had genuinely implemented the SMS provisions were much more encouraging. Saksvik found that over 50 per cent of those who had implemented IC had done assessments, implemented measures, made action plans, and gone through a revision of the IC system—all encouraging indicators that IC was indeed creating new organizational activities, new roles, and higher levels of OHS activity.[37]

A case study of IC in four sectors (newspaper, steel company, sawmills, and local communities) also reported generally disappointing early results, particularly (but far from exclusively) for small enterprises, to whom the objectives were not well communicated.[38] This study is consistent with individual observations of inspectors themselves that while IC continues to work well for off-shore oil (where injuries have been reduced considerably),[39] and for other high risk and sophisticated industrial sectors such as chemicals, and for already successful enterprises, performance overall has so far been disappointing.

The Norwegian results have been reciprocated in Sweden.[40] In these circumstances, with evidence that the success of legislated SMSs has been very uneven, and confronted with the many obstacles raised above, to which there seems no ready solution, it seems premature to recommend mandatory provisions applying to all employers and applying to all types of activity. For most jurisdictions, the resource implications of such a move would also be daunting. Rather, there are strong reasons for adopting an incrementalist approach, for experimenting with less interventionist and intrusive strategies, for working out means of overcoming the weaknesses while building on the strengths, before reassessing, at some future date, the merits of prescriptive legislation.[41]

A halfway house: incorporating systems into codes of practice

The main alternative to either pure voluntarism on the one hand, or mandatory legislation on the other, is to encourage the use of SMSs through the use of incentives. There is also some evidence that positive implementation of SMSs is 'linked with a requirement by demanding customers for the documentation of quality of products and working procedures'.[42] We explore these options in some detail below. However, before doing so, there is one other possibility which must be considered, namely that of a 'halfway house', which would encourage rather than mandate the adoption of SMSs by giving them legislative recognition.

In its most moderate form legislative recognition would entail a provision whereby employers are given the option of using a SMS as one means of demonstrating their compliance with their general duty of care. Such a provision already exists in many jurisdictions in which the existing general duty requirement to ensure health, safety, and welfare[43] is already specified to include (but is not limited to) providing a safe system of work. Provided an agency and/or the courts are prepared to interpret this broadly, then recognition of systems-based approaches can be brought about by administrative practice rather than legislative change.[44]

However, reliance on the general duty requirement, without more, provides very little guidance to enterprises as to what would suffice as a SMS. The most obvious way of articulating appropriate SMSs is through a code of practice. For example, the UK Management of Health and Safety at Work Regulations 1992 require employers to have effective arrangements in place for managing health and safety, and the HSE has published guidance on health and safety management.[45] Such a code should specify in some detail the elements the SMS must contain, and might also include other requirements.[46] Adoption of a SMS in compliance with a code could be regarded, prima facie, as evidence of compliance with the legislation.[47] Because of the difficulties confronting small firms in adopting SMSs, it might, however, be limited to enterprises over a certain size. But what precisely would be the status and implications of a code of practice on SMSs? It is here that the real problems of this approach become apparent. There are essentially two possibilities.

First, a code might serve as an *alternative to* other means of discharging the general duty requirement, in which case those satisfying the systems code would in effect be exempted from complying with other codes and regulations. The difficulty is that a SMS is process-based, and that implementation of a system does not guarantee outcomes. In effect, it would mean that a systems-based enterprise might, at least in some circumstances, be permitted to operate at standards lower than those of other enterprises.

Second, a code of practice on management systems might simply be a means of meeting the general duty requirement. However, since that requirement is further articulated in outcome and process standards all of which must continue to be satisfied, it might be argued that this approach is not a substantial advance on the status quo—or at least will not be an advance once the short-term goal of substantially performance (outcome)-based standards, is achieved. Consistent with this view, it is pointed out that enterprises already have very considerable flexibility in choosing how to attain prescribed outcomes. There is nothing to prevent them adopting a systems-based approach to achieve these outcomes, and a code of practice might provide useful guidance on how to do so. However, its advantages would be limited to that of evidence that the enterprise has discharged its duty of care in the circumstances of the case.[48]

On the other hand, a systems-based code might play an important role in changing perceptions. It sends out a signal that a SMS containing elements specified in the code is an approach favoured by the agency and the courts. Specifically, adoption of a SMS in compliance with a code could be regarded, prima facie, as evidence of compliance with the legislation.[49] This result would be consistent with the existing approach to codes of practice in most jurisdictions,[50] which are of evidentiary value but which are not prescriptive. If this approach were taken it should be made clear (to avoid the difficulties involved in the first position above) that the 'application of an approved code does not . . . constitute legal proof of compliance', that an employer would still have to demonstrate 'that the system used is relevant to the situation, and that health and safety risks had been addressed to the extent reasonably practicable',[51] and that adherence to the code does not absolve the enterprise from compliance with all other existing standards (for example, performance and specification standards). That is, what is required are *outcomes* and a management system is simply one way (albeit often a very good one) of achieving that result. Even so, the introduction of a code in these circumstances would, at the very least, have educational and persuasive value, and might serve to reinforce the demands of OHS managers for resources for safety systems.[52]

The first approach identified above could result in a substantial 'writing down' of the general duty requirement and in a lowering of the standards expected of enterprises adopting a SMS. For these reasons, it should not be supported. In contrast, the second approach may have persuasive value. However, we do not claim that it alone will be sufficient to bring about substantial change in approaches to workplace health and safety. Indeed, we believe much more might potentially be achieved through various types of incentives.

Incentive-based approaches

A potentially effective means of inducing enterprises to adopt a SMS is by means of incentives. Incentive-based approaches have considerable attractions. They can influence behaviour without direct intervention in the affairs of enterprises, they encourage them to seek out the most cost-effective (and often innovative) solutions to problems, they decentralize decision making to enterprises who often have better information on how to solve a problem than government, and they reduce government's enforcement costs. However, the effectiveness of incentives will depend largely upon the design and appropriateness of the particular mechanisms adopted.[53] There are a considerable range of credible options, including:

- offering administrative benefits, such as easing off on regular inspections for enterprises who agree to adopt a SMS (or blitzing recalcitrants who choose not to);
- supplying a logo or other positive publicity to enterprises who adopt a management system;
- making a SMS a precondition for granting of self-insurer status;
- giving 'up front' bonuses under the workers compensation system to firms which adopt SMSs;
- making implementation of a system a condition for tendering for major government contracts, or providing bonuses to systems-based firms;

- using subsidies to 'kick-start' a systems-based approach in firms which, by reason of their size, economic circumstances, or other factors, would otherwise be unlikely to adopt such an approach;
- reducing penalties if prosecutions take place (addressed in Chapter 5); and
- using a court order to implement a SMS in addition to any other penalty (addressed in Chapter 5).

In the following section we review the main features of, and strengths and limitations of, the various types of incentives available, before evaluating such incentives more generally, and the implications for regulatory strategy.

Administrative incentives

Administrative incentives might take one of two forms: those directed at best performers, and those directed at worst performers. Turning first to the former, it is clear that regulatory agencies have considerable discretion in enforcing the legislation for which they are responsible. Such agencies might choose to exercise that discretion in ways which provide regulated entities with incentives to adopt a SMS. Specifically, they might choose to adopt an enforcement strategy whereby companies which voluntarily adopt a SMS, audit its effectiveness, disclose the results to the agency, and take effective action to redress deficiencies, are subject to less frequent inspection and lighter regulation than those who do not.[54]

An alternative but related strategy would be for companies to be put on notice that if they fail to implement appropriate internal mechanisms (such as a SMS), then tough penalties may be imposed where violations occur which such internal mechanisms would have prevented or (put more softly) that the adoption of a management system may be viewed as a mitigating factor in exercising enforcement discretion.[55] In Holland, for example, companies which do not have an efficiently operating EMS at their disposal will in the Government's view be sooner considered for intensified enforcement activities by enforcement authorities relative to companies which are trusted to have an efficiently operating EMS.[56]

By invoking such a strategy, an agency could provide an incentive to regulated enterprises to adopt a SMS voluntarily. Just how important this incentive will be in practice, depends largely upon the credibility and effectiveness of a regulator's general enforcement policy and its capacity to escalate to the top of an enforcement pyramid invoking severe penalties for serious breaches.[57] These are matters to which we return in subsequent Chapters. Here, it should simply be noted that if enforcement is a rare event, even for recalcitrants, and if penalties themselves are perceived to be insignificant, than the proposed strategy would be untenable.[58]

Despite the advantages of incentives, it has been argued that they are so 'fraught with legal and policy obstacles' that most companies 'would not support or participate in an incentives-based auditing program'.[59] However, this view is contrary to the empirical evidence. There are precedents for the success of such a policy in OHS in the USA,[60] and in the environmental audit systems of some European countries.[61] Provided the practical obstacles identified above are overcome, there is no reason why such a policy should not be equally successful elsewhere.

One legitimate problem is that only those with a substantial initial commitment to improved OHS are likely to respond to the incentives offered. Yet it is generally those with the worst OHS records, and with least interest in voluntary change, who might achieve the largest improvements in safety performance. What is also needed, therefore, is a set of incentives targeted directly at this group. Specifically, an agency might adopt an administrative strategy whereby 'worst performance' enterprises which do not have an efficiently operating OHS management system at their disposal are sooner considered for intensified enforcement activities relative to companies which are trusted to have an efficiently operating EMS.[62]

In its more sophisticated form, government might adopt a version of incentive known as the Maine 200 Scheme. The basis of Maine 200 is the use of national injury data to target inspection sites. Under the programme, the 200 most dangerous workplaces in Maine (representing only 1 per cent of employers and 30 per cent of employees but representing 45 per cent of claims) are each sent a letter informing them that they are among the 200 most dangerous workplaces in the state and offering them a choice to either work in partnership to improve safety or face stepped-up enforcement. Partnership, for our purposes, may be taken to imply a willingness to adopt and implement a SMS. Firms that accept the invitation to partnership are assisted to develop strong health and safety programmes and also given the lowest priority for inspection.

Under Maine 200, the OSHA reviews these management plans and then inspects the company to check implementation of the safety plan. Organizations that do not comply are subjected to traditional 'wall-to-wall' inspections. Random inspections continue for companies not in the top 200 in order to maintain their incentive to comply with their legal obligations under the OSH Act.

The early evidence suggests that the Maine programme is extremely promising:

> . . . in two years, the employers self-identified more than 14 times as many hazards as could have been cited by OSHA inspectors (in part, because OSHA's small staff could never have visited all 1,300 workplaces involved). Nearly six out of ten employers in the program have already reduced their injury and illness rates, even as inspections and fines are significantly reduced.[63]

However, a detailed analysis by John Mendeloff sounds a note of caution, noting that being one of the 'top 200' is not necessarily a sign of poor safety performance, since large companies can have a high absolute number of injuries.[64] Moreover, in terms of its impact on work injuries, the picture is complicated[65] and the scheme itself is resource intensive.[66]

Nevertheless, so successful is the Maine 200 regarded as being that the Clinton Administration is now expanding the most successful features of the programme nationwide. These successful elements include: 'using workplace-specific data to help identify high-hazard workplaces, providing information to employers about effective safety and health programs, offering employers a choice in how they wish to work with the OSHA, ensuring management commitment and worker involvement, and modifying enforcement policies for high-performance employers'.[67] Aspects of the Diamond Project, conceived by the Workers Compensation Board of British Columbia in 1997, also build on the success of Maine 200, in recognizing the need to identify (through claims management experience), and work with, the worst firms in any particular industry sector and to bring them up to, and ultimately beyond, the minimum legal standard.[68]

Logos and public relations

Although strict rationalists might doubt the value of awards, logos, and other symbolic rewards, there is evidence, examined below, that such programmes can have considerable influence on enterprise behaviour. These programmes enable general managers and others with primary responsibility for OHS to provide tangible evidence, in a form acceptable to boards of directors, that their policies are working. As one of our respondents put it:

. . . its all about the shingle. How do I prove that my business is better and more macho than the next guy? I get a logo that shows I'm worth five stars for safety. That shows its working and the Board love it. The problem is they care more about the logo than the system itself, yet it's only if senior management is totally committed to the system that it will work effectively.

We return to the implications of this last comment, below.

One of the earliest and best-known schemes using this general approach is the USA OSHA's Voluntary Protection Program (VPP), which recognizes and rewards the voluntary efforts of employers and employees to go beyond the minima prescribed by law.[69] Although the VPP is not geared *explicitly* to reward a systems-based approach, it illustrates well how a programme based on the award of a logo or star, for exemplary OHS management, might work.

The VPP is designed to encourage more employers to provide outstanding worker protection. Management, labour, and the OSHA establish a co-operative relationship at a workplace, and in order to be awarded a 'star', management agrees to operate an effective safety programme that meets an established set of criteria (including corporate commitment and planning, a written programme and management system, hazard assessment, hazard correction and control, safety and health training, employee participation, and safety and health programme evaluation). This implies mechanisms to identify responsible individuals within the organization.[70] In turn, employees agree to participate in the programme and work with management to ensure a safe and healthful workplace.

In addition to the positive publicity and recognition in the community generated by VPP logos, the benefits include improved employee motivation to work safely, reduced worker compensation costs, and improvement of SMSs through internal and external review. The VPP has been operating for some years, and statistical evidence suggests that participant sites have 60–80 per cent fewer lost workday injuries. In 1992, of the 104 companies in the programme, 6 had no injuries at all. Overall, the sites had only 46 per cent of the injuries expected, or were 54 per cent below the expected average for similar industries.[71] More recent studies show lost workday case rates for VPP companies ranging from 35 per cent to 92 per cent below the national average.[72] To what extent these figures can be attributed to the VPP is unclear, given that VPP companies are initially selected on the basis of their good safety records. Nevertheless, they have been widely regarded as vindicating the VPP and, in 1997, the Workers Compensation Board of British

Columbia conceived the Diamond Project which has many similar features.

The experience of the Cooperative Compliance Program (CCP), conducted in California in the early 1980s, a precursor to the VPP, is also significant. The CPP established a three-way relationship between unions, management, and the OSHA by authorizing labour-management safety committees on seven large construction sites to assume many of the OSHA's responsibilities, while it ceased routine compliance inspections and pursued a more co-operative relationship with these companies.[73]

A critical success factor of such programmes is a set of formal criteria for determining which voluntary efforts to recognize, thus enabling all who meet the criteria to be publicly recognized. Suitable criteria include: compliance with existing regulations; a focus on a plant or facility rather than merely on a single process or product; an emphasis on management control and planning systems; appropriate employee training; and a focus on monitoring and measurable results (enabling self-evaluation and agency oversight).[74]

With all logo-based and public relations schemes there exists a tension between the desire of firms to convince the public of their OHS credentials (requiring some disclosure) and their desire to avoid revealing OHS problems or to avoid disclosing commercial confidential information. As a result, there remains a danger that such schemes will degenerate into shallow public relations exercises by industry.

For example, in the European Community (EC), in the related area of environmental protection, the Eco Management and Audit Scheme (EMAS) includes provision for what are, in effect, 'eco-audit logos' for firms which meet certain environmental benchmarks assessed by independent verifiers. Some commentators have expressed concern at the winding back of public disclosure provisions (following intense industry lobbying), fearing that the limited publicly available material (an 'environmental statement') may become merely a sanitized summary of a firm's performance. While the jury is still out on this issue, it is clear that incentive-based audit can only work in the public interest if the requirements necessary to obtain a logo are clearly stated, transparent, and effectively enforced.

Building incentives into Workers Compensation Schemes

In jurisdictions with workers compensation schemes which allow some enterprises to self insure, then regulators have an opportunity to apply considerable leverage, by requiring the implementation of a SMS as a prerequisite to the granting of self-insurer status. This approach is already being utilized by the New South Wales WorkCover Authority in Australia as a means of persuading one particular group of large enterprises to adopt a systems-based approach.[75] In order to assist firms to adapt to the systematic quality management approach which is now required of them, WorkCover has developed OHS management guidelines.[76] These have formed the basis of periodic review of industry self insurers since January 1994.

Specifically, the OHS systems of self-insuring firms are audited on a three-year cycle as part of their licensing process. The audit system contains the key components of:

- A Quality Occupational Health, Safety and Rehabilitation (OHS&R) System Model which has been aligned with International Quality Standards AS 3901. This allows the use of individually developed management systems created to suit the needs of a particular enterprise, offers the opportunity for an enterprise to integrate OHS with other areas of management, and provides a suitable framework for management; and
- A set of audit guidelines which self insurers are encouraged to use. However, as with the system model, the enterprise may use any audit methodology which meets the guidelines, demonstrates performance, and provides appropriate measurement against the model. Reporting requirements include self-audit plans and its summarized results.

An important element of the policy requires performance in any area of OHS&R which is below the statutory minimum to receive prompt attention and correction. A failure to address serious or continuing deficiencies will be detrimental to the ability of an enterprise to maintain a self-insurer licence.

Because this strategy has only been in operation for a relatively short time, no firm conclusions can yet be drawn about its success. Anecdotal evidence, however, suggests that a significant number of enterprises with self-insurer status still have an undistinguished record on safety. If this approach is to maintain its credibility, it may be necessary to suspend self-insurer status for those enterprises which, after an appropriate phase-in period, have still not implemented credible and effective SMSs.

It would also be possible to provide incentives for the large majority of enterprises who are not self insurers, through a bonus scheme for those who adopt a systems-based approach to OHS. For example, rebates might be given on workers compensation premiums for employers who apply enterprise SMSs or who agree to implement the recommendations of safety management audits.

It might be thought that such an approach is unnecessary if a system of differential premiums is in operation, providing in itself sufficient incentives for improved safety performance, whether through a systems-based approach or otherwise. Unfortunately, this is not entirely correct, because a substantial component of workers compensation premiums is based on factors over which the employer has no current capacity to control.[77] Accordingly, there is a case for much more direct incentives in the form of up-front payments for those who commit themselves to a systems-based approach.[78]

Making a management system a prerequisite for tendering for certain contracts

Extending the self-insurer principles described above, New South Wales WorkCover, in conjunction with a construction policy steering committee, has provided the construction industry with specific guidelines which address the objective of proactive OHS&R management.[79] Subsequently, tenders for major construction projects have only been accepted by the government from contractors that have an accredited OHS management system which conforms to the guidelines. From July 1995, in particular, a condition of contract has been the preparation of an acceptable site-specific project OHS&R Management System and the audit of the contractor's OHS&R systems as outlined above.

A particular and intractable problem with making a SMS a prerequisite for tendering is that there is only a very limited opportunity to test the system or to verify its effectiveness or credibility *in operation*. All that is possible is to insist on the creation of a paper system, prior to the tender itself, and to have the systems subsequently audited by a government authority or its nominee to ensure that the programme is in fact being adhered to. However, by this time it is too late to withdraw the contract. There is therefore a considerable risk that enterprises will view this requirement as just another formality to satisfy in order to bid for the contract, and that it will produce the cynical manufacture of largely meaningless paperwork rather than genuine changes in enterprises' approaches to OHS.

This outcome is especially likely in the case of the construction industry, where teams are put together for individual projects which largely disband after the project is over. The incentives for training the workforce in a systems approach, for altering perceptions at grass-roots level, and for attempting to achieve cultural change within the workplace, are accordingly far less than in more stable work environments. Moreover, since the names and structures of companies themselves change frequently, a strategy of penalizing, in subsequent tenders, those who previously adopted mere 'paper systems', may be only partially effective.

Even so, it is arguable that, at least in some circumstances, an insistence on a SMS as a prerequisite to tendering may achieve far more than token compliance and paper systems. This approach at the very least identifies for management a potentially different approach to safety, and gives them a road map of what is required. How many of them choose to use that map remains to be seen, although early indications are that both contractors *and subcontractors*[80] are having their OHS standards raised by this approach. It may be that with further experience many of these problems can be overcome. For example, it may be possible to track key members of management to see whether their past commitments to apply a SMS have been honoured, and if not to disallow future tenders involving those individuals as leading members of project teams and thereby provide disincentives to 'repeat players' to renege on their commitments. The policy's effectiveness in the construction industry is currently being assessed. At present, all that can be said is that analogous experience in the QA area shows that purchasing policies are indeed effective in encouraging continuous improvement on the part of suppliers.[81]

Subsidizing management systems

One approach to encouraging the adoption of SMSs, which is understandably supported by industry itself, is that participating enterprises should be provided with government subsidies. In theory, there are objections to using subsidies for such purposes. In economic terms, it is important to achieve 'efficient' resource allocation by ensuring that firms 'internalize externalities' thereby ensuring that the costs of workplace injury and disease are borne by the person or enterprise that causes them, rather than passed on to the taxpayer. However, in practice, we know that the start-up costs of developing a management systems approach are considerable, and that *in the short term* such systems are beyond both the financial and technical means of many SMEs. Since in practice the majority of the costs of workplace injury and disease are borne by either the taxpayer or by the victims and their families,[82] it may well be in the public interest to provide seed money to 'kick-start' a systems-based approach in smaller enterprises.

The most appropriate vehicle for providing such subsidies may be the workers compensation system (in those jurisdictions where it exists), thereby avoiding the implication that firms are being paid to do what the law requires them to do anyway. In contrast, under workers compensation such subsidies can be seen as an investment in reducing premiums in future years.[83] Certainly such an approach could be adopted on a trial basis to see how well it worked and whether the costs exceeded the benefits. Models already exist in related fields such as environmental protection.[84]

Supply side pressure

In the longer term, it may be that enterprises will have a further incentive to adopt a systems-based approach, over and beyond anything that government directly can provide. That is, it may be that other major enterprises will refuse to deal with them unless they adopt such a system. We can already see the beginnings of this approach in respect of EMSs. Many European customers in some industries now insist on dealing only with ISO 9000/14000 accredited companies. Accordingly, accreditation for implementing a systems approach can offer a competitive advantage and serve to strengthen marketing credibility internationally. Enterprises increasingly find pressure exerted upon them to adopt such standards, through customers, competitors, and shareholders. If this happens in the area of OHS (which, at the time of writing, still remains in doubt), then a happy coincidence of private sector and public incentives may result in the very widespread adoption of SMSs.

Evaluating incentive-based approaches

The main advantages of incentive-based schemes are that they encourage firms to exceed the bare minimum required by law, and that some firms which would not otherwise participate in a voluntary scheme are persuaded to do so. If appropriately designed, incentives have the very considerable advantage of changing behaviour without direct intervention or mandating such change. As such they have much greater political acceptability and generate far less resentment and resistance on the part of enterprises themselves.

Nevertheless, much depends on how individual incentive-based schemes are designed. For example, since management systems are expensive to set up initially, a major challenge is to devise incentives that are sufficiently attractive to overcome the initial resistance of enterprises to adopting them. This issue arises in a particularly acute form in respect of schemes such as the USA OSHA's VPP, whose requirements are so onerous (being very demanding and reserved for companies with exemplary OHS records) that only the very best firms have an opportunity to qualify. It is arguable that a more modest programme should be devised for firms unable or unwilling to meet these standards but who might nevertheless respond to other incentives and voluntary initiatives.

A further and crucial issue concerns strategies that rely on targeting particular enterprises (for example, Maine 200). For such strategies to work effectively, it is important that the regulatory agency has accurate information upon which it can rely in order to identify the target firms. In New South Wales, one consequence of merging OHS and workers compensation under WorkCover has been the opportunity to utilize a wider range of information to identify problem issues and workplaces.[85] Nevertheless, there is reason to doubt whether those sources of information are anything like adequate for effective targeting, an issue to which we return in another context in the next Chapter.

Perhaps the most serious problem with incentive-based approaches is that which was raised in a more acute form in respect of mandatory approaches earlier: namely, that to the extent that enterprises feel pressured to adopt them purely in order to get some form of reward, then this may produce an instrumentalist response, where the enterprise does just enough to jump through the hoops necessary to satisfy the regulators, but is not actually committed to the system or the benefits it might bring—the problem of 'paper systems' referred to earlier. As Frick puts it: 'Systems Control [i.e. government regulation of IC by the enterprise] can (at best) enforce what it regulates, which are the procedures of IC. Compared with other systems of quality control, IC is—and can, with the inherent limitations of regulation, only be—simply specified and audited, while its OHS goal is complex. Legal compliance with SC regulation of IC is thus easy, without having to manage the quality of OHS . . . *Only management can ensure that the procedures of IC are actually used to improve OHS.*'[86]

This problem cannot be lightly dismissed. Some regulators themselves express similar reservations about the capacity of management systems to influence those who are uninterested in adopting them. However, there may be at least a partial resolution, if one distinguishes, conceptually, between volunteers (those who recognize the benefits of management systems and embrace them willingly), recalcitrants (those who do see costs, rather than benefits, from the adoption of management systems, and who adopt them only as a matter of expedience if the incentives offered are sufficiently powerful), and incompetents (i.e. those who could benefit from improved OHS but fail to take the initiative to achieve such improvements through ignorance, lack of in-house competence or organizational capability to understand or implement such improvements).[87]

Volunteers will embrace the systems route with full commitment, and the greatest potential for the success of SMSs lies with this group. Many of them would introduce SMSs even in the absence of incentives, but the incentives may tip the balance in favour of this approach at the margin. Recalcitrants/the wilfully disobedient are most likely to take an instrumentalist approach, doing the least possible to meet the regulator's requirements, and, if needs be, producing mere 'paper systems' in order to get the incentives offered. Unless the regulator develops a sophisticated capacity to distinguish paper and real systems, this latter group are unlikely to produce significant improvements in OHS performance as a result of adopting OHS management systems. However, the third group, the incompetents, may achieve much better results, under an incentive-based approach, provided they can be persuaded of the benefits of this approach and of how, in practical terms, to implement it. As Frick argues,[88] this may involve:

- first, enforcing requirements (or in our view, providing incentives) and then trying to achieve internal compliance;
- second, reducing resistance. This involves demonstrating that IC is uncomplicated and compatible with overall management; and
- third, increasing motivation. This involves arguing that systematic IC is: efficient (that the money employers invest in OHS would be better guided by IC); ethical (IC is necessary to fulfil ambitions not to hurt employees); and profitable (improved OHS can be a sound investment, with the help of IC).

We have argued earlier that there are many investments that employers could rationally make in terms of OHS which they do not at present make—which we explain in terms of bounded rationality.[89] If so, then there may be much to be gained by providing incentives (coupled with persuasion and education as above, and the utilization of accounting systems which enable enterprises to evaluate their own health and safety performance[90]) to employers to motivate corporate decision makers to embrace a systems-based approach to OHS. First, the incentives (and OHS-oriented accounting systems) may help overcome limited corporate horizons and bounded rationality (bringing OHS forcibly to their attention, challenging a 'business as usual' mentality, and, through the system, locking in the changes). Second, the incentives may, at least in some cases, make it rational to embrace a systems approach where previously it might not have been. Finally, there is at least the possibility of turning the recalcitrants into volunteers, by means of persuasion and education along the lines that Frick suggests. However, it is unlikely that compulsion alone can overcome the 'paper systems' problem.

There will of course remain uncertainties about how enterprises will respond to incentives, particularly public relations-based incentives (such as logos) or administrative incentives (such as a promise to 'ease off' on regular inspections). These are empirical questions, and the most responsible path is to experiment on a limited basis, and to proceed incrementally, before devoting substantial resources or making a major change in policy direction. Our proposed incentives-based Track Two approach described in Chapter 5 adopts precisely such an incrementalist approach. At least initially, only the best enterprises, in OHS terms, are likely to participate in a systems-based approach so that one will be dealing with volunteers in circumstances which have the best chances of success. Only to the extent that a systems-based approach demonstrates its success with this group (and if necessary with modifications) will it be extended to others.

However, even at this stage there is reason for optimism as to the viability of this approach. Regulators report that in Australia the experience is that many larger companies, which have experienced poor OHS performance, have been very receptive to advice as to how to set up a systems-based approach, and have subsequently done so, even in the absence of direct pressure.

More substantive empirical evidence about the effect of incentives exists in the closely related area of environmental protection. There, the evidence is very largely positive, though much depends upon the design of the particular incentive programme. However, in broad terms, where the incentives for industry to participate are clear and substantive, participants see economic benefits, competitive advantage or the flexibility to choose means to achieve specified goals, then prospects for success are good.[91]

Implications for regulatory and administrative policy: Towards a two-track approach

Although we have argued that the adoption of a systems-based approach has considerable advantages over traditional regulation, it would not be appropriate to encourage or insist upon such an approach in all circumstances. As we have indicated, such an approach has the greatest potential where it would enable enterprises to make economic gains—where, in short, there is a coincidence between what is good for safety and what is good for private profit. Even here we have indicated that there may be resistance to adopting a systems-based approach, arising either from ignorance, incompetence, or irrationality, or from the gap between short-term and long-terms gains.

In respect of the latter, the extent to which a firm may be willing to sacrifice short-term profit for long-term dividends will depend on a number of factors. First, it is clear that those firms which are economically marginal cannot afford the luxury of a long-term view. Such firms tend also to be more heavily reliant on old, inefficient plant and, as a result, more commonly expose their workers to higher risks of injury and disease. In those industries whose technology does not change fast, and where it is hardest to justify investing the capital in new plant, there is little economic rationality in voluntarily moving beyond compliance. Industries such as waste management and transport, with low profit margins and a 'bottom line' mentality based on squeezing the last drop of short-term profit in a highly competitive market, will similarly see little attraction in a systems-based approach unless almost immediate dividends can be demonstrated.

By contrast, in circumstances where companies have substantially higher profit margins and a rapidly changing technology, they are in a far better position to take OHS initiatives yielding only long-term dividends. Moreover, larger firms, in comparison to their small and medium-size rivals, often have the technological capacity and the economies of scale to make system-based improvements both technically feasible and economically realistic.

Certainly the culture of the individual enterprise can also be important but only within economic limits. As Haines, drawing from her own extensive empirical research, points out: 'In principle, while cultural influences may be capable of arising independently, in fact they tend to have effect only if economic circumstances have generated "breathing space" within which they can work.'[92]

There are also other reasons why systems-based approaches are less appropriate (though not necessarily wholly inappropriate[93]) for most SMEs. In particular, the costs associated with developing an individual system for each workplace, and the lack of specialized resources, may seem prohibitive for many small employers.[94] Current OHS management systems may also be too complex and time-consuming for a small business which is often not a member of a trade association. The orientation of such firms is often towards compliance rather than towards best practice and it appears that these firms benefit most strongly from direct contact with government agencies.[95]

Again, systems-based approaches lend themselves more readily to some industry sectors than to others. For example, in respect of the control of major hazardous facilities, a systems-based approach is particularly appropriate. As one senior British regulator put it: 'visual inspection is a thing of the past in high hazard, large, complex facilities. You can't walk around a chemical plant and see much. All there is shiny tanks and pipework.' In consequence, the HSE rely very largely upon auditing the management system and on interviewing personnel based on that system.

In contrast, the construction industry has many features that make the introduction of a systems-based approach problematic. These include the fact that standard employment is daily hire or for the length of the project, the large number of small employers, the lack of expertise in OHS within the industry, the fact that risk assessment often falls on external consultants so that employers have less involvement and take less responsibility, the fact that construction is project-based, involving differing and multiple teams of subcontractors with no long-term relationship, and the lack of opportunity for employers and employees to develop mutual relationships of trust.[96]

It follows that a systems-based approach will suit some enterprises, and some circumstances, far more than others, and that it will not necessarily be a preferred path to improved OHS performance for all enterprises. Indeed, probably the majority will not choose this approach voluntarily, and many may not respond to incentives. Even if one were to contemplate mandatory implementation of a management system, it is questionable to what extent this approach could be imposed on small or medium-sized companies, or across all industry sectors.

One means of recognizing the limits of a SMS approach would be to confine it to enterprises in certain sectors (or of certain sizes) which lend themselves most readily to this approach. A sector-specific approach would be consistent with aspects of the UK approach where, although there is no legislative requirement to adopt a systems-based approach (beyond that implied in implementing the EU Framework Directive), the HSE in effect regulates different sectors differently. Chemicals and major hazardous substances, and hospital trusts, are essentially expected to have SMSs in place and are audited on the basis of these systems, while other sectors, including construction, are inspected by traditional means. Not coincidentally, it was the technically advanced and economically well resourced North Sea oil companies, where quality control generally was already well advanced, that provided the first, and successful, model for the incorporation of SMSs through regulation.

However, while more companies in some sectors are likely to success-fully embrace a management systems approach than in others, this is not in itself a reason for confining systems-based regulation to these sectors, because it would unnecessarily preclude some enterprises in other sectors from taking advantage of Track Two. Evaluations of the Norwegian IC approach also suggest that successful enterprises are able to cope with the requirements of IC, irrespective of industry sector.[97] Australian evidence suggests that even some enterprises in the construction sector (which the UK prefers to address through traditional regulation) are capable of successfully adapting to this approach,[98] and the anecdotal evidence of UK inspectors suggests that even a limited number of SMEs may be able to successfully adopt this approach, provided the quality and leadership of management is high.

In conclusion, not all workplaces are alike and not all employers are similarly motivated. Accordingly, SMSs are unlikely to work 'across the board', and a 'one fits all' approach to regulation would be both ineffi-cient and ineffective. As we saw earlier, even in a small and homog-eneous country such as Norway, this approach is already experiencing serious problems, which would likely be exacerbated in larger and more diverse jurisdictions. From this we conclude that two very different types of regulation will be necessary: one for those enterprises which adopt a systems-based approach, and another for those which do not. We explore the implications of this conclusion for the design and adminis-tration of regulation in the following Chapters.

NOTES

1   Industry Commission, Australia, *Work, Health & Safety: Inquiry into Occupational Health & Safety*, Volume II Appendices, Report No. 47 (AGPS, Canberra, 1 Sept. 1995), p. 189.

2   See N. Gunningham, 'Beyond Compliance: Management of environmental risk', in B. Boer, R. Fowler, & N. Gunningham (eds.), *Environmental Outlook: Law and policy* (ACEL, Federation Press, 1994).

3   N. Gunningham, *Safeguarding the Worker* (Law Book Company, Sydney 1984), Ch. 11. See also S. Dawson, P. Willman, A. Clinton, & M. Bamford, *Safety at Work: The limits of self-regulation* (Cambridge University Press, 1988); and H. Genn, 'Business Responses to the Regulation of Health and Safety in England' (1993) 15 *Law and Policy* 219–33.

4   A. Rappaport & M. Flaherty, *Multilateral Corporation and the Environment* (Centre for Environmental Management, Tufts University, 1991).

5   Of course, some commitment to OHS priorities will have short-term pay-offs. Many accidents can be prevented at low cost (for example by improvements in good housekeeping) while in other circumstances the potentially very high costs of workplace injury and disease to the employer may also provide a compelling reason, even in the short term, for making a major investment in OHS.

6   B. Fisse & J. Braithwaite, *Corporations, Crime and Accountability* (Cambridge University Press, Sydney, 1993), Ch. 3.

7   See below pp. 91 and 107 (bounded rationality).

8   See O. Saksvik, *The Norwegian Internal Control Regulation*, SINTEF IFIM, Institute for Social Research in Industry, Trondheim, Workshop on Integrated Control/Systems Control, Dublin, 29–30 Aug. 1996.

9   S. Tombs 'Law, Resistance and Reform: "Regulating" Safety Crimes in the UK' (1995) 4 *Social and Legal Studies* 343–65 at 344.

10   For example, 1990 data suggests that only 25% of workplaces have a comprehensive approach to OHS (indicated by the presence of workplace health and safety committees and/or health and safety representatives, written health and safety policies, a person with primary health and safety responsibilities, and appropriate record keeping systems). See B. Pragnall, *Occupational Health Committees in NSW: An analysis of the AWIRS data*, ACIRRT, Working Paper No. 31, Sydney, Mar. 1994.

11   D. Else, *Enhanced Cohesion and Co-ordination of Occupational Health and Safety Training in Australia*, Report to the Minister for Industrial Relations, Commonwealth of Australia (AGPS, Canberra, March 1992).

12   P. James, 'Reforming British Health and Safety Law: A framework for discussion' (1992) 21(2) *Industrial Law Journal* 87; D. Dawson, et al., 'Is Your Safety Policy Accurate?' (1984) Aug. *Health and Safety at Work*, 51–4; and P. Barrett & P. James, 'Do Safety Policies Assist Workplace Safety?' (1982) July *Health and Safety at Work* 24–2.

13   Also see Dutch approach where, although the internal EMS is voluntary, it is intended in future to be underpinned by a statutory requirement for companies in specified categories to produce an annual environmental report, which will be compulsory. This 'relates to reports to the competent

authority on the company's present and future emissions (planned emission reductions) including waste. The obligation applies to emissions which can be regulated and therefore assessed in the framework of the environment alliance' (T. C. R. van Someren, J. van der Kolk, K. ten Have, & P. T. Calkoen (KPMG Environment/IVA), *Company Environmental Management System, Interim Evaluation 1992,* Summary on behalf of Ministerie van Volkshuisvesting Ruimtelijk Ordening en Milieubeuheer, The Hague/ Tilburg Jan. 1993). Arguably, the information needed to produce the latter will be of such a nature and degree of detail as to make it rational for companies to develop an EMS as a means of discharging their obligation to provide the environmental report.

14   M. Aalders & T. Wilthagen, 'Moving Beyond Command and Control: Reflexivity in the regulation of occupational safety, health and the environment' (1997) 19(4) *Law and Policy* 415–44.

15   It is significant that over half of the chemical releases, and over half of the injuries reported in an OHSA survey of chemical producers and chemical users, 'occurred in facilities that had implemented state of the art process safety management practices. However, that survey indicates that no company had all process safety management elements in place' (N. A. Ashford et al., 'The Encouragement of Technological Change for Preventing Chemical Accidents: Moving firms from secondary prevention and mitigations to primary prevention'. A Report to the US Environmental Protection Agency, Centre for Technology, Policy and Industrial Development at MIT, Cambridge, Mass., July 1993, I-7 and Appendix A).

16   Ibid. I-8.

17   B. Fisse & J. Braithwaite, *Corporations, Crime and Accountability* (Cambridge University Press, Sydney, 1993), Ch. 3.

18   It has been estimated (Industry Commission, Australia, *Work, Health & Safety: Inquiry into Occupational Health & Safety* Volume I Report, Report No. 47 (AGPS, Canberra, 1 Sept. 1995), p. xix) that the costs of work related injury and disease in Australia are in the order of $20 billion a year, and that these are borne approximately 40% by employers, 30% by workers and their families, and 30% by the community in social benefit and health subsidies. Since a substantial proportion of these costs are externalized onto workers, their families or the state, there are good arguments, in conventional economic theory, for action to internalize them onto the party who causes these costs, or, where more than one party is involved, onto the 'least cost avoider' (see N. Gunningham, *Safeguarding the Worker* (Law Book Company, Sydney 1984), Ch. 12). See also Industry Commission, Australia, *Work, Health & Safety: Inquiry into Occupational Health & Safety,* Volume II Appendices, Report No. 47 (AGPS, Canberra, 1 Sept. 1995), pp. 99–105. Only if the administrative costs of intervention exceeded the benefits gained, would this intervention be unjustified. The costs of intervention to enforce a systems-based approach are not quantifiable at this stage but are unlikely to substantially exceed those arising from conventional forms of OHS regulation (which are already perceived as socially and economically justified).

19   Again, the arguments in favour of state intervention have been rehearsed
      at length elsewhere and will not be repeated here. See further N.
      Gunningham, *Safeguarding the Worker* (Law Book Company, Sydney, 1984).
20   K. Kaasen, 'Post Piper Alpha: Some reflections on offshore safety regimes
      from a Norwegian perspective' (1991) 9(4) *Journal of Energy and Natural
      Resource Law*, 281 at 286. The Cullen Report (1990) specifies that The
      Safety Case should demonstrate that certain objectives have been met
      including: (1) that the safety management system of the company (SMS)
      and that of the installation are adequate to ensure that: (a) the design; and
      (b) the operation of the installation and its equipment are safe; (2) that the
      potential major hazards of the installation and the risks to personnel
      therein have been identified and appropriate controls provided; and (3)
      that adequate provision is made for ensuring, in the event of a major
      emergency affecting the installation, (a) a temporary safe refuge and (b)
      their safe and full evacuation (Lord Cullen (Chairman), *Piper Alpha Inquiry*
      (HMSO, London, 1990)).
21   Cullen found that existing safety regulations were unduly restrictive, im-
      posing 'solutions' rather than 'objectives' which appeared to be technologi-
      cally out of date (Lord Cullen (Chairman), *Piper Alpha Inquiry* (HMSO,
      London, 1990), pp. 390–1 (para. 17.1)).
22   Regulation 4 provides that: 'every employer shall make and give effect to
      such arrangements as are appropriate, having regard to the nature of
      his activities and the size of his undertaking for the effective planning,
      organisation, control, monitoring and review of the preventive and protec-
      tive measures.'
23   See P. Lindoe, *Quality Management Systems: Between regulation and self-regulation*,
      Workshop on Quality Management, Kristiansand, Apr. 1997.
24   Provisions of the National Board of Occupational Safety and Health on
      Internal Control of the Working Environment (AFS, 1992:6).
25   T. Larsson, *Systems Control Development in Sweden*, SAMU, Institute for Social
      Science, Uppsala, Workshop on Integrated Control/Systems Control,
      Dublin, 29–30 Aug. 1996.
25(a) C. Woolfson, J. Foster, & M. Beck ' *"Paying for the Piper": Capital and Labour
      in Britains' Offshore Oil Industry*' (Mansell Publishing, 1997). The authors
      explain that there is also concern that this regime may have generated
      increasing pressures not to report lost time accidents (ie data manipulation
      which may in itself distort statistical assessments of its success. However,
      others have provided more positive evaluations. We explore the empirical
      evidence relating to mandatory approaches further in the next section
      below.
26   Dutch Lower House Parliament (1993), Letter from Minister, p. 10.
27   K. Frick, *Enforced Voluntarism—purpose, means and goals of systems control*, Na-
      tional Institute for Working Life, Solna, Workshop on Integrated Control/
      Systems Control, Dublin, 29–30 Aug. 1996.
28   J. A. Sigler & J. E. Murphy (eds.), *Corporate Lawbreaking & Interactive Compli-
      ance: Resolving the regulation-deregulation dichotomy* (Quorum Books, US, 1991),
      p. 158.

29   V. Blewett & A. Shaw, 'Quality Occupational Health and Safety?' (1996) 12(4) *Journal of Occupational Health Safety: Australia and New Zealand*, 481–7 at 485–6.

30   Dutch Lower House Parliament (1993), Letter from Minister, see above pp. 74.

31   Dutch Interim Report 41 also argued that blanket regulations make it more difficult on the other hand to apply legal tests, giving rise to problems of enforcement. If laws are very general how do we credibly make them stick? This questions whether government can ever credibly test if a system is in fact genuine (M. Aalders, 'Regulation and In-Company Environmental Management in the Netherlands' (1993) 15(2) *Law and Policy* 75).

32   Queensland Department of Employment, Vocational Education, Training and Industrial Relations, *Occupational Health & Safety*, Submission Number 79 to the Industry Commission, 30 Sept. 1994, p. 9.

33   This is currently the Danish view, which may be contrasted with that of Norway and Sweden, that IC can be applied to enterprises of all sizes.

34   O. Saksvik & K. Nytro, *Implementation of Internal Control of Health, Environment and Safety: An evaluation and a model for implementation* (SINTEF IFIM Institute for Social Research in Industry, Trondheim, Norway, 1995).

35   They identify the basics of such a system as concepts like mission statements, objectives, plans, strategies, feedback, and evaluation (see O. Saksvik & K. Nytro, 'Implementation of Internal Control of Health, Environment and Safety in Norwegian Enterprises,' Paper presented to *Seventh European Congress on Work and Organisational Psychology*, Gyor, Hungary, Apr. 1995; and O. Saksvik & K. Nytro, *Implementation of Internal Control of Health, Environment and Safety: An evaluation and a model for implementation* (SINTEF IFIM Institute for Social Research in Industry, Trondheim, Norway, 1995)).

36   P. Lindoe, *Quality Management Systems: Between regulation and self-regulation*, Workshop on Quality Management in Service VII, Kristiansand, Apr. 1997, p. 9.

37   O. Saksvik, *The Norwegian Internal Control Regulation*, SINTEF IFIM, Institute for Social Research in Industry, Trondheim, Workshop on Integrated Control/Systems Control, Dublin, 29–30 Aug. 1996.

38   K. Flagstad, *The Functioning of the Internal Control Reform*, Doctoral Thesis, NTH, Trondheim, 1995.

39   The Act of Petroleum Activities 1985, and accompanying regulations, established a regime for safety of off-shore oil. The success of IC in this 'elite' industry, has been documented in P. Lindoe, *Internal Control — Conflicting Interests between Bureaucratic Control and Participatory Co-operation*, Doctoral Thesis, University of Trondheim, Norwegian Institute of Technology, 1992; and in P. Lauredsen, Working Paper, Stavanger College, Rogaland Research, Stavanger, 1995.

40   Labour inspectors and the Board of Occupational Safety and Health suggest that IC has, in general, been favourably received in larger private and in public enterprises. However, even here, problems remained in involving middle management and local supervisors in problem solving

activities at local level: the essence of effective systems control (T. Larsson, *Systems Control Development in Sweden*, SAMU, Institute for Social Science, Uppsala, Workshop on Integrated Control/Systems Control, Dublin, 29–30 Aug. 1996). Among small businesses the results have been less encouraging (C. Frostberg, *Internal Control of the Working Environment*, Analysis and Planning Division, National Board of Occupational Health and Safety, Sweden, December 1994, p. 5). The latter do not have the knowledge, assets, or funds to meet the requirements and often find the rules difficult to understand, and the procedures overburearcratic (National Board of Occupational Health and Safety, *Internal Control of the Working Environment in Small Enterprises: A monitoring project of the occupational health and safety administration, 1995–1997* (Stockholm, March 1997)). Case studies of 15 small firms by Johansson concluded that changes in regulation had very little impact on the motivation to deal with work environment issues among small firm heads (B. Johansson, *The motivation for working environment activities among heads of small firms* (TULEA, Lulea, 1994)).

41 There may be limited exceptions, in exceptionally hazardous industries, where a legislative requirement may be justified by the exceptionally hazardous nature of a particular industry. Extremely hazardous facilities are a case in point.

42 P. Lindoe, *Internal Control* (Stavanger College, Rogaland Research, Stavanger, 1995).

43 See, for example, Health and Safety at Work Act 1974, UK, s. 2.

44 This approach has already been adopted in New South Wales, where WorkCover has stated that: 'a firm can attempt to demonstrate that it has met its duty by reference to (and demonstrated compliance with) its safety system' (New South Wales Government, *Occupational Health & Safety*, Submission Number 397 to the Industry Commission, 1994, p. 7).

45 See HSE (UK), HS(G)65 (HMSO, London, 1991).

46 For example, the Industry Commission, Australia, *Work, Health & Safety: Inquiry into Occupational Health & Safety*, Volume 1, Draft Report (AGPS, Canberra, 12 Apr. 1995) recommended that 'safety management systems be approved by the regulatory agency on conditions that: sufficient consultation had taken place between the employer and his or her employees; the safety management system addresses all the risks to health and safety at the workplace in question; the safety management system fulfils the duty of care for the employer in question; and the safety management system meets relevant exposure and process requirements'. There are some considerable problems with the concept of agency endorsement which we return to later, and certainly the Commission itself substantially qualified these views in the final report. See further Ch. 3 below on Industry Commission recommendation that system be approved by regulatory agency.

47 See Industry Commission, Australia, *Work, Health and Safety: Inquiry into Occupational Health and Safety*, Volume Report I, Report No. 47 (AGPS, Canberrra, 1 Sept. 1995). This result would be consistent with the existing approach to codes of practice in most jurisdictions, which are of evidentiary value but which are not prescriptive (South Australian Government,

Occupational Health & Safety, Submission Number DR275 to the Industry Commission, 11 May 1995, p. 13).

48   The Industry Commission, Australia, *Work, Health and Safety: Inquiry into Occupational Health and Safety*, Volume I Report, Report No. 47 (AGPS, Canberra, 1 Sept. 1995), pp. 86–7, emphasizes the obstacles to adopting a system for those companies which already face regulatory overload.

49   See Industry Commission, Australia, *Work, Health and Safety: Inquiry into Occupational Health and Safety*, Volume I Report, Report No. 47 (AGPS, Canberra, 1 Sept. 1995).

50   But compare Queensland Workplace Health and Safety Act 1995, s. 38.

51   Industry Commission, Australia, *Work, Health and Safety: Inquiry into Occupational Health and Safety*, Volume I Report, Report No. 47 (AGPS, Canberra, 1 Sept. 1995), p. 86.

52   Moreover, such a code would be read subject to the defence contained in s. 53 of the OHSA of 'reasonable practicability'. As a result, the code would have considerable flexibility in that, for example, while a substantial degree of sophistication and complexity would be reasonable for a large firm, far less would be required of a small enterprise.

53   See Note: Encouraging 'Effective Corporate Compliance' (1996) 109 *Harvard Law Review*, 1783–800 for an examination of means of dealing with the issue of incentives including the exercize of prosecutorial discretion.

54   For example, under the US OSHA's Voluntary Protection Program, successful applicants are no longer subject to routine inspection.

55   See further the discussion on sentencing guidelines in Ch. 5.

56   Documents, Lower House (Parliament) 1988–9, 20633, nr. 3, *Internal Company Environmental Management*, Summary of the note for Dutch Lower House, Netherlands, English translation, 30 Aug. 1989, p. 3.

57   See further pp. 121–9 below.

58   See Note: Encouraging 'Effective Corporate Compliance' (1996) 109 *Harvard Law Review* 1783–800 at 1794.

59   United States Environment Protection Agency (USEPA), *Environmental Auditing Policy Statement*, 51 Fed. Reg. 25004, 9 July 1986. The main objection, as the USEPA has argued, is that a reduction of enforcement efforts or inspections for those who perform environmental audits would eliminate the current incentive for them to perform effective audits and correct deficiencies.

60   See J. Rees, *Reforming the Workplace: A study of self-regulation in occupational health and safety* (University of Pennsylvania Press, 1988); and J. Pendergrass & J. Pendergrass jun., 'Beyond Compliance: a call for EPA regulation of voluntary efforts to reduce pollution' (1991) 21 *Environmental Law Reporter*.

61   Dutch Lower House Parliament, Letter from Minister, 1993.

62   M. Aalders, 'Regulation and In-Company Environmental Management in the Netherlands' (1993) 15(2) *Law and Policy* 75.

63   B. Clinton (President) & A. Gore (Vice-President), *The New OSHA: Reinventing Worker Safety and Health*, National Performance Review, White House, Washington DC, May 1995, p. 4.

64   J. Mendeloff, *A Preliminary Evaluation of the 'Top-200' Program in Maine,*

Report to the Office of Statistics, Occupational Safety and Health Administration, US Department of Labor, Washington DC, Mar. 1996; and C. Needleman, 'OHSA at the Crossroads: Conflicting Frameworks for Regulating Occupational Health and Safety in the United States, Paper delivered to *Policies for Occupational Health and Safety Management Systems and Workplace Change Conference*, Amsterdam, 21–4 Sept. 1998.

65 As Needleman points out: 'Between 1991 and 1994, the total number of workers' compensation claims in Maine fell sharply, by almost 41%. However, the Maine 200 program cannot claim all the credit. Even though the decline was more rapid among participating firms, most of the drop preceded the start of the Maine 200 program and probably reflected changes in Maine's economy and workforce. Also, during the program's first year, there were major changes in Maine's worker compensation laws that almost certainly account for some of the decline in claims. Labor advocates note that the program itself created disincentives for workers to report work injuries for compensation, so that a drop in claims does not necessarily mean fewer actual safety problems' (C. Needleman, 'OHSA at the Crossroads: Conflicting Frameworks for Regulating Occupational Health and Safety in the United States, Paper delivered to *Policies for Occupational Health and Safety Management Systems and Workplace Change Conference*, Amsterdam, 21–4 Sept. 1998, p. 13).

66 C. Needleman, 'OHSA at the Crossroads: Conflicting Frameworks for Regulating Occupational Health and Safety in the United States,' Paper delivered to *Policies for Occupational Health and Safety Management Systems and Workplace Change Conference*, Amsterdam, 21–4 Sept. 1998.

67 B. Clinton (President) & A. Gore (Vice-President), *The New OSHA: Reinventing Worker Safety and Health*, National Performance Review, White House, Washington DC, May 1995, p. 4.

68 K. M. Rest and N. A. Ashford, *Occupational Health and Safety in British Columbia* (Ashford Associates, Cambridge, Mass., 1997).

69 J. Pendergrass & J. Pendergrass jun., 'Beyond Compliance: A call for EPA regulation of voluntary efforts to reduce pollution' (1991) 21 *Environmental Law Reporter* 10305.

70 See further Chs. 4 and 5 below.

71 OSHA, *Voluntary Protection Program (VPP)* (Washington DC, 1994).

72 C. Jeffress, Assistant Secretary, Occupational Safety and Health Administration, *Statement Before the Subcommittee on Oversight and Investigations*, Committee on Education and the Workforce, US House of Representatives, 8 May 1998.

73 J. Rees, *Reforming the Workplace: A study of self-regulation in occupational health and safety* (University of Pennsylvania Press, 1988). See also C. Jeffress, Assistant Secretary, Occupational Safety and Health Administration, *Statement before the Subcommittee on Oversight and Investigations*, Committee on Education and the Workforce, US House of Representatives, 8 May 1998.

74 J. Pendergrass & J. Pendergrass jun., 'Beyond Compliance: A call for EPA regulation of voluntary efforts to reduce pollution' (1991) 21 *Environmental Law Reporter* 10305.

75   WorkCover Authority (NSW) (Working Party of Representatives of the NSW Self Insurers and the WorkCover Authority), *A Quality Approach to Occupational Health and Safety and Rehabilitation for Self Insurer: Quality OH&S system model and system audit guidelines,* Internal WorkCover Report (undated).

76   Ibid.

77   These include disease cases (whose origins are in the past), stress cases (which are only partly connected to the mainstream safety programme), and manual handling (many of which will be pre-existing injuries exacerbated by work).

78   For example, in South Australia, the Safety Achiever Bonus Scheme provides financial incentives for larger employers to reduce claim numbers and costs by application of a systematic approach to managing prevention. See L. Owens and K. Lantis, 'Safety Management Systems and Incentives' *Journal of Occupational Health & Safety: Australia—New Zealand* (1996) 12(5) 597–601.

79   NSW Construction Policy Steering Committee (undated).

80   Personal communication, G. Mansell, NSW WorkCover Authority, 9 Feb. 1996.

81   Occupational Health and Safety Authority of Victoria, *Occupational Health & Safety,* Submission Number 176 to the Industry Commission, 25 Oct. 1994, p. 7.

82   Industry Commission, Australia, *Work, Health & Safety: Inquiry into Occupational Health & Safety,* Volume I Report and Volume II Appendices, Report No. 47 (AGPS, Canberra, 1 Sept. 1995), p. xix says approximately one-third each.

83   See discussion of 'upfront' payments at p. 84–6 above.

84   These sorts of proposals have already been taken up by the Australian Federal Department of Primary Industries and Energy (DPIE) which in July 1991 introduced a scheme to subsidize companies performing energy consumption audits.

85   In addition to workers compensation data, under s. 27 of the New South Wales Occupational Health and Safety Act 1983, employers must notify WorkCover of any accidents resulting in an employee being unable to perform normal duties for seven days or more, and other dangerous occurrences likely to threaten health and safety of persons at work. This data is used to target 'particular industries, geographical areas and even individual workplaces'. This replaces the old system where the inspector's activities were largely confined to accidents or random visits to workplaces.

86   K. Frick, *Enforced Voluntarism—purpose, means and goals of systems control,* National Institute for Working Life, Solna, Workshop on Integrated Control/Systems Control, Dublin, 29–30 Aug. 1996.

87   Jensen suggests by implication that this group is a large one, pointing to evidence that many firms are essentially reactive in relation to OHS, failing to make regular assessments of their OHS problems, but rather 'awaiting problems to be presented to them in the form of accidents or complaints' (P. L. Jensen, *Internal Control in Denmark,* Technical University of Denmark, Lyngby, Workshop on Integrated Control/Systems Control, Dublin, 29–30 Aug. 1996).

88 P. L. Jensen, *Internal control in Denmark*, Technical University of Denmark, Lyngby, Workshop on Integrated Control/Systems Control, Dublin, 29–30 Aug. 1996.

89 See section on bounded rationality at pp. 91 and 107.

90 Within the European Commission, DG V/f/5 commissioned a report, 'The Economic Appraisal of European Union Health and Safety at Work Legislation', which outlines a framework which can be used by individual enterprises for this purpose. In essence, it is a much simplified personnel accounting system which focuses on the costs of sick absence, the costs of personnel turnover, and the costs of preventive measures.

91 T. Davies & J. Mazurek, *Industry Incentives for Environmental Improvement* (Global Environmental Management Initiative, Washington DC, 1997); and D. Beardsley, *Incentives for Environmental Improvement: An Assessment of Selected Innovative Programs in the States and Europe* (Global Environmental Management Initiative, Washington DC, 1997).

92 F. Haines, *Corporate Regulation: Beyond 'Punish or Pursuade'* (Clarendon Press, Oxford, 1997), p. 31.

93 The Victorian SafetyMAP, for example, is not intended to be restricted to large companies. Similarly, in Holland, the government is of the view 'that environmental management should become an integral part of corporate management practices'. The government's objective is that in 1995 nearly all 10–12,000 companies causing major or moderate pollution or presenting special hazards should have an effective integrated EMS tailored to the nature, size, and complexity of the company concerned. The government also assumes that by 1995 the estimated 250,000 companies causing limited pollution will have made clear progress towards introducing an adequate partial EMS geared to the nature of the company in question. In Norway, the IC approach is intended for all firms, including small ones, although regulators informally acknowledge that they have far lower expectations of small firms in terms of their adopting and implementing systems-based approaches.

94 See further, Industry Commission, Australia, *Work, Health & Safety: Inquiry into Occupational Health & Safety*, Volume I Report, Report No. 47 (AGPS, Canberra, 1 Sept. 1995), p. 88.

95 Submission by Worksafe Australia to the Industry Commission, Australia (1995).

96 See Construction Forestry Mining and Energy Union—Mining and Energy Division (CFMEU), *Occupational Health & Safety*, Submission Number 153 to the Industry Commission, 28 Oct. 1994.

97 P. Lindoe, 'Self-regulation of Occupational Health and Safety in Norway: A revitalisation of the Working Environment Act', Stavanger College, Rogaland Research, Stavanger, undated.

98 Personal communication with Mr G. Mansell, NSW WorkCover Authority, Dec. 1997.

4 Two Paths to Enlightenment: A Two-Track Approach to Regulation

Most regulatory agencies have very considerable discretion in how they choose to operate. This Chapter explores ways of using this administrative discretion to best advantage, to develop innovative, efficient, and effective regulatory strategies. In particular, we examine the implications of adopting a 'two-track' approach to regulation. We argue that enterprises should be offered a choice between a continuation of traditional forms of regulation on the one hand (Track One), and the adoption of a SMS on the other (Track Two). The latter will put primary responsibility on employers and workers themselves to find optimal means of reducing occupational injury and disease subject to government and third party oversight, and will involve a partnership between the agency and the enterprise, from which both sides will benefit. However, even in respect of conventional regulation (Track One), we argue that there is scope for considerable effectiveness and efficiency gains, through appropriate regulatory redesign.

Track One: Towards efficient and effective regulation

This section, and indeed the remainder of this Chapter, examines the question of how to design efficient and effective forms of regulation for those enterprises which, either through choice or necessity, do not adopt a SMS. Here, the issue is how to improve, rather than how to replace, traditional forms of regulation and inspection. The duty holders who will need be regulated under Track One include not just many employers, but also manufacturers, suppliers, managers, directors, workers, third parties, and others who hold a duty of care.[1] Noting the changing nature of the work relationship, they will also include increasing numbers of contractors, subcontractors, and self-employed. The very large majority of small employers, in particular, will fall into this category. Indeed, Hopkins has argued that: 'A visit by an inspector is almost the only way to get to small employers. Moreover, such a visit is likely to be relatively effective since small employers are more impressed than are many larger employers by the authority wielded by government inspectors.'[2] While we will argue that Hopkins's conclusion is too pessimistic, he is undoubtedly correct in pointing out that the range of tools capable of dealing with small employees is very limited.

Perhaps the most important context in which regulatory policy must be designed is that of regulatory resources which are severely limited and which are likely to remain so for the foreseeable future. For example, in the UK, in 1996 there was an average of one factory inspector to 1,000 worksites (the equivalent figure for 1984 being one inspector to 420 worksites). In 1991 about half the worksites had not been inspected for three years and nearly 70,000 had not been inspected for eleven years.[3] Similarly, on average, in Australia, there is a 22 per cent chance of a workplace being visited by the OHS inspectorate in any year.[4] Inspections per number of workers in continental Europe are, by and large, equally thin on the ground.[5] Since the emphasis is on inspections of larger workplaces, small employers have only a very small chance of inspection in any one year. This was highlighted in a 1995 HSE (UK) publication which acknowledged how rarely many small firms were visited and committed the HSE to 'contact' (as distinct from inspect) all the firms it is aware of at least once every six years.[6]

We also recognize that much of OHS inspectors' time is necessarily taken up in reactive tasks such as investigating accidents, which, for practical and political reasons, is both necessary and important, and in administration.[7] In these circumstances, it is crucial that those scarce resources which are still available for discretionary and proactive tasks are used in the most effective and efficient manner so as to get 'the biggest bang for the regulatory buck'.

Below, we review the empirical evidence about regulatory enforcement internationally, we examine recent and important developments in regulatory theory, and we build on both of these to argue a case for the substantial modification of current regulatory practice in any jurisdiction.

Allocating resources

Here, we examine three issues which are crucial to the deployment of regulatory resources under Track One and to the achievement of efficient and effective enforcement practice. These are: (1) what principles should govern the allocation of scarce enforcement resources *among different sizes of firms*, and how could an agency improve its allocation under current laws with current resources? (2) what should be the nature of inspection? For example, 'should inspections be made more intensive, or should more firms be inspected with shorter inspections?'[8] and should there be a greater focus on audits rather than upon traditional inspections? and (3) how should inspections be targeted? For example, should inspections be random or targeted to worst performers?

As regards the first issue, most of the empirical evidence comes from the USA and from studies of the behaviour of the OSHA. These studies provide convincing evidence that additional OSHA enforcement would prevent more injuries in large and mid-size plants than in small plants.[9]

However, care must be taken in applying the results of these studies since they are based on marginal shifts in resources from the current OSHA base allocation of resources between plants of different size, which of course may be different from the base of other enforcement agencies. There will come a time, with all agencies, where further redistribution of resources from small to medium and large enterprises, will become inefficient, and this will be at a point far short of shifting all enforcement away from smaller plants.

The broader conclusion is that agencies should assess their current distribution of resources between small, medium, and large firms, the likelihood being that a strategy that focuses substantially on the latter will be more efficient than the converse—though it may also be that a further distinction has to be made between larger firms (which, by reason of having more employees in one place, are easier to inspect than lots of small firms), and many of the very largest firms, who, by virtue of being close to industry best practice already, are not sensible targets for inspection. This is consistent with the further finding of Gray and Scholz, that enforcement actions against firms with 100 to 500 employees had greater specific and general deterrence effects than those against smaller *or larger* firms.[10]

In determining the optimal strategy, there may also be a tension between efficiency and equity considerations. For example, it would arguably be unfair to spend a disproportionate amount of time at large plants, and to largely ignore small plants, even if this proved to be the most efficient strategy. It is also important to remember that, under the two-track policy we advocate, some large firms will be regulated on an entirely different basis under Track Two, in a manner that involves substantial self-regulation. The result should be that many of the largest firms are no longer the subject of conventional routine inspection (that is, by virtue of adopting Track Two regulation, they become a very low inspection priority) but that, of those firms that remain within

Track One, those of between 100 and 500 employees become a higher inspectoral focus.

The second issue concerns what *type* of inspection is likely to be optimal in terms of achieving changes in employers' behaviour. For example, should an agency conduct a relatively small number of intense and detailed inspections or a much larger number of superficial ones? Here, more may depend upon the consequences of inspection than upon their length. In particular, it may be *even relatively superficial inspections, if accompanied by formal (albeit limited) enforcement action*, will have a substantial impact on behaviour and will often prove to be the most efficient strategy. The most important finding, coming from a major empirical study by Gray and Scholz,[11] using USA data on injuries and OSHA inspections, is that even relatively small fines still achieved a change in employer behaviour.

Such a conclusion would come as a surprise to proponents of strict economic rationality models, which predict that penalties will only change behaviour if the risk of detection x anticipated penalty is greater than the cost of fixing up the problem (i.e. employers will only respond if expected penalties are high enough to offset compliance costs).[12] The authors' own very credible explanation of their data is that it supports a bounded rationality view of corporate compliance:

... the bounded rationality model of corporate compliance[13] emphasises the fact that regulatory law plays an important but limited role in guiding human behaviour. Regulation presumably evolves when existing motivations lead to socially undesirable behaviour. To change such behaviour, regulations must not only overcome the set of existing motivations but must also call the attention of citizens to the particular behaviour in need of changing.

According to the bounded rationality approach, even the best-intentioned citizens have difficulty learning of the multiple demands that socially desirable laws impose on them amid the 'boomin'-buzzin' complexity of modern life.[14] Legal demands compete with a cacophony of political and commercial advertising and other forms of persuasive communications that attempt to change the citizen's habitual behaviour.

Citations by an enforcement agency help this process of legal change by interpreting legal duties in concrete situations, pointing out the discrepancies between the cited behaviour and (the agency's interpretation of) legally mandated social responsibilities. For a minority of individuals, avoiding future penalties may be the only motivation capable of inducing compliant behaviour. But for most citizens, a legal citation is likely to focus the citizen's attention on a set of habitual behaviours that the citizen may have not fully integrated with the citizen's own beliefs about relevant social responsibility. Like other forms of persuasive communication,[15] a citation is most likely to succeed in changing behaviour if it: (1) gains attention; (2) points out discrepancies between behaviour and normative beliefs; and (3) suggests compliant behaviour that is more consistent with beliefs and social obligations.[16]

Although citations and immediate fines are not an integral part of many enforcement systems outside the USA, the broader lesson may be that *the very fact of an inspector's visit (albeit superficial), coupled with some form of enforcement action* (for example, an improvement or prohibition notice or an 'on the spot' fine), may have a significant impact on behaviour and consequently on injury levels, even in circumstances where compliance costs will likely exceed the economic benefits to the employer of compliance. Essentially, this is because such action may serve to refocus employer attention on safety and health problems they may previously have ignored or overlooked.[17]

Amongst the main policy implications of the above analysis are that a regulatory agency can achieve a considerable impact by a regular programme of inspections reinforced by some degree of formal enforcement action sufficient to bring the problem forcibly to the employer's attention *even if the latter is not substantial,*[18] *and even in the absence of full or 'wall-to-wall' inspections.* However, this is *not* an argument against stronger penalties. On the contrary, while there are reasons of *specific* deterrence for conducting more frequent inspections even with only modest enforcement action, there are compelling reasons of *general* deterrence in favour of substantially higher penalties,[19] which we address later.[20]

Another innovation, under the USA 'reinventing government' initiative, is that of 'focused inspections' for employers with strong and effective safety programmes. The OSHA's intent is to work with targeted industries to identify the most serious hazards in those industries. These hazards will then be given focused attention during inspections, as a means of encouraging the adoption of effective safety and health programmes.

This scheme has been piloted on the construction industry. Under it, if an OSHA inspector conducts an inspection and finds an effective safety programme operating on-site, then the remainder of the inspection will be limited to the top four hazards that kill workers in the construction industry.[21] Under the scheme: 'if these hazards are well controlled, the inspector closes the inspection and leaves the site. Conversely, where a safety and health program has not been established or is ineffective, OSHA conducts a complete site inspection.'[22] The scheme is to be extended to a number of other industries.

In other jurisdictions, such a scheme might be applied as an incentive to larger companies to establish a SMS, or it might be applied to smaller companies who, even without a SMS, might be able to demonstrate a quality of safety performance sufficient to justify an inspection limited to priority hazards—thereby both providing them with an incentive to reduce workplace injury and disease and enabling regulatory resources to be redeployed elsewhere.

It must be acknowledged that regulatory resources are so limited that a 'wall-to-wall' inspection in never a serious option, and that such a policy of differential inspections for enterprises with different levels of safety performance would need to be modified—the realistic choices being between partial inspections on the one hand, and *very* partial inspections on the other. Even if this reality detracts from the degree of incentive that can be offered to employers, it in no way limits the value of focused inspections as a rational way to deploy inspectoral resources.

Another important trend is away from conventional inspection towards a greater emphasis on OHS audits. For example, the UK HSE's Accident Prevention Advisory Unit now devotes about one-third of its resources to auditing the OHS performance of large firms. Within Australia, the Comcare's Planned Investigation Program also takes this approach, with investigations focusing on legislative compliance by an employer's OHS management system, policies, and practices, and including identification of potential hazards at the workplace and systems of work to ensure that any risks from these hazards are controlled. Two-thirds of investigations are now based on this approach rather than on conventional inspection.[23] All of these approaches are still in their infancy, and while early results are promising, they are not conclusive. The potential benefits of an audit-based approach are further examined below.

The third issue concerns whether, and how, to target inspections. The empirical evidence suggests that focusing on plants with the highest level of risk, as indicated by consistently high injury rates, may provide the greatest incentives for those plants to comply voluntarily and therefore lead to more efficient enforcement.[24] The success of the Maine 200 experiment, referred to earlier, also suggests the importance of concentrating on the most serious threats to OHS and that redeploying enforcement efforts in this manner will do most to protect workers. Similarly in the UK, the HSE is now giving priority to inspecting companies who create the greatest risks—using a relatively sophisticated weighting of risk described below,[25] which includes a particular emphasis on risk assessment and managerial competence and commitment. A focus on the industries where subcontracting is prevalent might also be rational, given the high incidence of OHS problems associated with this practice.[26]

We conclude there is much to be gained in identifying and targeting inspections to those employers who have serious OHS problems which they are unwilling to address voluntarily, and in focusing on worst cases and major hazards rather than inspecting primarily on a random basis across an industry group.[27] However, it is rational to retain *some* random component in order to keep other firms 'on their toes' because they might still be inspected.

Initiatives taken by the HSE in the UK, to expand the range of inspection techniques and, in particular, to improve the targeting of inspection, have also become increasingly sophisticated, although it is too early at present to evaluate the results. In essence, the HSE is seeking to concentrate its resources on those activities which create the greatest risk and to use techniques geared to securing maximum impact.[28] A particular focus is on the enhancement of the inspection rating system (which substantially influences the frequency of inspection) and on the development of a risk-based workload formula to assist management decisions on the deployment of resources between employment sectors. The application of this formula has resulted in a shift of resources away from traditional manufacturing and from agriculture, to service industries, to construction, and to chemical manufacturing. Consistent with the recommendations of the 1994 review of OHS legislation in the UK, enforcement is also designed to be proportionate, consistent, and transparent.[29]

Finally, it must be noted that although effective targeting can play an important role in an efficient and effective enforcement policy, it is heavily dependent on the depth and accuracy of an agency's statistical database and other information sources. Only with adequate data collection and interpretation can a targeted inspection programme realize optimal results. An over-reliance on traditional sources of information such as workers compensation data for these purposes may be dangerous, given the significant limitations of this data. In particular, there are dangers that it may be manipulated by employers for claims management purposes,[30] that it will not reveal information about hazards, with long latency periods, and that it tells us insufficient about particular industry sectors or groups of disadvantaged employees.[31] There are further problems in that significant numbers of workers are not covered by workers compensation and many injured workers do not claim for various reasons.[32]

It follows that there is a need to explore alternative sources of information and to establish databases which provide more accurate profiles of individual firms, hazards, and industries. For example, injuries that have resulted in medical treatment, or even lost time injuries,[33] might provide a better indicator than workers compensation data, although much will depend on the circumstances of a particular industry.[34]

Strategies for inspection and enforcement

Legislation that is not enforced seldom fulfils its social objectives, and effective enforcement is vital to the successful implementation of OHS legislation. This Chapter now examines the question of how best an inspectorate should go about the enforcement task, in terms of a policy of rational enforcement designed to achieve maximum compliance under Track One.[35]

As a context for what follows, it should be remembered that inspectors in the large majority of jurisdictions have very extensive investigatory powers, powers to institute prosecutions, both against individuals and corporations, and powers to issue administrative notices against a wide range of duty holders. They also make use of less formal warnings and other strategies.

In practice, the full range of formal powers may not be used in anything like equal measure. This is because most inspectorates (though less so the USA OSHA[36]) have a very considerable degree of administrative discretion and can choose between (or incorporate some mixture of) two very different enforcement styles or strategies: those of deterrence and compliance. The deterrence strategy emphasizes a confrontational style of enforcement and the sanctioning of rule-breaking behaviour.[37] It assumes that those regulated are rational actors capable of responding to incentives, and that if offenders are detected with sufficient frequency and punished with sufficient severity, then they and other potential violators will be deterred from violations in the future. The deterrence strategy is accusatory and adversarial. Energy is devoted to detecting violations, establishing guilt, and penalizing violators for past wrongdoing.

In contrast, a compliance strategy emphasizes co-operation rather than confrontation, conciliation rather than coercion.[38] As described by Hawkins:[39]

Compliance strategy seeks to prevent a harm rather than punish an evil. Its conception of enforcement centres upon the attainment of the broad aims of legislation, rather than sanctioning its breach. Recourse to the legal process here is rare, a matter of last resort, since compliance strategy is concerned with repair and results, not retribution. And for compliance to be effected, some positive accomplishment is often required, rather than simply refraining from an act.

Bargaining and negotiation characterize a compliance strategy. The threat of enforcement remains, so far as possible, in the background. It is there to be employed mainly as a tactic, as a bluff, only to be actually invoked where all else fails, in extreme cases where the regulated remains uncooperative and intransigent.

These two enforcement strategies are ideal types, hypothetical constructs unlikely to be found in their pure form. They can best be regarded as two polar extremes on a continuum. Any enforcement agency is likely to employ some mixture of both. The extent to which one strategy predominates over the other determines where on the compliance-deterrence continuum an agency's approach can be located.[40]

Experience suggests that neither a pure deterrence, nor a pure compliance, strategy will achieve the best results in terms of improved OHS performance. For example, the USA approach to enforcement is considerably more punitive, deterrence-oriented, litigious, and confrontationist, than that of either Australia, the UK, or the Scandinavian countries. There is evidence that an over-reliance on deterrence in the USA has been counter-productive and has produced a 'culture of regulatory resistance' amongst employers.[41]

In part, the USA's approach to enforcement is a product of that country's history and in particular of the hostility that business commonly shows towards any state intrusion into its affairs. In this environment of conflict between business and government, a legalistic deterrence-oriented approach is at least understandable, even if its efficacy may be questioned. In countries and cultures where such antagonism does not exist to anything like the same degree, a confrontational strategy of enforcement makes little sense.[42]

However, there are also dangers in adopting a pure 'compliance'-oriented strategy of enforcement, which can easily degenerate into intolerable laxity,[43] and fail to deter those who have no interest in complying voluntarily. As one of our respondents put it:

Ten per cent of firms are engaged in OHS best practice. Thirty per cent are well intentioned, but haven't got a clue, which is where the information side of things is important. These people are quite willing to do it . . . thirty per cent have their head in the sand, hoping that it will all go away. With these people you need targeted enforcement—if you don't do things you will suffer. Then there are the people at the bottom—you have to enforce with them.[44]

Unless these figures are wildly inaccurate (and there is broad agreement among regulators that they are not[45]), they suggest that a credible enforcement strategy must include a significant deterrent component. Significantly, a rise in the incidence of fatalities and major injuries in the UK, in the first half of the 1980s, occurred at a time when inspection and enforcement activity fell significantly, due to a combination of increased workload and staff cuts.[46] The fact that deterrence does not apparently work across the board,[47] and that it is not necessary in all cases, is not an argument against the need for criminal sanctions, but rather for targeting those sanctions to circumstances and actors where deterrence is most likely to be effective (and for avoiding the use of sanctions where it is likely to be counter-productive).

Most contemporary specialists on regulatory strategy point to the severe limitations of both pure deterrence and pure compliance strategies, and argue, on the basis of considerable evidence from both Europe and the USA, that a judicious mix of compliance and deterrence is likely to be the optimal regulatory strategy.[48]

But how can such a mix best be achieved, and does current administrative and enforcement practice approximate it? In considering alternative strategies, it is important to remember that there may be a range of situations where improving OHS makes good business sense in that it is likely to improve profit and productivity and possibly provide other benefits also (for example, improved public image, improved industrial relations). However, this is far from always being the case. Policy makers must also confront those circumstances where there is a direct conflict between safety and profit[49] or where regulatees, through irrationality or incompetence, do not recognize the other benefits that improved OHS may provide. In either of these latter situations, voluntary initiatives alone are unlikely to provide an adequate response and stronger forms of direct intervention in industry's affairs may be justified.

It follows that regulators must invoke enforcement strategies which successfully deter egregious offenders, while at the same time encouraging and helping the majority of employers to comply voluntarily. Thus good regulation means invoking different strategies depending upon whether on not business has a self-interest (or perceives itself as having a self-interest) in improving OHS outcomes. However, the dilemma for regulators is that it is rarely possible to be confident in advance of which classification a regulated firm falls into and still less to distinguish rational economic actors (who consciously calculate costs and benefits in terms of their self-interest) from the irrational or incompetent (who may have self-interest in improved OHS but do not recognize or act upon it).

If the regulator assumes all firms have a self-interest in improving OHS, or will behave as good corporate citizens, it may devise a regulatory strategy that stimulates voluntary action but which is incapable of effectively deterring those who fall into the second category and who have no interest in responding to encouragement to voluntary initiatives. On the other hand, if regulators assume all firms face a conflict between safety and profit, or for other reasons that they will require threatening with a big stick in order to bring them into compliance, then they will unnecessarily alienate (and impose unnecessary costs on) those who would willingly comply voluntarily, thereby generating a culture of resistance to regulation.[50]

Towards a pyramidal enforcement response

A solution to this dilemma confronting Track One regulation, suggested by John Braithwaite, is the regulatory enforcement pyramid.[51] Similar approaches have been suggested by Rees, in terms of flexible enforcement,[52] by Scholz, in terms of co-operative enforcement,[53] by Nonet and Selznick, in terms of responsive enforcement,[54] and by Carson and Johnstone,[55] in terms of a graduated enforcement response. Here, we focus specifically on Braithwaite's model, which, for reasons indicated below, has particular advantages.

Under this model, regulators start at the bottom of the pyramid assuming virtue—that business is willing to comply voluntarily. However, they also make provision for circumstance where this assumption will be disappointed, by being prepared to escalate up the enforcement pyramid to increasingly deterrence-oriented strategies.

As Braithwaite puts it:

At the base of the pyramid, regulators assume and nurture virtue—corporate responsibility . . . When virtue fails, regulatory strategy shifts through escalating deterrent responses. When deterrence fails, strategy shifts again to an incapacitative response.[56]

For example, an enforcement pyramid might begin with the provision of advice and formal directions, move to the issuing of improvement

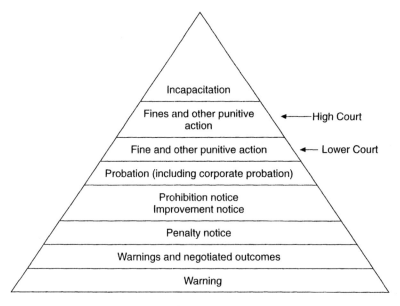

Figure 2

and prohibition notices, and on-the-spot fines, and then escalate to prosecutions with increasingly serious consequences, as represented in Figure 2.

Braithwaite points out:

A paradox of the pyramid is that the signalled capacity to escalate regulatory response to the most drastic of measures channels most of the regulatory action to the cooperatives base of the pyramid. The bigger the sticks at the disposal of the regulator, the more it is able to achieve results by speaking softly. When the consequence of firms being non-virtuous is escalation ultimately to corporate capital punishment, firms are given reason to cultivate virtue.[57]

In essence, by adopting a 'tit for tat' strategy, by co-operating with firms until they cheat, 'regulators avert the counter productivity of undermining the good faith of socially responsible actors. By getting tough with cheaters, actors are made to suffer when they are motivated by money alone; they are given reason to favour their socially responsible, law-abiding selves over their venal selves.'[58] Braithwaite's conclusion is that pyramidal forms of responsive regulation 'hold out the possibility of nurturing the virtuous citizen, deterring the venal actor and incapacitating the irrational or dangerously incompetent actor'.[59]

Four general points may be made about this framework for regulatory enforcement under Track One. First, the strategy involves both carrots and sticks, and it will almost invariably be better to start with the carrots. Second, it is absolutely essential to the credibility and thus to the success of the strategy that regulators are prepared to climb to the top of the pyramid where action at the lower end fails to achieve results. This is an issue concerning which there is a large gap between theory and practice. Third, there may be practical limits to the implementation of a 'pure' pyramidal approach, which necessitate modifications in practice. Fourth, the penalties at the top of the pyramid must be sufficiently serious and effective that they do, indeed, serve as an effective deterrent or to incapacitate the offender. We deal with this fourth point in Chapter 6, and with the first three issues immediately below.

The importance of carrots

The essential reasons for preferring carrots to sticks is that people and organizations usually respond better to incentives, they are less demanding of enforcement resources, and they avoid unnecessary antagonism between regulator and regulatee. Not only do people respond better to rewards than to punishments[60] but those who comply willingly will do so with far more commitment and effect than those who do so reluctantly, under threat of penalty: volunteers almost always behave better than conscripts.

The enforcement pyramid set out at Figure 2 is illustrative of the sorts of carrots and sticks that an agency might wish to invoke as part of an escalating strategy of enforcement but it is not intended to be exclusionary. Many different mechanisms might be utilized under this general approach.

For present purposes, we note that the present 'advice' role of most inspectorates is a useful but essentially modest one. It may include the provision of information, assistance in finding professional help in respect of a particular problem, and technical assistance. However, given that inspections are quite rare events, the opportunities for effective advice are limited. Moreover, there may be some reluctance on the part of inspectorates to giving such advice 'for fear it may be relied upon by the employer in a subsequent prosecution'.[61] As an alternative, it might be valuable for an inspectorate to suggest the use of a consultant to assist the duty holder in resolving a problem, and to agree to give them a period of grace in which to do so.

There may also be opportunities to assist small employers in particular, for example by developing guidance material collectively where they face a limited range of control measures, or by providing both subcontractors and those who engage them with material on the OHS hazards they face and how to deal with them, as well as on their broader legal responsibilities. Advisory standards, codes of practice, and guidance notes on the *use* of subcontractors will be particularly important.

Here, specific strategies will vary greatly between industry sectors. For example, the UK HSE now complements routine inspection with mail shots (targeting specific health and safety information to specific sectors by post and inviting contact with an inspector); seminars on specific topics; publicity campaigns; and use of intermediaries to target information (for example, trade associations, trade unions, small business and citizen advice bureaux, training centres, etc). Similarly, the Farmsafe strategy, which was developed by Worksafe Australia, the Department of Primary Industries and Energy, and the agriculture industry, created regional farm safety action groups for this purpose.[62] Such strategies may stimulate awareness, and encourage innovation and commitment to OHS.

Most valuable at the base of the enforcement pyramid may be those approaches which actively encourage duty holders to regulate themselves, and which give them a positive incentive to do so. Of course, the most developed strategy designed to achieve this result is the one we advocate under Track Two below. However, this approach is in effect limited to the minority of enterprises which are ready, willing, and able to embrace a SMS approach. There may be many SMEs for whom such an approach, at least in the short term, is not viable, but who are open to other forms of 'partnership' and who will respond to other incentives at the bottom end of the pyramid.

For such enterprises, there are a number of innovative alternatives which are well worth exploring. We considered a number of incentive-based approaches earlier in Chapter 3, and there is no reason why similar incentives should not be offered under Track One, though here such incentives would be designed to encourage OHS initiatives which are less demanding than commitment to a SMS.

For example, the VPP program, described earlier, might be modified so as to encourage firms to go beyond the minima prescribed by law, by recognizing and rewarding excellence in OHS programmes provided by employers, but in a less demanding form which did not require the implementation of a full SMS.

Again, some of the changes to enforcement practice, recommended under the Clinton Administration's 'reinventing government' initiative, could be used to give incentives to enterprises to demonstrate a greater commitment to OHS than mere compliance with prescribed standards, even if this falls short of introducing a SMS. Under this initiative:

> OSHA will provide special recognition including the lowest priority for enforcement inspections (which, given remaining priorities, means that inspections will be quite rare) and the highest priority for assistance, appropriate regulatory relief . . . For those firms that are well intentioned but have room for improvement, OSHA will offer a sliding scale of incentives depending upon the degree to which the employer demonstrates real efforts to find and fix hazards.[63]

Particular attention could be given to encouraging OHS self-audits, which, while far less comprehensive than a full systems-based approach, are nevertheless very valuable in encouraging self-regulation, in alerting enterprises to OHS problems, and in prompting corrective action. Here, government can play an important role in providing the tools which enable firms to audit themselves, and in rewarding such behaviour through the sorts of incentives examined earlier in Chapter 3, including mechanisms to recognize achievement.[64]

More broadly, inspectorates in Scandinavia and the UK, amongst others, are having significant success through a general strategy which involves emphasizing risk assessment as a crucial part of an enterprise's OHS responsibilities—including SMEs. Under this approach, an inspector, in addition to identifying specific hazards and breaches of individual legislative requirements, emphasizes the importance of risk assessment and the need for the enterprise to *manage* OHS and the work environment. A critical element here is to encourage and facilitate mechanisms that prevent the enterprise from sliding backwards (a common scenario being a marked improvement in OHS performance in the wake of an inspection followed by a gradual deterioration in performance). Here, one tool with particular merit is the action plan. The inspector insists the enterprise identifies its hazards, assesses the risk, sets priorities, and then establishes an action plan setting out the actions to be taken to address the risks identified and prioritized. Since the expectation is the action plan will be revised periodically (an inspector will not accept the same plan on subsequent visits), this is an important vehicle through which to encourage change. It also means that the inspector can begin a subsequent visit by comparing performance against the previous action plan as a means of assessing the enterprise's bona fides and competence. In the case of small enterprises, however, the strategy, while valuable, has important limitations. As one British inspector put it: 'you try to sell it to them but if they don't grab the bait and run with it you've had it because you won't be back for another seven or eight years'.

Mindful of the severe resource constraints which inevitably limit the opportunities for regulating small enterprises, the USA has introduced an incentives-based innovation which promises effective and resource-efficient ways to promote voluntary compliance in small firms. While the scheme is directed to environmental protection rather than OHS, it could readily be translated to the latter. The scheme involves an agreement signed by the State of Minnesota and the Printing Industry of Minnesota Inc. (PIM), a state-wide trade association, designed to significantly increase the use of environmental audits by printing companies in Minnesota, many of whom are relatively small operations. It is based on the premise that a clear distinction should be made between companies that have adopted detailed company environmental compliance and pollution prevention policies and those that take little or no positive initiatives in respect of environmental protection. It seeks 'to better differentiate the good from the bad actors by increasing the incentives to voluntarily comply with environmental laws and to pursue pollution prevention initiatives'.[65] By encouraging voluntary compliance, the scheme will enable regulatory resources to be redeployed and refocused on those who are not responsive to voluntary initiatives.

Under the scheme, PIM established a separate corporation, PIM Environmental Services Corporation, to provide auditing services to PIM members. However, it was also necessary to provide some incentive to companies to engage in such audits.[66] Government regulators were reluctant to provide a total amnesty from prosecution for breach of regulations, simply because a firm had engaged in such an audit, for fear of damaging the integrity of their enforcement programme. However, an agreement was reached whereby an auditing company which discovers environmental violations, and corrects them promptly, will have this fact taken into account when regulators decide whether to initiate any enforcement action, whether an enforcement action should be civil or criminal in nature, and what penalties to impose.[67] Thus 'a company which conducts an auditing program in good faith and makes appropriate efforts to achieve environmental compliance is likely to mitigate the consequences of any violations it discovers'.[68]

As Humphrey points out,[69] the PIM audit agreement demonstrates how the use of audit programmes might be expanded to thousands of SMEs in other industries that similarly do not have the resources or the motivation to conduct safety, health, or environmental audits but who might be given incentives and encouragement to do so. Such a policy might result not only in voluntary initiatives that substantially improved environmental performance but also in a far more flexible and cost-effective response on the part of participants than is likely to be achieved through traditional 'command and control' government regulation.

Moving to the 'next rung' on the enforcement pyramid, options include the issuing of formal directions, improvement and prohibition notices, and on-the-spot fines. Improvement and prohibition notices have already more than demonstrated their very considerable worth in countries such as the UK and Australia, and have proved to be perhaps Robens's most successful innovation. They are preventative in nature, and allow action to be taken swiftly without the necessity of going to court. Appeals are rare and infrequently successful. In contrast to the cumbersome and time-consuming nature of the traditional prosecution process, these orders offer a quick and simple mechanism capable of being used on the spot to deal with serious hazards immediately they are detected. Moreover, such orders are particularly flexible in that they do not necessarily specify how an employer may come into compliance, thereby leaving her or him free to choose the least-cost method and avoid unnecessary expense.[70]

On-the-spot fines also have considerable potential. The research of Gray and Scholz, cited earlier,[71] suggests the very considerable value even of a slight slap on the wrist, such as on-the-spot fines provide, in focusing attention on OHS issues. Early anecdotal evidence in New South Wales (where such fines were recently introduced) also supports this conclusion.[72] Similarly, the broader international research on administrative penalties generally[73] also suggests that such penalties can provide credible deterrence at a very modest administrative and legal cost.[74] Provided that use of these fines does not become a substitute for more serious enforcement action in serious or repeat cases,[75] and provided they do not serve to trivialize OHS offences through misuse, then they have considerable value. The latter possibility might be reduced by doubling the current level of penalty for employers but coupling that with some form of due process.[76]

In discussing the lower part of the pyramid, carrots have been empha-sized more than sticks, and even the sticks have been small ones, in-tended more as a 'tap on the shoulder' than to provide a punitive sanction. Those sticks have also focused on specific rather than on general deterrence. Improvement and prohibition notices, for example, create no incentive to comply with the law until the illegal behaviour has been detected and the preventative order issued. The importance of such strategies, in serving to focus attention on OHS, and in many cases producing voluntary action, should not be underestimated. Nevertheless, there is strong evidence, referred to above, that it will not be sufficient across the board and that there will remain a substantial minority of recalcitrants, for whom stronger action is necessary.

Climbing the pyramid: Theory and practice

From here on, further escalation up the enforcement pyramid is uncom-promisingly deterrence-oriented. The regulator will have already, and unsuccessfully, attempted to persuade the duty holder to comply volun-tarily, and further action at the lower end of the pyramid is unlikely to achieve results. Such escalation will most likely be necessary when deal-ing with the intransigent, with the irrational, and, most important, with the rational calculators, who believe that it is not in their self-interest (usually financially defined) to comply voluntarily, and who are only likely to respond when the costs outweigh the benefits. This latter group are strategic 'game players'. They are only likely to change their behav-iour where the severity and likelihood of punishment make it rational for them to do so.

Unfortunately, the administrative practice of many OHS agencies falls short of providing an effective deterrence strategy against this latter group. Lack of resources, the cost of prosecution, perhaps the unwilling-ness of the judiciary to impose substantial penalties, and a cultural resistance to prosecution amongst some inspectorates,[77] may be factors which help explain this response.

Moreover, our empirical research suggests that even where the aggre-gate data for the use of the various enforcement mechanisms (notices and prosecution) confirm the existence of what *appears to be* an enforcement pyramid, the reality may be far more complex. For example, in many jurisdictions, statistics suggest that, from a macro-perspective, enforce-ment practices in New South Wales conform to a pyramid structure. What the data fails to reveal, however, is that *there is little evidence that such*

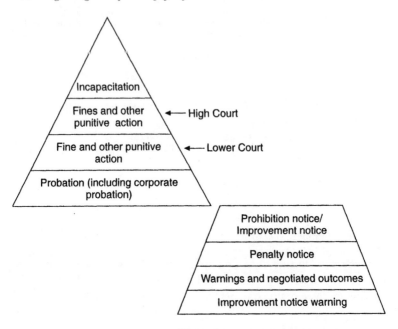

Figure 3

a pyramid is actually used by inspectors as a framework for an escalated enforcement response against individual employers. And where inspectors do in fact use what resembles an enforcement pyramid, this enforcement response is informal and unstructured. We found no evidence that inspectors routinely begin their dealings with employers by advising and persuading, and then escalating their enforcement response systematically using improvement, prohibition, and penalty notices, with prosecutions being conducted against employers who failed to comply in the face of these lesser enforcement measures.[78]

Instead, the data seems to suggest that the enforcement pyramid is in fact split (see Figure 3). Where routine enforcement and programmed inspections are conducted, the data would appear to show that there is a fairly inconsistent and unstructured enforcement pyramid beginning with advice, education, and persuasion, and ending with the issuing of improvement, prohibition, and/or penalty notices. Prosecutions in these matters are rarely, if ever, conducted. On the other hand, the vast majority of prosecutions (around 90 per cent) were in response to workplace injuries, where there was very little evidence that the inspectorate had first attempted to use enforcement mechanisms at the bottom end of the pyramid. In other words, where serious injuries or fatalities occur, the enforcement response begins at the top end of the pyramid, with prosecution in the magistrates' court or Industrial Court, depending on how serious the incident and its consequences are. In short, the evidence suggests that in fact there are two incomplete, and separate complementary pyramids: the bottom half for OHS matters not resulting in injury or death; and the top half in relation to events resulting in serious injury or death.

There are serious dangers with this approach. Agencies are both failing to take advantage of the considerable benefits that a pyramidal enforcement strategy can deliver, and leaving themselves with no escalating deterrent to address the recalcitrant minority. The result is that the 'rational calculators', in particular, have every encouragement not to comply with their legal obligations. Where there is a conflict between safety and costs (as, for example, in occupational disease cases there almost invariably is), and where the regulated industry behaves as a rational economic actor, with a focus on short-term outcomes, then an emasculated enforcement pyramid will fail to deter. The regulated enterprise, knowing that even if it is detected no serious enforcement actions will be taken unless there is a serious injury (and in this case their general approach to safety will be a secondary consideration anyway), is unlikely to undertake expensive remedial action.

The limits of a 'pure' pyramidal approach

Recently, commentators have pointed to limitations in the practical workings of the enforcement pyramid. Haines[79] reminds us that OHS regulators operate side by side with other processes, for example, provisions for the compensation of ill and injured workers, and procedures for coronial inquests. These different processes each shape organizational behaviour, and operate simultaneously, but not always in the same direction, or to the same ends. While the OHS regulator may adopt measures to promote trust and co-operation, an organization may respond distrustfully to the possibility of compensation claims, or coronial inquests. She also observes that while the theory behind the pyramid is that there be an immediate 'de-escalation' of enforcement measures in response to shifts by the regulated organization to a more co-operative position, in practice 'de-escalation of penalty may be far more difficult than previously anticipated'.[80] The initial escalation of penalty may produce long term defensive behaviour in the regulated organization.

We suggest that a drawback of Braithwaite's pyramid, particularly in the context of under-resourced OHS enforcement agencies, is that the approach is 'bottom heavy'. Most OHS enforcement agencies do not have the resources to have anything more than sporadic contact with employers, and contact with other types of duty holders (for example, manufacturers and suppliers of plant, equipment, and substances for workplaces) may be even less frequent. The shortage of regulatory resources restricts the opportunities for a genuine 'tit for tat' strategy. Most agencies will not have the opportunity to begin enforcement measures in every case at the bottom of the pyramid and gradually respond to non-compliance by working their way through to the top, and *in some cases* it may not be appropriate to begin at the bottom of the pyramid. In short, we query whether the resource constraints on OHS agencies, and the range of OHS offenders and offences which agencies are likely to confront, would allow agencies *always* to begin with advisory and persuasive measures, and to presume that there will be a uniform rate of escalation up the pyramid.

Our empirical data suggests that in recent times OHS agencies have tended to target their enforcement programmes to selected industries or hazards.[81] This adds the further complication that entry to the lower levels of the pyramid will often be concentrated on employers in those targeted areas at the expense of others. In short, given the resources of most OHS enforcement agencies, the Braithwaite model coupled with targeted enforcement strategies will result in slow movement up the pyramid in the selected industries, and very low-level enforcement elsewhere. We also fear that an enforcement strategy that always begins at the bottom of the pyramid, always presuming virtue, will result in an enforcement profile heavily weighted towards persuasive and advisory measures, so entrenching the widespread perception, which we referred to in Chapter 1, that OHS enforcement has largely been decriminalized.

Further, no matter how geared towards proactive inspection an OHS enforcement agency might like to be, in practice a significant proportion of its time will be taken up with 'reactive' inspections, in response to reports of workplace injuries, disease, and death, and complaints from workers or members of the public. In many of these instances the agency will be investigating serious contraventions, often where great harm has already resulted, and resort to measures at the bottom of the pyramid may not be feasible.

Rather than a blanket policy of co-operation with duty holders until they demonstrate a lack of good faith in their compliance efforts,[82] we suggest that agencies will need to develop enforcement strategies, based on the pyramid, where, depending on factors such as the duty holder's record of exposing its employees to hazards, its previous record of compliance, the quality of its OHS management systems, and the circumstances of the contravention, there are different *points of entry* to the pyramid, and different *routes* (sometimes amounting to 'leapfrogging') up the pyramid. The enforcement agency will not assume virtue in every case. Rather, the OHS agency might instead distinguish a presumption of what we might call 'proactive virtue', where an enterprise has clearly taken measures (such as introducing a verified SMS) to manage OHS risks systematically and proactively. In these cases the OHS agency will designate the enterprise to be a Track Two enterprise, and we discuss the implications of this later in this Chapter. Where there is no such SMS in existence, the OHS agency might presume 'compliant virtue', in the sense the enterprise is deemed to be 'persuadable' and co-operative, and will respond positively to information and advice from the OHS inspectorate. In these cases, provided there are no factors (see below) which undermine the presumption, enforcement measures might begin at the bottom of the pyramid.

However, there will also be clear circumstances when OHS agencies are able to decide that the presumption of compliant virtue is not justifiable, based on previously acquired knowledge, or easily ascertainable information which can be gathered on a workplace visit. In these cases we envisage a 'grid' of suitable enforcement responses, where the actual enforcement response (level of entry, and the route taken up the pyramid) depends on what the inspectorate knows, or can easily find out, about:[83]

- the characteristics of the organization, such as the size and nature of the enterprise, and its OHS record (previous OHS management measures, its history of compliance and previous violations, and injuries and diseases suffered by employees, as reported in the organization's compulsory injury and disease records, or available through workers compensation records);
- the duty holder's level of control over the OHS measures it can introduce, which will be determined by the size of its business, and its position and relative strength in the industry;[84]
- the employer's good or bad faith in trying to meet its OHS obligations, as evidenced in the employer's knowledge of its OHS obligations,

the quality of the OHS management systems in place (including the employer's adoption of OHS action plans, compliance programmes, and self-audit or audit by an outside organization, as discussed earlier in this Chapter), and its level of co-operation with the inspectorate;

• the knowledge that the duty holder has of the duty that was contravened, based on earlier complaints by workers, the extent to which the duty holder's OHS practices fall short of those regularly carried out in the industry, previous enforcement measures taken against the duty holder by the inspectorate, and the degree to which the duty holder has intentionally disregarded its obligations; and

• the severity of the defects in the system of work, and the potential or actual harm to employees and others.

In short, the OHS enforcement agency will need to develop a litmus test of the duty holder's attitude to its OHS obligations. Rather than basing these judgments on a detailed examination of the duty holder's actual compliance with the minutiae of OHS regulatory requirements, the inspectorate will need to form judgments about the business culture of the duty holder, and whether the presumption of trust and co-operation is warranted. Does the duty holder take OHS seriously, so that it can nurtured by advice, encouragement, and persuasion, or is the duty holder recalcitrant, or attempting to avoid the substance of its OHS obligations? A key element in such a judgment is the quality and effectiveness of the duty holder's existing system of OHS management, or lack of such a system. This emphasis on inspecting management systems is consistent with our observations earlier in this Chapter of the emerging trend in many North American, European, and Australian jurisdictions for inspectorates to focus on the inspection of OHS management systems, rather than concentrating on assessing compliance with specific detailed regulatory requirements.

For example, we would envisage that an OHS agency's enforcement strategy might state that the agency's initial response might be purely advisory (i.e. entry at the lowest level of the pyramid) where the duty holder is co-operative, there are at least signs of a rudimentary OHS management system, and the detected contraventions are relatively minor or do not have the potential to lead to an immediate injury, disease, or death. The inspector would be seeking to assist the duty holder to improve the development and implementation of its OHS management system, as discussed in the previous sections of this Chapter. Where the duty holder failed to respond to these advisory measures, and, in particular, was tardy in development of an OHS management system which included the removal of the identified hazards, the inspector would resort to an administrative sanction. Where there was no response to either persuasion or the administrative sanction, a prosecution at the lower level of the court hierarchy might be launched (see Figure 2 above, and Chapter 6). Where non-compliance was accompanied by demonstrated recalcitrance, the prosecution might be initiated in the upper level court.

In order to maintain the integrity of the pyramid, it is important that enforcement measures be escalated in such a manner, rather than following the traditional practice of allowing non-compliance to persist for an unreasonable period of time. While there is a danger that initially prosecutions resulting from escalated enforcement measures might be numerous, as duty holders become aware of, and understand, the regulator's resolve to ensure compliance, escalation from the bottom to the top of the pyramid will not often be required.[85]

The agency might *begin* with administrative sanctions, such as an instruction to remedy a contravention within a specified time (for example, an improvement notice), where contraventions discovered during a workplace visit have resulted, or have the potential to result, in a relatively mild form of injury or illness (for example, a sprain or minor fall); the duty holder already has a fairly bad record of injury and disease; and the duty holder's OHS management system is poor. Here the OHS enforcement agency is expressing scepticism about the good faith of the duty holder, but is prepared to give the duty holder the benefit of the doubt, provided compliance is prompt. Where compliance is slow, the inspector might escalate to an on-the-spot fine in cases where there has been some, but not complete, compliance, or to a prosecution in the lower level court where there is no compliance. Where an on-the-spot fine fails to trigger adequate compliance, a prosecution should be initiated in the upper level court.

Where the risk is more significant, and more immediate, the appropriate point of entry would be to take measures to stop the dangerous work (by issuing, for example, a prohibition notice), unless the duty holder was able to respond immediately to advice, and removed the hazard while the inspector was at the workplace. Where a notice was issued, non-compliance would result in escalation to a prosecution in a lower level court where the duty holder had previously had a good OHS record, and had at least a rudimentary OHS management system. Where both conditions were not satisfied, prosecution for non-compliance with the notice would be in the higher level court.

Inspectors might initiate enforcement measures near the top of the pyramid in cases where there is a workplace fatality, or great potential for a fatality or series of very serious injuries or diseases, in a workplace where the employer has a bad OHS injury and claims record, and where there is a clear failure to follow widespread industry approaches to OHS management. In a case like this there is little basis upon which a presumption of compliant virtue can be justified, and the integrity of the pyramid will only be undermined if enforcement measures are initially confined to advice and warnings.

It might be argued that here we are endorsing the 'split pyramid' which we criticized earlier in this Chapter. The 'split pyramid', however, results where OHS prosecutions are initiated principally in response to workplace injuries and fatalities, and where other forms of contravention rarely result in prosecution. As we show later in Chapter 6, here we are arguing that a variety of factors, only one of which is the nature of the injury which actually results, determines the level of entry to the pyramid. We envisage that prosecutions will sometimes be initiated after lower level enforcement measures have not been effective to remedy high-risk situations, even though an injury has yet to result. We also suggest that prosecutions (together with appropriate administrative sanctions to call a halt to dangerous work) might be initiated immediately where there is a deficient OHS management system and an extremely high level of exposure to risk, even though an injury has, fortuitously, yet to materialize. Finally, we are arguing that prosecutions should be initiated where the trust placed in duty holders voluntarily to remedy relatively minor breaches has been met with intransigent attitudes, and compliance has not been forthcoming even in the face of escalated enforcement measures.

We also note at this point that the pyramid should encompass enforcement measures against individual managers, directors, and other corporate officers, rather than focusing solely on the corporate duty holders. We examine the rationale for taking enforcement measures against individuals in Chapter 6. As we discuss in Chapter 6, the OHS agency should have a well-publicized OHS enforcement strategy which clearly sets out the points of entry to the enforcement pyramid, and the likely rates of escalation.

In summary, while a *modified* pyramidal approach to enforcement is likely to be optimal, such an approach can only work where it is consistently applied and where there is a willingness to escalate to the top of the pyramid in cases where compliance is inadequate. Despite the difficulties of determining in advance whether a duty holder is compliant and persuadable, or recalcitrant, the OHS enforcement agency must be willing to enter the pyramid at the highest levels where there are serious breaches which demonstrate that the presumption of compliant virtue is ill-founded, even if no injury has yet materialized. What is most seriously lacking under present arrangements is not only an escalated enforcement response against individual employers, and a willingness to pursue action towards the top of the pyramid, but also a credible range of sanctions at the top end (including penalties against individuals). We return to the issue of deterrence and sanctions in Chapters Six and Seven.

Controlling contractors and suppliers under Track One

In Chapter 1 we noted the proliferation of SMEs, and the growing use of contractors and subcontractors. All of these kinds of enterprises have a high incidence of OHS problems. SMEs commonly lack sophistication and understanding of OHS problems and their motivation to solve them is often compounded by economic marginality, and a preoccupation with short-term profit. As Mayhew, Quinlan, and Bennett point out, contractors and subcontractors, in addition to these factors, confront an:

. . . economic/reward structure [which] encourages enhanced competition for contracts between different subcontracting groups, the paring of profit margins, and work intensification—all of which have been known to increase injury ratios. [Moreover] the inherent disorganisation associated with the outsourcing of labour increases as the numbers of subcontractors on a site multiply, and hence, the overall standard of communication and co-ordination normally diminishes. Thus responsibility for OHS becomes fragmented and attenuated.[86]

We also noted the difficulties of controlling the health and safety performance of SMEs or contractors/subcontractors by traditional means.[87] Inspectoral resources are simply too limited to develop long-term relationships with such enterprises, and even establishing contact with them presents difficulties. Subcontractors, in particular, are 'a largely invisible category as far as most OHS agencies are concerned'.[88] Although there have been some encouraging developments in terms of providing information and guidance to such enterprises[89] there will remain very considerable limits as to what can be achieved by such approaches.

One complementary approach with considerable promise is that of using supply-chain pressure as a means of informal market control over the practices of such entities. In the related area of environmental protection, supply-chain pressure has been an increasingly common means of achieving indirect changes in the behaviour of buyers and suppliers.[90] Within the chemical industry, the product stewardship code of practice under the Responsible Care self-regulatory programme is perhaps the most sophisticated example of this approach.[91]

In the case of OHS, it has been largely the use of SMSs that has prompted many large employers to contemplate extending their influence in a similar fashion. That is, the application of such systems has been extended not only to those who are employees, but also to all others with which the firm has substantial dealings, including, in particular, suppliers, contractors, and subcontractors. It has been possible to use the considerable leverage which large companies have over their smaller suppliers and contractors (whose very survival commonly depends on maintaining their contractual relationship with one or two large companies) to insist that those suppliers and contractors conform to the requirements of their larger contracting partner's SMS (and with certain other requirements); with the threat that they will no longer be doing business if they fail to do so. This development has had enormous significance in influencing the behaviour of many small enterprises.

As one senior UK regulator put it:

> . . . you have got to get the requirement to improve health and safety into the bloodstream of the organisation, into its culture. Information isn't enough because many contractors will ignore it. Regulation often isn't enough because inspections are such rare events in the lives of small companies. But when a contractor finds they can no longer get work from the local authority because it requires a system, or they cannot get work in certain areas of the country or as a contractor on a chemical installation without having a system of accreditation then their attitude changes dramatically. What was previously a very low priority rapidly becomes a high one.

To be effective, the extension of SMSs to non-employees should include a number of characteristics. First, care should be exercized in the selection of contractors, so that only those satisfying certain safety criteria (for example, retraining, accreditation, or otherwise demonstrating a clear commitment to OHS) will be shortlisted. Second, a term of any contract should be that the contractor satisfies certain requirements, including the implementation of specified features of a SMS. In sophisticated contractor control systems this may extend to issues such as orienting contractors to the site, emergency procedures, training requirements, and hazard prevention and control.[92] Third, the contractor must be required to measure, monitor, and report progress in implementing the system. Fourth, there must be the right to monitor and audit the contractor's system, in order to overcome the familiar temptation to adopt mere 'paper systems'. It is regular monitoring and review which ensures that the contractor is implementing the system correctly. This last requirement is crucial because the successful implementation of OHS management systems is more than a technical task, it is also socially determined. In particular, subcontractors, who are not socially integrated into the employer's workforce, are far less likely to comply with the technical requirements of a system than core employees. On the other hand, if they are adequately monitored, and the consequence of persistent noncompliance is being struck off the tender list for future work, then this problem may be overcome.

What is also critical to the success of this approach is that it should be engaged in comprehensively rather than superficially. For example, Quinlan rightly points out that an increasing number of employers are including subcontractor provisions as part of their OHS system, but that: 'the provisions are often confined to a rule-based compliance model, sometimes backed up by induction, training, performance auditing and sanctions. These provisions do not assess outsourcing decisions (including factoring in the cost of additional internal controls that may be required) or fully comprehend the organizational implications of subcontracting.'[93] What is needed is an insistence on more comprehensive programmes which form an integral part of the employer's OHS management system.[94]

In some industries, this development has occurred more or less spontaneously, with the recognition that good OHS practices can often make good business sense, both in terms of profit and corporate image. However, it is in those industry sectors with a high public profile and where the link between improved health and safety and profitability is most direct (for example, chemicals and hazardous facilities) that most has been done to insist that contractors and suppliers comply with the requirements of SMSs.[95] In industries such as these, considerable improvements in the performance of contractors and subcontractors may be achieved even in the absence of outside intervention.

However, in many other sectors, very little action has so far been taken to influence the behaviour of contractors or subcontractors by these means. For example, in the hospitality and construction industries (where the use of contractors is extremely common) the link between spending on health and safety and profitability is a tenuous one, and is widely perceived as such. For industries such as these, much more is likely to be gained by legislation and by the use of incentives.

Turning first to the legislative approach, much has been achieved through the use of deeming provisions in legislation. Specifically, in the construction industry many of the problems that result from the traditional approach of making each employer responsible only for the behaviour of their own employees (leaving many gaps where no one is responsible) can be overcome by specifying, for example, that the principal contractor will have overall responsibility for everything on the site and imposing duties on everyone who is a part of the construction procurement process. Thus, in the UK, the aim is to influence the behaviour not just of contractors and subcontractors but also designers, planners, supervisors, and implementers of projects.[96]

In this sector, again to quote a senior regulator: 'there is a limit to what morals and ethics and business advantage will persuade people to do. But a lot can be achieved by legislation and through fear of a prosecution. It's fear of penalties that drive people on. If you make something a legal requirement it achieves a response. Make it a law, with penalties, and most people will comply. Here, the law is the kick to get the cultural changes going.' For similar reasons, it is arguable that a licensing system should be introduced in relation to the contracting out of particularly hazardous services such as asbestos removal or the treatment or removal of hazardous material or wastes.

Turning to the role of incentives, a requirement that the employer's management system be extended and applied to contractors/subcontractors could appropriately be built into many of the specific incentives systems we proposed earlier. For example, such a requirement might be made a requisite for tendering for government contracts. By so doing, Track Two enterprises (most of whom are likely to be large companies with considerable leverage) would be prevailed upon to bring about important changes in the OHS behaviour of those with whom they contract: changes which would be much harder to achieve through a traditional direct legislative requirement. Government might also impose similar conditions in terms of its own outsourced labour, which might also have a ripple effect, 'encouraging industry associations and smaller organisations to establish similar measures'.[97]

In essence, we may be witnessing the first stages of a gradual transformation in the roles of the state and the market in addressing OHS issues. It is clear that there are substantial limitations to the state's capacity to regulate directly the behaviour of certain sectors and certain economic relationships (most notably subcontracting and small business). In part, in some sectors, the market can compensate for these inadequacies (see the role of supply-chain pressure in sectors where large companies perceive a self-interest in policing the behaviour of non-employees with whom they deal). Yet, in other sectors, the market, unaided, in unlikely to bring about a socially acceptable result.[98] Rather, the perceived self-interest of employers is in outsourcing on the cheapest possible terms, irrespective of the OHS record of those with whom they deal. In these latter circumstances, the state still has an important role to play, but while on some occasions this may involve regulating directly (as in the construction industry) on many others, as we have seen, the state's most effective role is in providing incentives which encourage third parties (large employers) to act as surrogate regulators.[99]

NOTES

1 In dealing with Track One enterprises, it must be remembered that they are likely to include a disproportionate number of SMEs, incapable, by virtue of their limited size, lack of sophistication, or economic circumstances, of embracing Track Two. Studies have shown that management in smaller workplaces generally prefers a regulatory system which clearly outlines the requirements and prescriptions of the legislation to be complied with. In practice most small employers rely on the health and safety inspectors to tell them what the relevant regulations are and to advise them on how to comply and are likely to continue to do so. If so, then 'an entirely different set of strategies may need to be identified for channelling OHS information to small business' (Australian Council of Trade Unions (ACTU), *Occupational Health & Safety*, Submission Number DR336 to the Industry Commission, 1 June 1995, p. 10).

2 A. Hopkins, 'Patterns of Prosecution', in R. Johnstone (ed.), *Occupational Health and Safety Prosecutions in Australia: Overview and Issues* (Centre for Employment and Labour Relations Law, The University of Melbourne, 1994), p. 177.

3 D. Walters (ed.), *The Identification and Assessment of Occupational Health and Safety Strategies in Europe*, Volume 1: The National Situations (European Foundation for the Improvement of Living and Working Conditions, Dublin, 1996), p. 191.

4 Industry Commission, Australia, *Work, Health & Safety: Inquiry into Occupational Health & Safety*, Volume I Report and Volume II Appendices, Report No. 47 (AGPS, Canberra, 1 Sept. 1995), p. 103.

5 See *Policies on Health and Safety in Thirteen Countries in the European Union*, Volume II: The European Situation (European Foundation for the Improvement of Working Conditions, Dublin, 1997) at p. 45.

6 HSE (UK), *Health and Safety in Small Firms* (HSE, 1995, c150, 12/95), p. 9.

7 Inspectors in the UK HSE report that as a result of increases in their administrative responsibilities they are now involved in inspections only an average of one day a week, compared to two or three in the past.

8 W. B. Gray & J. T. Scholz, 'Analysing the equity and efficiency of OSHA enforcement' (1991) 3(3) *Law and Policy* 185.

9 Their data indicates that a 1% increase in inspection hours for each category of plant would reduce total injuries by 296 in mid-size plants and 249 in large plants, compared with only 100 in small plants (see W. B. Gray & J. T. Scholz, 'Analysing the equity and efficiency of OSHA enforcement' (1991) 3(3) *Law and Policy* 185 at 193). Their conclusions on this point are consistent with those of previous studies cited in their article.

10 W. B. Gray & J. T. Scholz, 'Analysing the equity and efficiency of OSHA enforcement' (1991) 3(3) *Law and Policy* 185.

11 Ibid.

12 We discuss this further, Ch. 7 below.

13 W. B. Gray & J. T. Scholz, 'OSHA Enforcement and Workplace Injuries: A behavioural approach to risk assessment' (1990) *Journal of Risk & Uncertainty* 283.

14 J. T. Scholz, 'Cooperation, Deterrence and the Ecology of Regulatory Enforcement' (1984) 18 *Law and Society Review* 179.

15 R. B. Cialdini, *Influence: Science & Practice*, 2nd edn. (Scott Foresman & Co., Glenview, Ill., 1988).

16 W. B. Gray & J. T. Scholz, 'Does Regulatory Enforcement Work? A Panel Analysis of OHSA Enforcement' (1993) 27 *Law and Society Review* 177 at 198.

17 Such action will be particularly effective if an inspector makes clear it is part of a 'pyramidal enforcement strategy', described below at pp. 155–62.

18 There is also some evidence that longer inspections have much greater deterrent effect than superficial inspections which check only the plant's injury records. However, it would be dangerous to generalize from this to conclude that shorter inspections, generally, are less effective. See W. B. Gray & J. T. Scholz, 'Analysing the equity and efficiency of OSHA enforcement' (1991) 3(3) *Law and Policy* 185.

19 This study only focused on the specific deterrent effect of regulations on individual plants inspected. However, OHS enforcement also has a general deterrent effect on un-inspected plants.

20 Strong penalties should be invoked (consistent with the enforcement pyramid

considered below) against those recalcitrants who, whether through intransigence, incompetence, or irrationality, are unwilling to comply, even when 'tapped on the shoulder' by the regulator. Without such a deterrent, it is most likely that the minority of recalcitrants will continue to ignore their legal and moral responsibilities. As indicated in Chs. 3 and 4, such penalties should, wherever practicable, be directed against individuals as well as, or instead of, against the corporation.

21 These are falls from heights, electrocution, crushing injuries, and being struck by material or equipment.

22 B. Clinton (President) & A. Gore (Vice-President), *The New OSHA: Reinventing Worker Safety and Health*, National Performance Review, White House, Washington DC, May 1995, p. 4.

23 Comcare Australia, *Analysis of Data from the 1994–95 Planned Investigation Program* (AGPS, Canberra, 1996). Similarly, in Australia, Queensland's enforcement programme now has a random audit component, in conjunction with an emphasis on enterprise self-audit.

24 J. T. Scholz, 'Cooperation, Deterrence and the Ecology of Regulatory Enforcement' (1984) 18 *Law and Society Review* 179; and J. T. Scholz, 'Cooperative Regulatory Enforcement and the Politics of Administrative Effectiveness' (1991) 85(1) *American Political Science Review* 115.

25 See below p. 110.

26 See below p. 129.

27 The Queensland experience, where targeting is based on workers compensation data, incident and accident reports, complaints, and emerging health and safety issues, is also of note in this regard. See Submission by Department of Employment, Vocational Education, Training and Industrial Relations, Queensland, *Occupational Health & Safety* (1994) to the Industry Commission Australia, (1995).

28 Enterprises are rated under a variety of headings and a weighted sum calculated as a risk estimate. This score is then used to prioritize inspections. Relevant factors include: health hazards and risks; safety hazards and risks; welfare standards; public hazard potential; and confidence that management will maintain or improve standards and the number of years since the premises were last inspected.

29 See further Health & Safety Executive (UK), *Review of Health and Safety Legislation* (London, HMSO, 1994).

30 A. Hopkins, *Making Safety Work: Getting Management Commitment to Occupational Health and Safety* (Allen & Unwin, Sydney, 1995). However, note that changes to the WorkCover scheme have restricted the opportunities for claims management.

31 See South Australian Government, *Occupational Health & Safety*, Submission Number 147 to the Industry Commission, 18 Oct. 1994, p. 33.

32 Research undertaken in Australia by the Queensland Workplace Health and Safety Council revealed that workers compensation statistics are poor measures of the status of workplace injury and disease and their use for targeted purposes is extremely limited. See: (a) Workers Compensation Data: A Poor Indicator of Workplace Injury; and (b) Disease Division of Workplace Health and Safety Queensland 1994.

33 Lost time injury frequency rates, in themselves, are proving to be less than reliable. See E. A. Emmett, 'Regulatory Reform in Occupational Health and Safety in Australia', Paper presented at *The Institute of Public Affairs Conference on Risk, Regulation and Responsibility*, Sydney, July 1995, pp. 621–2.

34 For example, neither lost time, injuries nor medical treatment injuries would be appropriate for measuring safety performance in the construction industry, where there is little security of employment, where injury may result in a worker being laid off, and where reporting of such injuries is unreliable. Within Australia, the Queensland approach, where considerable work has gone into developing alternative database sources, may be particularly useful. This suggests the following possibilities:

• using workers compensation data subject to recognized limitations by developing a database coded to Workers Compensation National Data Set requirements;

• using a random audit programme to collect compliance data which can predict levels of compliance in particular industries or workplaces and decide on targeting priorities;

• monitoring industry, workplaces, or work practices which have been identified in external research as likely to put persons at risk of development of injury or disease;

• establishing data collection systems in conjunction with general practitioners, hospitals, and other health providers; and

• conducting specific purpose samples (e.g. a rural fatalities study).

35 See further C. Diver, 'A Theory of Regulatory Enforcement' (1980) 28(3) *Public Policy* 257 at 264; and Industry Commission, Australia, *Work, Health & Safety: Inquiry into Occupational Health & Safety*, Volume I Report, Report No 47 (AGPS, Canberra, 1 Sept. 1995), p. 415.

36 Because of fears of regulatory capture, and corruption, OSHA inspectors were given far less discretion than their counterparts in most of Western Europe and Australia, and are required, when they identify a violation, to issue a citation.

37 See further Ch. 5.

38 See B. M. Hutter, 'Regulating Employers and Employees: Health and Safety in the Workplace' (1993) 20(4) *Journal of Law and Society* 452.

39 K. Hawkins, *Environment and Enforcement* (Oxford University Press, Oxford, 1984), p. 4.

40 See B. M. Hutter, 'Regulating Employers and Employees: Health and Safety in the Workplace' (1993) 20(4) *Journal of Law and Society* 452.

41 E. Bardach & R. Kagan, *Going by the Book: The problem of regulatory unreasonableness* (Temple University Press, Philadelphia, 1982).

42 S. Kelman, *Regulating Sweden, Regulating America* (MIT Press, 1981); and D. Vogel, *National Styles of Regulation* (Cornell University Press, 1986).

43 N. Gunningham, 'Negotiated Compliance—A Case of Regulatory Failure' (1987) 9 *Law and Policy* 69.

44 For empirical support for this justification, see J. Hodges, 'Eight Years of Robens Style Legislation in Queensland—What have we learnt?', Paper presented to *Productivity, Ergonomics and Safety Conference*, Ergonomics Society of Australia, 24–7 Nov. Canberra, 1996.

45 See also I. Ayres & J. Braithwaite, *Responsive Regulation* (Oxford University Press, Oxford, 1992), who make a very similar claim.

46 P. James, 'Reforming British Health and Safety Law: a framework for discussion' (1992) 21(2) *Industrial Law Journal* 87 at 97.

47 We return to a more detailed discussion of deterrence theory in Ch. 5.

48 For a comprehensive review of the literature see R. Kagan, 'Regulatory enforcement', in D. Rosenbloom & R. Schwartz (eds.), *Handbook of Regulation and Administrative Law* (Dekker, New York, 1994). See also I. Ayres & J. Braithwaite, *Responsive Regulation* (Oxford University Press, Oxford, 1992), and the discussion below.

49 See further N. Gunningham, *Safeguarding the Worker* (Law Book Company, Sydney 1984), Ch. 11.

50 E. Bardach & R. Kagan, *Going by the Book: The problem of regulatory unreasonableness* (Temple University Press, Philadelphia, 1982).

51 I. Ayres & J. Braithwaite, *Responsive Regulation* (Oxford University Press, Oxford, 1992), pp. 35–41.

52 J. Rees, *Reforming the Workplace: A study of self-regulation in occupational health and safety* (University of Pennsylvania Press, 1988).

53 J. T. Scholz, 'Cooperation, Deterrence and the Ecology of Regulatory Enforcement' (1984) 18 *Law and Society Review* 17.

54 P. Nonet & P. Selznick, *Law and Society in Transition: Towards responsive law* (Harper and Row, New York, 1978).

55 W. G. Carson & R. Johnstone, 'The Dupes of Hazard: Occupational health and safety and the Victorian sanctions debate' (1990) 26 *Australian and New Zealand Journal of Sociology* 126.

56 J. Braithwaite, 'Responsive Business Regulatory Institutions', in C. Cody & C. Sampford (eds.), *Business, Ethics and Law* (Federation Press, 1993), p. 88.

57 Ibid.

58 I. Ayres & J. Braithwaite, *Responsive Regulation* (Oxford University Press, Oxford, 1992).

59 J. Braithwaite, 'Responsive Business Regulatory Institutions', in C. Cody & C. Sampford (eds.), *Business, Ethics and Law* (Federation Press, 1993), p. 88.

60 For example, there is evidence that trust fosters the internalization of regulatory objectives by regulated managers. In the case of nursing home regulation, nursing homes experience improved compliance after regulatory encounters in which facility managers believe they have been treated as trustworthy. See J. Braithwaite & T. Makkai, 'Trust and Compliance' (1994) 4 *Policing & Society* 1.

61 Australia Post, *Occupational Health & Safety*, Submission Number 86 to the Industry Commission, 1994.

62 Department of Industrial Relations, *Occupational Health & Safety* (Portfolio Response), Submission Number DR338 to the Industry Commission, 31 May 1995.

63 B. Clinton (President) & A. Gore (Vice-President), *The New OSHA: Reinventing Worker Safety and Health*, National Performance Review, White House, Washington DC, May 1995, p. 3.

64 For example, in Queensland, the Division of Workplace Health and Safety provides audit documents that enable enterprises to self-audit and to illustrate how risks can be identified and controlled, thereby encouraging all

organizations to take greater internal responsibility for risk management. By publicizing in advance the key audit criteria (which the inspectorate also use in random audits) workplace compliance may be facilitated even without an inspector's visit (see Queensland Department of Employment, Vocational Education, Training and Industrial Relations, *Occupational Health & Safety* Submission Number 79 to the Industry Commission, 30 Sept. 1994). Similarly, the Victorian SafetyMAP, developed by the Health and Safety Organisation, is designed to provide a framework to facilitate continuous improvement in OHS, and includes self-audit as one of its core components (see Victorian Government, *Occupational Health & Safety*, Submission Number 382 to the Industry Commission, 1994).

65 H. Humphrey III, 'Public/Private Environmental Auditing Agreements: Finding better ways to promote voluntary compliance' (1994) 3 *Corporate Conduct Quarterly* 1–4 at 2.

66 See also 'Note: Encouraging Effective Corporate Compliance' (1996) 109 *Harvard Law Review*, 1783–888 at 1795.

67 See further the discussion of sentencing guidelines in Ch. 4.

68 H. Humphrey III, 'Public/Private Environmental Auditing Agreements: Finding better ways to promote voluntary compliance' (1994) 3 *Corporate Conduct Quarterly* 1–4 at 3.

69 Ibid.

70 See, further, D. Dawson, P. Willman, A. Clinton, & M. Bamford, *Safety at Work: The limits of self-regulation* (Cambridge University Press, Cambridge, 1988).

71 W. B. Gray & J. T. Scholz, 'Does Regulatory Enforcement Work? A Panel Analysis of OHSA Enforcement' (1993) 27 *Law and Society Review* 177. See also A. Hopkins, *Making Safety Work: Getting Management Commitment to Occupational Health and Safety* (Allen & Unwin, Sydney, 1995), p. 90 for anecdotal evidence in support of the same point.

72 Personal communication, J. Cox, WorkCover Authority, 9 Feb. 1996. See also A. Hopkins, 'Patterns of Prosecution', in R. Johnstone (ed.) *Occupational Health and Safety Prosecutions in Australia: Overview and Issues* (Centre for Employment and Labour Relations Law, The University of Melbourne, 1994), p. 90.

73 R. M. Brown, 'Administrative and Criminal Penalties in the Enforcement of Occupational Health and Safety Legislation' (1992) 30(3) *Osgoode Hall Law Journal* 691.

74 See summary and evidence cited by Industry Commission, Australia, *Work, Health & Safety: Inquiry into Occupational Health & Safety*, Volume I Report and Volume II Appendices, Report No. 47 (AGPS, Canberra, 1 Sept. 1995), Volume I, p. 118.

75 On the contrary, in NSW, the introduction of on-the-spot fines coincided with an increase in prosecutions.

76 For example, an appeal mechanism.

77 For example, see W. G. Carson, *The Other Price of Britain's Oil* (Martin Robertson, UK, 1982). In their submissions to the 1995 Industry Commission inquiry, most agencies explicitly opposed a deterrence-oriented approach on the basis either that it was likely to be counter-productive, or that a compliance strategy, alone, would be appropriate in almost all circum-

stances. Australia-wide, if a prima facie breach of the OHS legislation is detected, there is only a 6% chance of a conviction and fine by the courts.

78 Our empirical research was based in New South Wales, and, in a previous study, elsewhere in Australia. While we have no detailed evidence, informal discussions with regulators suggest that our conclusions are likely to have much wider application.

79 F. Haines, *Corporate Regulation: Beyond 'Punish or Pursuade'* (Clarendon Press, Oxford, 1997), at p. 221.

80 F. Haines, *Corporate Regulation: Beyond 'Punish or Pursuade'* (Clarendon Press, Oxford, 1997), at p. 221.

81 See European Agency for Safety and Health at Work, *Priorities and Strategies in Occupational Safety and Health Policy in the Member States of the European Union* (European Agency for Safety and Health at Work, Bilbao, Spain, 1997), pp. 42–5; and S. A. Shapiro & R. S. Rabinowitz, 'Punishment versus Cooperation in Regulatory Enforcement: A Case Study of OSHA' (1997) *Administrative Law Review* 713 at 171.

82 For an analysis of the strategies available to the OHS agency, see S. A. Shapiro & R. S. Rabinowitz, 'Punishment versus Cooperation in Regulatory Enforcement: A Case Study of OSHA' (1997) *Administrative Law Review* 713–62 at 724–9.

83 For an example of the types of issues which might be taken into account in determining an appropriate response, see the United States Department of Labor's Occupational Health and Safety Administration Field Inspection Reference Manual, 26 Sept. 1994, particularly Directive CPL 2.80 (Handling of Cases to be Proposed for Violation-by-Violation Penalties), and the four factors upon which civil penalties under s. 17 of the OSH Act are to be assessed.

84 See F. Haines, *Corporate Regulation: Beyond 'Punish or Pursuade'* (Clarendon Press, Oxford, 1997), especially Ch. 2, pp. 7–10.

85 I. Ayres & J. Braithwaite, *Responsive Regulation: Transcending the Deregulation Debate* (Oxford University Press, Oxford, 1992), pp. 40–1.

86 C. Mayhew, M. Quinlan, & L. Bennett, *The Effects of Subcontracting/Outsourcing on Occupational Health and Safety*, Executive Summary, Industrial Relations Research Centre, University of New South Wales, Sydney, Australia, 1996, p. 137.

87 There is evidence that subcontracting is associated 'with increased economic competition, work disorganisation, regulatory failure, and a divided workforce. These factors had adverse OHS effects on all workers in these industries but especially marginal or peripheral workers' (C. Mayhew, M. Quinlan, & L. Bennett, *The Effects of Subcontracting/Outsourcing on Occupational Health and Safety*, Executive Summary, Industrial Relations Research Centre, University of New South Wales, Sydney, Australia, 1996).

88 C. Mayhew, M. Quinlan, & L. Bennett, *The Effects of Subcontracting/Outsourcing on Occupational Health and Safety*, Executive Summary, Industrial Relations Research Centre, University of New South Wales, Sydney, Australia, 1996, p. 61.

89 See above p. 110.

90 N. Gunningham, 'Environmental Management Systems and Community

Participation: Rethinking Chemical Industry Regulation', forthcoming (1998) 16(2) *UCLA Journal of Environmental Law and Policy*.

91 See N. Gunningham, 'Environment, Self-Regulation, and the Chemical Industry: Assessing Responsible Care' (1995) 17(1) *Law and Policy* 57.

92 See further C. Mayhew, M. Quinlan, & L. Bennett, *The Effects of Subcontracting/Outsourcing on Occupational Health and Safety*, Executive Summary, Industrial Relations Research Centre, University of New South Wales, Sydney, Australia, 1996.

93 M. Quinlan, *The Development of OHS Control Systems in a Changing Environment: An international perspective*, School of Industrial Relations and Organizational Behaviour, University of NSW, Workshop on Integrated Control/Systems Control, Dublin, 29–30 Aug. 1996.

94 For an excellent study of how a safety system can fail in the absence of these conditions, see C. Wright, 'A Fallible Safety System: Institutionalised Irrationality in the Offshore Oil and Gas Industry' (1994) Feb. *The Sociological Review* 79–103.

95 On the role of North Sea oil companies in playing this role, see P. Lindoe, *Internal Control* (Stavanger College, Rogaland Research, Stavanger, 1995).

96 See the Construction Design Management Regulations 1994/5.

97 C. Mayhew, M. Quinlan, & L. Bennett, *The Effects of Subcontracting/Outsourcing on Occupational Health and Safety*, Executive Summary, Industrial Relations Research Centre, University of New South Wales, Sydney, Australia, 1996.

98 See also A. Hopkins & L. Hogan, 'Influencing Small Business to Attend to Occupational Health and Safety' (1998) 14 (3) *Journal of Occupational Health and Safety—Australia and New Zealand* 237–44, arguing that the majority of small businesses are routinely in contact with national supplier companies and with accountants, bankers, and insurance brokers and that the OHS authorities could use these points of contact to construct OHS leverage.

99 In bringing about such a transformation, organized labour may also play a significant role (though it has not so far done so) by negotiating, into collective agreements, similar conditions which oblige employers to take responsibility for the OHS problems of contractors and subcontractors.

5   From Adversarialism to Partnership: Track Two Regulation

We have argued that the best strategy is to reward the efforts of good OHS performers, and that negative strategies such as deterrence should be confined to those situations where rewards are unlikely to be effective. The evidence suggests that the best results are likely to be achieved by those who adopt a SMS. By implication, those doing so commit themselves to continuous improvement, to going beyond compliance with existing regulation, and to establishing a safety culture within the enterprise. Such an approach, in its present form, is intended to achieve a partnership between business and government. We will argue that this partnership should be extended to workers as well.

All three parties will benefit from this arrangement, provided it is implemented effectively and applied to the range of enterprises best suited to it (described above). Since primary responsibility is placed on employers themselves to find optimal means of reducing occupational injury and disease, they gain the benefits of considerable autonomy, flexibility and cost-effectiveness. In particular, this approach enables them to prioritize and to tackle their worst safety hazards first, and with the most efficient methods available. Governments also gain an opportunity to pass the principal responsibility back to those at the workplace to regulate themselves, leaving the agency free to redeploy its scarce regulatory resources elsewhere.[1] Finally, workers will also be beneficiaries, since a safety systems approach promises to deliver greater health and safety benefits than conventional regulation.[2] Indeed, this is ultimately the greatest gain for all three parties.

Earlier, we suggested the use of various incentives to encourage enterprises to adopt a systems-based approach, while recognizing that not all enterprises would have the willingness or, in some cases, the capacity to do so. In this Chapter we explore the questions of how to design a regulatory system appropriate for those who choose to adopt a systems-based approach under Track Two, and of how regulators should respond to this development. How best, can regulators move from adversarialism to partnership?

From inspection to audit

Primary responsibility to provide and update a SMS lies with the employer. However, there will be a need for that system and its implementation to be monitored, otherwise there is a risk that, either deliberately or through ineptitude, the system will fail to achieve its stated goals. How then should this task be addressed, and does this imply a redefinition of the appropriate role of the inspectorate and the nature of inspection?

The emphasis under the traditional regulatory approach, described under Track One in the previous Chapter, is on inspection, and on compliance with specification, performance, and, more recently, process standards. The main aim is to bring enterprises up to the minimum level of performance prescribed by law, and inspection, coupled with enforcement, will continue to play an important role in achieving this.[3] Indeed, in the case of small employers, inspection may be one of very few credible strategies available.[4] There is little point trying to conduct an audit based on the firm's own documentation, when there is little, if any, such documentation available.

In contrast, under Track Two, the emphasis will be on the performance of the system, and on the effectiveness of the company's internal monitoring of that system, in achieving levels of OHS performance which go 'beyond compliance' with existing standards. That is, audit and oversight of the system itself will take priority over conventional inspection and enforcing the rules, because the former will be more effective than the latter in ensuring that the system is operating effectively. Auditing, for example, 'is a much more detailed review of procedures beyond what can merely be observed at a workplace',[5] and provides an independent assessment of the validity and reliability of the management planning and control systems. What are the implications of a shift from inspection to audit, and of enforcing a systems-based approach?

The problem of resources

A core problem in enforcing a systems-based approach is that of resources. Checking whether an enterprise has genuinely and successfully adopted and implemented this approach not only requires greater and different skills on the part of the inspectors, but is also extremely demanding of inspectors' time. These demands may be particularly intense during the period when the new system is being introduced, where the dangers of implementation failure are greatest. The difficulties are exacerbated by the fact that inspectors have to deal not just with one uniform 'off the peg' system but with a number of different systems applied to the particular contexts of particular workplaces. Even after the system is in place, it will require change to accommodate to new hazards, with the result that the inspectorate has to address a moving rather than a static target.

Accordingly, if a systems-based approach is to be administratively viable, ways must be found to ease the burden on regulatory resources. Solutions to this problem are fundamentally important, for without them the entire Track Two regulatory approach may founder. There are a number of credible ways of addressing this problem. They fall into two broad categories: first, various strategies which encourage effective self-regulation on the part of enterprises that commit themselves to a SMS; and second, various forms of third party oversight. The former approach transfers a substantial part of the regulatory burden onto employers themselves, the latter onto third parties who can act as surrogate regulators.

Strategies to encourage self-regulation by individual enterprises

In terms of making SMSs work effectively, and in the public interest, there is a key role for utilizing self-regulation by individual firms,[6] for it is the enterprise itself that has the greatest capacity for making a systems approach work to optimal effect. Moreover, enterprises are more committed to rules they write and enforce themselves, and such rules can be tailored to match each enterprise's needs and functions.[7] Given extremely limited inspectoral resources, self-regulation is also the only way that the workplace can realistically be monitored on a continuous basis. However, experience suggests that pure self-regulation, without outside scrutiny or oversight, is rarely capable either of overcoming the gap between public and private interest, or of providing the credibility necessary for public acceptance.[8] For these reasons, those regulatory analysts who nevertheless see virtue in self-regulation have commonly advocated its use in conjunction with some form of government or third party oversight, or both.

Our own views have been influenced by this literature.[9] The model we propose, which seeks to devise self-enforcing, self-referencing SMSs, is also influenced by the literature on reflexive law. Reflexive law recognizes the limitations of command and control regulation, in terms of its limited impact, its rigidity, and its tendency to produce regulatory overload. Instead, reflexive law 'focuses on enhancing the self-referential capacities of social systems and institutions outside the legal system, rather than direct intervention of the legal system itself through its agencies, highly detailed statutes, or delegation of great powers to the courts . . . [it] aims to establish self-reflective processes within businesses to encourage creative, critical, and continual thinking about how to minimise . . . harms and maximise . . . benefits'.[10]

In effect, this is essentially the aspiration of OHS management systems, which seek to encourage internal self-critical reflection within enterprises about their OHS performance, and to establish processes and procedures that encourage self-reflective learning and thinking about reducing occupational injury and disease, rather than seeking to influence behaviour directly by proscribing certain activities.

At a practical level, the core issue is how to ensure that enterprises do indeed establish mechanisms that are self-monitoring, self-correcting, and self-improving. It is crucial that they do not, intentionally or otherwise, produce the trappings of self-regulation without delivering the promised outcomes in terms of a safety culture, a commitment to continuous improvement, and, as a result, improved OHS performance.

It will be argued that an effective model for the self-regulation of Track Two enterprises should contain the minimum components which we set out below. There will also be a need for oversight; either directly by a regulatory agency or indirectly by a third party but underpinned by a safety net of government regulation. Such oversight will be critical in circumstances where self-regulation fails to deliver on its promises. These aspects of our model are dealt with in subsequent sections.

Essential prerequisites for effective self-regulation

To ensure that SMSs do indeed deliver the results of which they are capable, it is essential that governments impose certain minimum requirements. First, in order to minimize the likelihood of abuse, Track Two should only be available to enterprises which can demonstrate an existing high level of OHS performance. Second, governments should require participating Track Two firms to commit themselves to a number of 'bottom lines'. In particular, each participating enterprise would be required to commit itself to implementing a SMS with prescribed minimum components. These would be identified in the code of practice on SMS proposed earlier. Minimum requirements under such a code might include:

- periodic review;
- education and training of personnel at all levels;
- documenting and monitoring activities and results;
- setting up procedures for investigation and corrective action in cases of system failure;
- establishing internal and external communication procedures concerning OHS practices;
- establishing operational procedures designed to maintain 'operational control';
- keeping a register of legislative, regulatory, and other policy requirements; and
- keeping a register of OHS incidents deemed significant at a participating site.[11]

The UK HSE's approach also encapsulates the central features of a successful health and safety management (Box 4 below). Thus while the details of a SMS would be unique to each individual firm, the minimum criteria which the system must satisfy would not.

Of the above criteria, some in particular deserve emphasis. Prime examples are corrective and preventive action. If a responsive and reflexive system is put in place, capable of reacting positively to failures, variations, preventive action ideas, and improvement suggestions, then the evidence suggests that there will be substantial pay-offs in terms of improved OHS. Accordingly, a participating enterprise could be required to commit itself to specified internal procedures designed to ensure that problems are identified and dealt with, so that the system does indeed become self-regulating. These procedures would include investigation of suspected failures of the system and breaches of legislation, followed by internal actions to remedy the situation.[12] It would also be required that employees report any suspected violation to appropriate officials within the organization and that a record be kept by the enterprise of any such report.[13]

It is also essential that there is provision for internal auditing, monitoring, reporting, and tracking systems, including continuous on-site monitoring by specifically trained personnel and independent checks on the status of the system.[14] For example, following Ayres and Braithwaite, the enterprise could be required to establish an independent inspectoral/

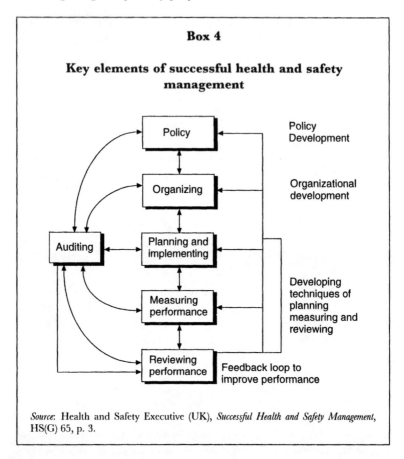

Box 4

Key elements of successful health and safety management

Source: Health and Safety Executive (UK), *Successful Health and Safety Management*, HS(G) 65, p. 3.

audit group, responsible for identifying problems, taking action, and reporting directly to a nominated member of senior management, who themselves would be required to 'sign off' on the system, undertaking to government that the system did indeed meet government criteria and that it was being implemented effectively.[15] Both the internal audit unit and the nominated manager might also be required to report to government any major breakdown in the system or any major incident involving serious risk to OHS. However, as a quid pro quo for such voluntary disclosure, regulators would routinely provide a 'period of grace' to rectify the defects, rather than using such notification as a basis for prosecution.[16]

Finally, borrowing from the proposed USA sentencing guidelines for environmental offences, monitoring would also be required to directly involve line managers, so that the organization as a whole is forced to internalize OHS compliance, and line managers in particular, are required to confront it on a daily basis.[17] This is because:

> In the day to day operation of the organisation, line managers, including the executive and operating officers at all levels, direct their attention, through the management mechanisms throughout the organisation (eg, objective setting, progress reports, operating performance reviews, departmental meetings) to measuring, maintaining and improving the organisation's compliance with [OHS] laws and regulation. Line managers routinely review . . . monitoring and audit reports, direct the resolution of identified compliance issues, and ensure application of the resources and mechanisms necessary to carry out a substantial commitment.[18]

An excellent example of how self-regulation can be made to work in practice is that of the VPP, described in Chapter 3. Under the self-inspection and hazard-correction requirement, an employer must describe its hazard assessment procedures in detail, show how hazard assessment findings are incorporated in planning decisions, training programmes, and operating procedures, and agree to provide to the OSHA its self-investigation and accident investigation records, its safety committee minutes, its monitoring and sampling results, and its annual safety and health programme evaluation. It also pledges to correct in a timely manner all hazards identified through self-inspections, employee reports, or accident investigations, and to provide the results of these investigations to its employees.

More recently, a number of USA and Canadian initiatives have expanded on this partnership approach.[19] Indeed, the Clinton–Gore Reinventing Worker Safety and Health initiative asserts that the 'New OSHA' will change its fundamental operating paradigm from one of command and control to one that provides employers with real choice between a partnership and a traditional enforcement relationship.[20] Consistent with this the OSHA is adopting 'common sense regulation' by 'identifying clear and sensible priorities, focussing on key building block rules, . . . emphasising interaction with business and labor in the development of rules . . . [and] insisting on results instead of red tape'.[21]

Compatible with this philosophy, the OSHA's Draft Proposed Safety and Health Program Standard contemplates a standard requiring employers to take a systematic approach to addressing safety and health hazards, including obligations to identify and prioritize all hazards in terms of their seriousness and to track progress in controlling them. Other elements of the programme include employee participation and an emphasis on flexible performance-based obligations.[22]

In Canada, there is an increasing emphasis on 'building relationships with employers and workers and sharing responsibilities for health and safety on the job'.[23] Perhaps the most notable of the Canadian initiatives is the Workers Compensation Board of British Columbia's Diamond Project, created in 1996, and modelled on the VPP.[24]

For 'best practice' companies, the requirements identified, under the VPP, above may not be particularly onerous. For others, they might appear, at least in the short term, to be unduly demanding. It may well be that (following a pilot phase in which only best practice companies are permitted to participate in Track Two), in order to attract sufficient numbers of enterprises to Track Two, these requirements might be relaxed, at least in the short term. If so, a full-fledged system might appropriately be introduced after a significant transitional period. SMEs, in particular, should be encouraged rather that discouraged from participating in Track Two.[25] Although, as we indicate below, it is important to balance the provision of incentives with a provision to deter abuse.[26]

Having said that, it must be emphasized that some at least of the above requirements should be attainable, even by less advanced enterprises. SMSs, because they are substantially based on principles of TQM, can readily be designed to be self-referencing and self-enforcing. Indeed, an essential element of quality management systems is their built-in capacity for self-correction, whereby 'plans are made, carried out, assessed for effectiveness, and improved in an ongoing, formal and systematic manner'.[27] Thus, when a problem is detected within the OHS system, it should be designed so that there is a prompt, efficient, and effective process which automatically intervenes to correct the deficiency in the system. Indeed, only if the deficient procedure or process is corrected will the TQM system be working successfully.

In part in recognition of the need to design more innovative forms of regulation to meet the needs of the worst, as well as those of the best, enterprises, the OSHA introduced its CCP in 1997, reducing the number of inspections for plants that establish health and safety management programmes, and targeted to facilities with the worst records. However, unlike the VPP, which met widespread industry support, the CCP is encountering considerable dissension and legal action from those

who feel threatened by increased OSHA scrutiny if they do not participate.[28] At present, it seems, unlikely that the initiative will be successful in
providing a model whereby some of the worst Track One firms might
ultimately transcend to Track Two.

The role of regulatory or third party oversight

Although self-regulation on the model described above might achieve
a great deal, much will also depend on the effectiveness of oversight mechanisms, for even the most credible self-regulatory mechanism
may succumb to the temptations of short-term self-interest, in the absence of outside forces capable of blowing the whistle and keeping it on
track.[29]

Where regulatory resources are not already greatly overstretched,
then oversight of systems-based self-regulation might best be conducted
by the regulatory agency itself. This is undoubtedly the most desirable
solution because it minimizes the potential shortcomings of competence,
independence, and integrity entailed in using third parties, described
below. Moreover, by retaining a central role for the inspectorate in
auditing Track Two firms, it reinforces its information and intelligence
gathering functions (enabling it to remain a centre of knowledge and
expertise on industry hazards and on the history of individual enterprises). These functions might be damaged if the oversight role is passed
on to third parties.

Oversight by the regulatory agency itself is the approach adopted in
Norway and Sweden under their systems of IC, described earlier.[30] In the
UK, where resources are more limited, this approach is relied upon in
selected areas, most notably chemicals and major hazardous installations. It is also relied upon in circumstances where a 'permissioning'
regime is in place (whereby a facility can only operate once its systems
have received approval). For example, under the UK off-shore oil
installation safety regime, referred to earlier, the written safety case is
subjected to a 'desktop' or 'paper' audit by the regulatory authority,
which, once it has been accepted, becomes a set of legal requirements
against which the operator is assessed. Further guidelines are issued by
the regulatory authority which also fulfil the function of a semi-structured assessment methodology for assessing safety cases.[31]

In countries such as these, with sufficient regulatory resources to fulfil
the oversight role (or, as in the UK, to the extent that this is partially so),[32]
the most important questions concern how the inspectorate's role should

be modified to accommodate to its new task (an issue we address be-low[33]). However, oversight and vetting of SMSs impose very consider-able burdens which some less well-resourced agencies have argued are impractical. For example, the South Australian government maintains that:

> . . . ensuring that all risks have been addressed and that the duty of care has been complied with, would require a very high degree of auditing by government. Even then, to ensure that health and safety is protected at any time within an enterprise requires a dynamic system that responds to changes in plant, equip-ment, substances or processes. Government could not audit and provide ap-proval with the necessary frequency to ensure the ongoing maintenance of health and safety. Nor is such extensive auditing by government a desirable objective. It is completely contrary to the spirit and intent of self-regulation.[34]

Moreover, for regulators to themselves audit enterprises adopting SMSs is to focus on a very resource-intensive activity which may well have the ironic result of transferring resources to the very best industries (i.e. those who volunteer for Track Two) and, by implication, away from those which are in most need of regulatory scrutiny (and assistance).

For the many jurisdictions that do not have anything approaching adequate inspectoral resources, there are compelling pragmatic reasons for seeking to pass off at least some of the oversight role on to appropriate third parties, in order to preserve scarce regulatory resources to address problems for which there are no alternatives to government regulation. The remainder of this section is concerned with precisely this issue: how best to use third parties as partial surrogates for government regulation, and the residual role of government regulators as an under-pinning where self-regulation or third parties fail do deliver their prom-ised outcomes.

The most obvious form of third party oversight would be by audit conducted by an independent OHS professional.[35] OHS audits can provide systematic, documented, periodic, and objective reviews of whether OHS requirements are being met or whether systems are being adhered to.[36] In particular, such audits involve 'the structured process of collecting independent information on the efficiency, effectiveness and reliability of the total SMS'.[37] The audit might examine a 'vertical slice' of an enterprise's OHS activities (for example, examining one specific aspect in each of the elements identified) and/or a 'horizontal slice' (examining one particular element of the safety management in detail) with the overall goal of assessing the validity and reliability of the man-agement planning and control systems.[38]

A third party audit of the SMS could be relevant on a number of occasions. For the purposes of accreditation, completion of a systems audit and compliance audit could generally be required, the precise scope of the audit being agreed in consultation with the enterprise concerned.[39] After the system is in place, periodic verification audits could be required to establish 'whether the SMS is doing what it is claimed to do in its extent and quality, and whether this is adequate as operated. Validation audits . . . focus on such matters as whether the right kinds of subsystems and components are being adopted, whether the correct types of monitoring are being done and whether appropriate subsystems are in place.'[40]

Since the costs of the audit would be required to be borne by the enterprise concerned, the system has very few costs for regulators. In effect, the cost of inspection is transferred to the regulated entity. While in strict economic theory this may be desirable (in terms of internalizing the costs of OHS) it is far less attractive to regulated enterprises themselves. Some may regard this as an acceptable trade-off for the greater flexibility, autonomy, and other benefits that a systems-based approach may provide. However, if third party audit implies the need for enterprises to be formally certified as in compliance, then the costs may be heavy and a significant deterrent to participating in Track Two, particularly for SMEs. Significantly, USA, Canadian, and German employer interests all opposed the development of an ISO standard on OHS management systems because of the likely cost of certification under such a standard. Accordingly, one important practical hurdle that must be overcome is to ensure that costs do not blow out as a result of the generation of a formal certification system and the excessive documentation that this might generate. This suggests the emphasis should be away from the mechanistic approach which characterized ISO 9000 product quality certification with its associated heavy documentation, regimentation, and use of statistics, and towards a more flexible, integrated, problem-solving style emphasizing simple, consultative, and team-based models.[41] While it is still too early to be confident of precisely how to avoid the serious pitfall of overburdensome and overexpensive third party certification, there are at least some alternative models available, which suggest cause for optimism. One is the experience of the Australian best practice approach which has provided an informal and flexible integration style.[42]

Another option, which would be much more acceptable to many employers, would be to follow the model of the European Eco

Management and Audit System (EMAS), and rely upon internal rather than external audits.[43] While this option is very appealing in terms of costs, it raises concerns about the independence and integrity of such audits.[44] It will be some time before it is clear whether the EMAS solution to this problem—a system of independent verification of a random sample of audits—will suffice to maintain not only the integrity of the audit system but also public (and worker) perception of that integrity.[45]

There is another serious difficulty in the strategy of utilizing third party auditors as surrogate regulators, namely the tension between the regulator's interests and those of the regulated enterprise. From the regulator's point of view, third party audits work best if the auditor's report is made accessible to the regulatory agency and does not remain confidential as between auditor and enterprise (which, after all, is footing the bill). However, such a requirement is likely to be unattractive to the enterprise itself, which may understandably fear that it is providing the regulatory agency with considerable information (and ammunition, in the event of a prosecution) which would otherwise not be available. There is thus a tension between the regulator's need to be reassured that it will be alerted to unsatisfactory audit results (enabling it to take corrective action) and an enterprise's reticence to adopt Track Two if required to make full disclosure of the audit report.[46]

If the regulator insists on full disclosure in every case, then Track Two may become insufficiently attractive for many enterprises to agree to participate.[47] Such an outcome would be counter-productive, and a solution must be found which is acceptable to both sides. The most satisfactory compromise might be one whereby only an overview or summary of the audit is ordinarily supplied to the regulatory agency by the auditor, indicating the conclusions, but not the details, of the audit. Thus the latter, including any specific identified breaches of the legislation, would remain confidential to the regulated enterprise. The fact that an audit itself is to be treated as a privileged document should be clearly indicated, either in enforcement guidelines or in the legislation itself.[48] However, as indicated below, where the audit summary indicates a generally unfavourable report, or major failings in individual aspects, this should be one circumstance triggering an inspection by the regulatory agency.

While this solution may serve to alleviate the fears of regulated enterprises, it does far less to assure the regulator that the audit system is working satisfactorily, that the auditors are operating in the public interest, and that they have not been captured by the client enterprise. To overcome these problems, and to ensure the integrity of the audit process, the government agency should have a right to spot-check (and verify) a random sample of full audits.

Even in this latter circumstance, the information gained from the audit report could not sensibly be used as a basis for enforcement action, for, if it were, it would provide a substantial and unnecessary disincentive to adopting a Track Two approach. Few companies would agree to participate in the scheme.[49] More appropriately, the agency might undertake to give participating firms a 'period of grace' to rectify problems revealed by the audit.[50] Indeed, any agency wishing to encourage the establishment of incentive-based management systems in conjunction with third party audits might be well advised to follow the stated policy of the USA Federal Aviation Agency (FAA):

I am announcing a major change in the FAA's enforcement policy. Simply stated, it is this: if you discover an inadvertent violation, correct the problem, report it promptly to the FAA, and put in place a permanent fix acceptable to FAA to make sure it will not happen again, the FAA will not penalize you . . . Period.

In other words, we want to encourage carriers to shift their resources from contesting punitive enforcement actions to making their operations safer. Internal evaluation promises to benefit the aviation industry and the FAA by allowing each of us to use our resources more positively, intelligently and effectively.[51]

However, even under these conditions, it must be acknowledged that only companies which are firmly committed to rectifying any OHS problems which an audit might identify are likely to participate under Track Two, because although regulatory agencies may provide a 'period of grace', having been alerted to the existence of a hazard, they are likely to insist on its removal in the long term.

In principle there is a further problem. Information passed on to the regulatory authorities as a result of audit may, in some jurisdictions, become publicly available through the operation of Freedom of Information legislation.[52] As a result, it could readily cause adverse publicity and damage the company's public image or even form the basis for third party legal action against the company. This latter problem, at least, could be overcome if the confidentiality of such audits were expressly preserved as between the firm and the regulatory agency. On balance, despite our general support for the principle of 'right to know', we believe such confidentiality protection is necessary in this context[53] in order to avoid deterring enterprises from adopting Track Two.

While it is argued that third party audits coupled with random spot checks of a minority of audits by the government regulator is the best approach, alternatives do exist. For example, rather than spot checks, the confidentiality of *all* audits might be preserved, with the integrity of the audit process being maintained through independent verification (either by the government regulator or a third party) of the audit summary or overview. This is the model adopted by the EU's EMAS. That scheme also provides for internal (but 'independent'[54]) in-house auditors to conduct the audit itself should the enterprise prefer this to the hiring of a third party.

On both counts, the European scheme lacks the degree of independence, and hence credibility, that is desirable. However, its political attractions, at least in gaining industry support, are obvious. Against this, there are the two serious objections to the European model touched upon earlier. First, the degree of scrutiny involved in an independent verification of a summary report falls far short of that provided by a spot check on the details of the audit itself. Second, an 'in-house' audit lacks the arms-length independence (and thus credibility) that only a third party or government regulator can provide.

Because the adoption of Track Two will involve substantial processual requirements, and significant development costs, it is likely to be limited to enterprises which have responded to one or more of the incentives identified in Chapter 2. It is important that, in return, the regulator not only provides the incentives described, but also provides a period of grace for the enterprise to rectify any breaches of legislation identified by the audit. Without these minimum requirements, together with the fact that audit implies the waiver of regular inspection, it is unlikely that Track Two will be sufficiently attractive for regulated enterprises to enter the partnership it implies.

A further problem with relying on third party audits is ensuring the professional integrity and independence of the auditors. For there remains the possibility of co-option of the auditors by the firm seeking accreditation—a hazard illustrated by the failure of financial auditors in the financial scandals of the 1980s. This problem is exacerbated by the fact that there are no universally agreed standards for carrying out safety audits.[55] Neither is there agreement as regards the professional requirements for auditors. However, generic audit standards do exist which may in time be further refined to apply specifically to OHS.[56]

There are a number of possible ways of dealing with these issues, none of them totally satisfactory. The two most important require-ments are random spot checks by the government regulator (audits of the audits) and a provision that the auditor be nominated by the regula-tor from a pool of accredited auditors, rather than by the regulated enterprise. Another is to follow the approach of the EU's EMAS and appoint third party verifiers who, in effect, themselves audit the auditors. Further possibilities include state regulation of auditors, an auditor accreditation scheme, peer review, civil liability, and the establishment of a set of national standards relating to the quality and scope of OHS audits.[57]

A final and important possibility is that the audit team be required to include a trade union or worker nominee of the relevant trade union (or of the work group), who might commonly be drawn from the workplace safety committee (where one exists) or be a work-place safety representative. Since, in most jurisdictions where there is provision for safety representatives, such worker representatives will have already undertaken an approved OHS training course and have, in most cases, considerable practical experience of OHS, their pre-sence on an audit team would make a useful practical contribution. It would also serve to maintain the credibility of the audit team, reduce the chances that it is being 'captured' by the regulated enterprise, build in a valuable 'whistle blower', and build the possibility that the partnership implied under a systems-based approach be extended to include workers and their representatives. A further attraction of this proposition is that it builds on an existing mechanism for worker partici-pation in OHS, which could readily be adapted for this additional purpose.

The inspectorate as backstop

Notwithstanding the important roles of self-regulation and (for some jurisdictions) of third party oversight, there will remain a basic function which the inspectorate itself must perform. As we have indicated, there will be temptations on those who self-regulate to cut corners and mini-mize costs in the short-term. Some enterprises, rather than genuinely implementing a systems approach in order to improve their OHS per-formance, may be tempted to simply devise cosmetic 'paper systems' to keep the regulators off their backs or to gain other perceived advantages. For example, they may seek to benefit from various incentives offered to Track Two enterprises while seeking to avoid the reciprocal obligations. How can enterprises be prevented or deterred from abusing Track Two in this way, and how will regulators or courts be able to distinguish between paper systems and the genuine article?

As regards this latter problem, many prosecutors in the USA have doubted 'both the utility of compliance plans and their own ability to distinguish serious efforts at compliance from merely cosmetic plans'.[58] The issue has arisen in the context of sentence hearings, because agreement to introduce corporate compliance plans (including a commitment to a systems-based approach) can lead to a sentence reduction in respect of environmental crime.[59] Here, the experience is that these plans are easily manipulated, with 'a virtual cottage industry of law firms cranking out compliance plans for their corporate clients (often with the mechanical uniformity of a cookie cutter)',[60] leading one to doubt that adoption of such plans would have much beneficial impact on corporate behaviour, at least until clear minimum criteria are prescribed (on which see further below).

At this stage, we do not have the experience of SMSs under a Track Two approach to know how seriously this approach might be abused. Early experience with monitoring the systems of self insurers and major tenders in the construction industry, suggests that inspectors/auditors can readily identify company system failures.[61] Nevertheless, given the obvious temptations and the experience of related areas, it seems likely that agency strategies to counter this problem will be essential to the successful operation of Track Two. This is particularly the case since third party oversight, while important, has its own limitations identified above, and cannot be relied upon, exclusively, to ensure the efficacy of Track Two.

When should the regulator intervene to ensure that the system is being complied with, and that an enterprise, through intent, inefficiency, or incapacity, is not failing to discharge its obligations under OHS legislation? As a practical matter (given the resources problems indicated earlier), it is essential that such agency intervention is within its budgetary and administrative capability.

The regulatory design that we propose involves a tiered regulatory response. First, it is designed to encourage Track Two enterprises to regulate themselves (as indicated above, one of the prerequisites to entry into Track Two will be that the SMS is self-referential and self-correcting). Second, comes third party oversight, both at the stage of accrediting the system when it is introduced, and through subsequent periodic audits (with a reserve power of inspection, but recognizing that Track Two enterprises would be a very low inspection priority). Thus the third party audit fulfils a substantial role as surrogate regulator. However, there is also need for a third tier, involving an underpinning of government regulation which 'kicks in' *as a back-up mechanism* in circumstances where there is reason to believe that tiers one and two have not delivered the required outcomes in terms of system-effectiveness and improvements in OHS.

What circumstances might be appropriate to trigger an inspection by the regulatory agency? We suggest four:

- if the worker representative, having exhausted internal procedures, complains to the agency that the audit was not, in his/her view, fair and accurate;
- if the third party audit report itself expresses serious reservations about the effectiveness of the SMS;
- if a regulator's verification of the third party audit (conducted randomly on a small minority of audits) suggests that the audit itself was not fair and accurate; and
- if there is a serious accident or incident, or a series of complaints from workers.

When one of these circumstances does trigger action on the part of the regulator, this will usually involve two tasks. The first will involve an audit of the SMS itself (i.e., a relatively quick and limited examination of the system). The regulatory burden in conducting this audit will be lightened by the fact that a Track Two enterprise, by virtue of having established a SMS, can be expected (or be required) to have created considerable documentation on the system's performance—in effect, a 'paper trail'—which will make an inspector's task much easier. We describe how such an audit might be conducted, and its general focus, below.

The second will involve a fuller audit, which seeks to determine whether individual performance and specification standards have been met, based not just on the system, but on the extent to which the enterprise is achieving the OHS *outcomes* required by legislation. We emphasize the need to focus on outcomes in these circumstances because systems themselves are process-based, and, although when well constructed and implemented, can be anticipated to provide required outcomes, cannot be guaranteed to do so.[62] We considered this point in more detail above when examining the appropriate relationship between specification, performance, and process-based standards, where we emphasized the need to insist that Track Two enterprises remain subject to the same outcome-based standards as other enterprises.

Our suggested approach raises a further question of what action to take with a Track Two enterprise whose SMS is failing. We recommend different responses depending upon whether the enterprise is identified as incompetent or recalcitrant. Incompetents can be given advice as to how to improve, but where, after a period, it seems that the system is not leading to improvements, then that firm should be given a period of grace in order to rectify breaches of the law (punishment is not an appropriate short-term response for firms which have genuinely attempted a Track Two approach but failed) and then returned to Track One.

However, in the case of those who deliberately abuse Track Two, a different response is required. If Track Two is to maintain its credibility, a betrayal of the trust implicit in the accreditation and self-regulatory arrangements must not only lead to loss of accredited status but also to rapid escalation up the enforcement pyramid. This is a necessary response to discourage foot-dragging by firms that claim to be regulating themselves, but who use this as a guise to avoid their legal obligations. The justification for such rapid escalation is that the privileges[63] bestowed by Track Two also imply obligations. Where industry is largely trusted to regulate itself, devising its own solutions in ways which best meet its own particular circumstances, so a betrayal of that trust must be seen to have serious consequences sufficient to deter the minority who might otherwise be tempted to do so (see Figure 4 below).

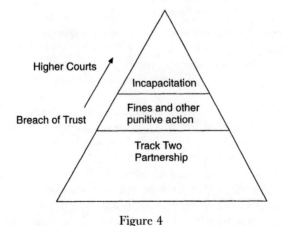

Figure 4

Having identified the circumstances in which a WorkCover inspection/audit would be triggered, we turn to the question of what precisely it should involve, and what implications this has both for the design of the SMS and for the skills required of government inspectors.

Track Two: From inspection to audit

Under Track Two, the inspectorate would take the internal management system rather than breach of regulations as their point of departure.[64] This will be the case irrespective of whether the inspectorate takes the principal oversight role (as in Sweden, Norway, and, partially, in the UK) or whether its role is as an underpinning to third party oversight, only being triggered in exceptional circumstances (as we proposed above for less well-resourced inspectorates).

In either event, the role of the inspectorate shifts in emphasis from that of conventional inspection of the premises and plant (which relies heavily on observation of site conditions) to that of audit of the management system. The latter implies a review of written documentation coupled with interviews with personnel at various levels of the organization to test the understanding of written practices and the extent to which they can and are being followed, together with inspection of selected areas of the plant based on interviews and the information that has come to light.[65] The result is that the inspectorate relies not only on its judgment of the management system, but seeks to validate that judgment by detailed examination of how well it has actually been implemented in specific contexts (for example, are subcontractors actually monitored and trained, to the extent the system claims?[66]). In essence, there are three steps: a desktop exercise of examining the system itself; the process of testing it on the ground to see whether it is being implemented at the enterprise; and finally, testing whether it actually achieves the results it claims to do.

For example, an opening question in a Track Two inspection/audit would be 'what is your management system in respect of OHS?' The inspector would then investigate how far the company's actual practice compares with that plan. Related questions (all of which a company with a management plan should be able to answer) are: how do you hold management accountable for OHS? What measurements do you use of OHS performance? How do you make OHS 'stick' as you go down the management chain? How do you set OHS goals and targets and how do you monitor performance against your plans?

Thus, under this approach, much of the audit can be based primarily on the paperwork and on interviews with personnel at all levels. (Can I see your quality manual? What are your targets with respect to . . . ? Show me your performance indicators regarding. . . .) If there is a disparity between what is expected to happen under the system, and the firm's actual practice, this is likely to be revealed by such an investigation. As one British inspector put it: 'you can tell if they're concealing the truth or disguising hazards—someone will depart from the script or simply not know what the paperwork claims they know'. If the firm seeks to fudge the system, to claim that it is doing far more than is actually the case, then someone will know. There will be the risk, which the enterprise cannot ignore, that somebody (most likely an individual worker or the relevant trade union) will 'blow the whistle'. Moreover, an OHS audit will also look at conformance with rules and standards to obtain evidence to test the effective implementation of the system.

Crucial to the success, both of the system and its audit, will be the development of appropriate performance measures, for it is these which determine whether OHS plans have been implemented and key objectives achieved. Provided such measures are in place then it should be possible for either a third party auditor or the inspectorate to audit the system effectively. In effect, the agency (or third party) will say, 'you draw up a plan and we will inspect you against it'. The performance and other indicators that will allow this to be done should be determined during the initial baseline audit during which the enterprise and the agency/or third party auditor agree on both performance indicators and benchmarks,[67] and the latter satisfies itself about their adequacy for their purpose. The outcome should be an agreed plan of how the enterprise intends to proceed and how to measure progress against the baseline.

For example, the enterprise could be required, through its documentation, to demonstrate that potentially hazardous conditions are recognized and corrected, that it has in place detailed hazard assessment procedures, effective self-inspection and hazard-correction procedures, and to show how hazard assessment findings are incorporated in planning decisions, training programmes, and operating procedures, and agree to provide the agency with its self-investigation and accident investigation records, its safety committee minutes, its monitoring and sampling results, and its annual safety and health programme evaluation. It might also be required to demonstrate how hazards identified through self-inspections, employee reports, or accident investigations are rectified, and to provide the results of these investigations to its employees.[68]

Identifying appropriate performance indicators has been one of the principal challenges for organizations applying TQM principles to their OHS management system.[69] More recent approaches represent a considerable advance, having the capacity to monitor the system both reactively and proactively, and including much more sophisticated principles of performance measurement and measurement techniques.[70] As a result, the problems an agency (and indeed the enterprise itself) is likely to encounter in identifying and utilizing appropriate performance indicators are likely to substantially diminish, both as our knowledge expands, and as appropriate and sophisticated 'off the peg' SMSs develop, enabling measurement and assessment to be based on relevant, appropriate, and consistent standards.[71]

The development of appropriate performance indicator techniques and of the regulator measuring performance against these will be crucial, because management systems in themselves are process rather than outcome-based mechanisms. That is, simply because the management system is functioning effectively does not in itself guarantee that the enterprise itself is functioning safely,[72] for reasons we identified earlier.

Of particular importance, in terms of ensuring a system's effectiveness, will be the practice of benchmarking: the process of measuring an enterprise's method, process, procedure, and results against those enterprises that consistently distinguish themselves in that same category of performance. The Australian Manufacturing Council (AMC), amongst others, has correctly identified benchmarking as the most important factor that distinguishes leaders from laggards in the pursuit of best practice environmental management.[73] As the AMC point out: 'Without the reference point that benchmarking provides, it is difficult for companies to embark on a comprehensive change program designed to achieve best practice.'[74] As Worksafe Australia has already demonstrated, government can play a positive role in setting industry benchmarks and fostering best practice.[75] Others, including the Workers Compensation Board of British Columbia under its[76] Worksafe Strategy are also paying increasing attention to planned data collection and analysis initiatives.

Among the additional benefits of systems-based audit is the very considerable gearing it provides in dealing with large, multi-site organizations, in which the same basic management system is likely to apply to all sites. As a result, a single audit of one site, which identifies a substantial flaw in the system as it operates on that site, is likely to have implications for all the company's other sites, and to result in remedial action, even in the absence of separate inspection of those sites.

The role of internal audits

In many circumstances, enterprises will also choose to conduct their own internal audits. Indeed, most systems require an enterprise to conduct internal audits as part of the overall system. Such audits offer considerable opportunities for self-examination, provide essential feedback to senior management on the workings of the system, and can often make a considerable contribution to the improvement of OHS generally, and to the efficiency of SMSs in particular.

As a matter of public policy such internal audits are highly desirable and to be encouraged. Nothing that we have said concerning the disclosure of third party audits should have application to internal audits. Indeed, it is fundamentally important either that the latter are treated as confidential and privileged documents, or, if the regulator is to have access to those documents, that industry is protected against self-incrimination in respect of them. At the very least, the regulatory agency should state clearly in enforcement guidelines that it would in no circumstances prosecute on the basis of evidence obtained from internal audits.

The need for highly skilled and flexible regulators

The shift from inspection to audit based on the SMS has a number of implications for regulators. One of the most important is that it requires regulators who are sufficiently skilled and adaptable so as to be able to supervise and oversee a broader approach, focusing not on the specific rules, but on the effectiveness of the system itself. That is, it requires professionally qualified and technically skilled inspectors of whom far more is required than has been the case in the past.

This shift is already happening in some jurisdictions. For example, in the UK, inspectors are trained in a diagnostic approach and increasingly examine a firm's policy, organization, and systems in the course of routine inspections of large organizations. In future, the inspectorate intends to 'adapt inspections to focus attention on health and safety management, for example by making more use of techniques for auditing management systems'.[77] Significantly, in the UK, graduate entry into the health and safety inspectorate is required, and both in the UK and Scandinavia there is a higher degree of professional skills than is common in some other inspectorates, notably the USA and Australia. Moreover, the inspectorate is specifically trained to examine quality systems.

Whether inspectors in countries such as the USA and Australia will readily adapt to a systems-based approach, and whether they have the technical skills to do so, remains an open question, although early experience with audit-based approaches is positive.[78] However, past experience of the transition from specification to performance standards suggests that there may be similar problems of adjustment in the move from conventional inspection to audit and oversight. For example, in The Netherlands many inspectors have strongly resisted a shift to a more systems-based approach. Wilthagen's research indicates that 'inspection and enforcement routines so far do not show a significant change. On the contrary, SC is to a large extent still a matter of semantics: inspectors get on with "business as usual".'[79] Similarly, there is a risk of heavy-handed inspectors mechanistically insisting on technical compliance with every detail of the system while missing the essential point, which is whether the system is performing well in practice.

The experience of Victoria, the only Australian jurisdiction where detailed empirical research has been conducted, is that inspectors with very modest technical skills, who had been accustomed to applying only specification standards, were most reluctant to enforce performance standards. In essence, it is far easier to ascertain whether a guard is on the machine, or a stair rail is of the requisite height, than it is to be sure that an employer failed to do what was 'reasonably practicable' in the circumstances of the case to measure outcomes.[80] They were also understandably concerned at the difficulties of obtaining successful prosecutions under the latter requirement, as contrasted with the relatively clear-cut nature of prosecutions under the former.[81] Recent reports from The Netherlands suggest a similar problem in that jurisdiction, not only with inspectors as mentioned above, but in addition with prosecutors and courts also showing reluctance to prosecute employers for malfunctioning of the company's internal system in the absence of specific breaches of concrete rules.[82]

As indicated above, these earlier problems have now very largely been overcome. However, one might anticipate a similar problem of transition from traditional inspection to auditing systems, given the skill levels required to implement the latter approach. However, if a greater skills and specialization base is indeed acquired (and only a retraining or reorientation of a limited part of the inspectorate is required to achieve this), there are considerable advantages to be gained. First, there are the predicted benefits to OHS performance that a systems-based approach can provide, but which are more certain with effective oversight to keep enterprises 'on track'. Second, a more skilled inspectorate is likely to gain far greater insight into a firm's OHS performance via its SMS than by conventional inspection methods. This in itself contributes to enforcement becoming more selective, and in known offenders being targeted more effectively.

The role of trade unions

The approach we advocate under Track Two is a tripartite rather than a bipartite one, and involves an important role for workers and the organizations that represent them.[83] Such worker involvement is essential both to the effective operation of Track Two and to its credibility. Workers have the most direct interest in OHS of any party: it is their lives and limbs that are at risk when the law fails to protect them. Moreover, workers often know more about the hazards associated with their workplace than anyone else, for the obvious reason that they have to live with them, day by day. In particular, the hazards at work need to be identified and evaluated, and workers' experience and knowledge is crucially important in successfully completing both of these tasks.

In principle, quality management approaches also recognize the importance of worker participation and of communication between workers and management to the successful implementation of a systems-based approach. Above all, they recognize that those who are genuine partners, and effective participants in a systems-based approach to regulation, will have the greatest commitment to it.

However, in practice, a management systems approach may fall far short of these ideals.[84] There is a serious danger that: 'such systems are used solely or mainly as management techniques used by company management to attain their own pre-set objectives'.[85] For example, Larsson has documented how the goal of improving OHS can be displaced by the goal of removing/making redundant those with the greatest vulnerability to injury, or those with long-term health problems, since the latter approach will provide a more immediate dividend in terms of short-term profit.[86]

Such a result is particularly likely if a management system is developed by the company from above without input from employees. This is indeed a potentially serious problem: the experience with implementing management systems in the case of North Sea oil installations is that this has in fact lead to poorer contact with the supervisory organ for employees and contractors and less genuine worker participation.[87] The risk that a systems-based approach may diminish genuine worker participation is also increased by the fact that regulators, in auditing the system, may themselves focus on top management (emphasizing the strategic level), with the unintended result that there is less dialogue with, and role for, worker safety representatives. The fact that it is much easier to measure concrete matters such as accidents, than levels of participation or co-operation (and that surrogate measures, such as numbers of hours of training, are vulnerable to manipulation), also increases the risk that the participatory dimension is downplayed.

The dangers of ignoring worker interests in the design of management systems are exacerbated by the reliance on models derived from the earlier generation of quality and EMSs. Yet there are important differences between systems which are applied to product quality and environmental protection on the one hand and OHS on the other. As a result, a management systems approach which has been applied to the former may not be appropriate to the latter. As the General Confederation of Spanish Workers' Unions has put it:

The company does not own the health of its employees at work; it cannot be treated as something to be hired out or transferred under a contract. There are no moral or legal grounds for seeing it as something that can be bought and sold. The same applies to accident prevention. Health and safety management cannot be left exclusively to the tender mercies of the employer, which is precisely why it is governed by labour law.[88]

Finally, there is an underlying tension for management between effective implementation of OHS management systems (which implies worker participation) and managerial prerogatives.[89] This has been encapsulated by Tombs as follows:

As occupational safety has come to be recognised as less about technical and more about organisational factors, there has emerged a growing recognition among regulators and on the part of some corporations that safety management requires some form of ongoing role for workers, both individually and through their collective organisation. This intervention can range from the ability of individuals to raise warning signals, which are taken seriously by managements, to the right of workers' groups to negotiate changes in plant, technology or work organisation. For corporations to act upon the recognition that safety cannot effectively be managed by 'managers' alone is in the interests of profitability if the negative economic effects of accidents and incidents are avoided; yet in this very recognition, managements also create a possible threat to their own right to manage.[90]

How, then, might the dangers of a management system approach be overcome, and how best might workers and trade unions function as a countervailing force, ensuring that systems function in practice consistent with their purported OHS objectives? We have already indicated a number of ways in which trade unions and workers might be involved under Track Two. In this section, we provide an overview of these various participatory measures, and of the relationship between them, in order to demonstrate that worker participation in this scheme is not token or *ad hoc*, but involves a genuine partnership between employers, government, *and workers*. We also suggest means by which these measures might become embedded in management practice.

In our view, for Track Two to be genuinely tripartite, and effective, then worker and trade union involvement need to be built in at each step. This means beginning with the standards setting process itself. As Lindoe puts it:

... the employees possess key competence and their 'tacit knowledge' can be replaced by neither management nor experts. They must gain real entrance to, and influence on, the arenas where both quality and internal control programs *are formulated*. The systems therefore *have to be developed* within a corporate culture providing the employees with a legitimate and natural role in the process.[91]

All standards development, and in particular the creation of a code of practice on SMSs,[92] should involve a tripartite process. Again, when an enterprise decides to adopt Track Two and to develop a management system, then the process of determining the objectives of the system itself should directly involve workers' representatives, as should the accreditation process. For example, this might also involve consultation by the third party auditor with a worker-nominated member of the workplace OHS committee or a worker nominee might become a member of the accreditation audit team. The role of the worker representative would be to satisfy themselves that the accreditation process was conducted fairly and properly. Such a person could play a similar role in respect of subsequent third party audits of the workplace, subject to undertakings of confidentiality in respect of details of the audit document which would normally have privileged status. Members of safety committees and safety representatives could also be directly involved in enterprise self-regulation and self-correction of system failures, by being involved in regular participative workplace inspections.[93] At a subsequent stage, workers would play important roles both as 'whistle-blowers' and in triggering agency inspections, thereby providing checks and balances, and acting as a countervailing force to any tendency to co-option of the regulator by the regulated enterprise. For example, as Frick points out, management system performance indicators are open to management manipulation, so the success of systems-based approaches (such as IC) depends 'on its alliance with the less powerful but more adaptive internal OHS actors, who know the reality behind the indicators. Especially employee participation can both be enhanced by the procedures which [systems control] *can* enforce, and fill these procedures with real content.'[94] In this context it may be particularly important to protect worker anonymity and protect against victimization. Indeed, the entire strategy of workers as 'genuine partners' and of direct worker participation can only work if adequate protection is provided by legislation against co-option and discriminatory action.[95] This includes both an adequate empowerment to stop an operation (i.e., a right to refuse hazardous work) and legal protection from employer victimization for OHS-related activities. Even if such rights are in place, experience suggests they are much more effective in industries where trade unions are strong and safety conscious than in the (now increasing) sectors where this is not the case.[96]

The most obvious way of ensuring that the above proposals are implemented, rather than simply paid lip-service to, would be to make such participation a prerequisite for Track Two participation, thereby embedding the principle of worker participation in a systems-based approach. The preservation or extension of largely existing worker rights (rights of warning, to withdraw to a place of safety, to information[97] training, and consultation, and a guarantee against reprisals by employers) would also be essential. Equally, from the union side, success requires a willingness to integrate OHS directly in negotiations with employers, and to shift from a largely representative role to one of co-ordinating (and training) workers to participate directly in this process.

Other recommendations made earlier also have implications for the role of workers and unions. For example, the requirement that Track Two companies issue a public statement summarizing the results of the OHS audit makes the audit process more transparent to unions. Similarly, there are now opportunities for unions themselves to take direct enforcement action, as discussed in Chapter 4. The proposal for a steep escalation of the enforcement pyramid, in the case of enterprises which betray the trust placed in them under Track Two, should also serve to reassure unions that Track Two does not involve any abdication of the regulatory agency's enforcement role but quite the contrary. Some of the new sanctions proposed, such as adverse publicity, could also be utilized directly by unions to their advantage.

Track Two might also be characterized as involving a shift from an inspection model to a disclosure model.[98] Workers often receive insufficient information to assess the risk to them from hazardous plant, equipment, and substances used at work, and this problem has most serious consequences in respect of exposure to diseases with long latency periods. This has prompted recommendations that OHS legislation should contain a right to know and a duty to tell,[99] and note that such provisions would also enhance the capacity of workers to participate in OHS issues and to more effectively oversee employer commitment. In particular, broader disclosure requirements would enable trade union and worker representatives to cross check the accuracy of audit summaries and raise questions if they doubt their veracity, and in other ways act as 'whistleblowers' where enterprises fail to live up to their commitments under Track Two. Information is also likely to be a crucial factor in building trust and co-operative action.

Finally, one particular problem under a genuinely tripartite approach is how to avoid situations where trade union or worker involvement results in demands for enforcement which are counter-productive to the success of a Track Two approach. Such a situation is most likely to arise where a firm which has in place an accredited SMS nevertheless has a serious incident or accident. Even the very best firms, with the very best systems in place, confront this possibility, even though it is less likely to occur in such firms than elsewhere. If the trade union response to such an incident is to attack the credibility of Track Two as a whole, then the possibilities for a co-operative tripartite systems-based approach are jeopardized. Government, in particular, will face considerable pressures to revert to a traditional regulatory approach.

Arguably the best way to avoid, or at least to minimize, the chances of this outcome is to involve trade unions so fully in the partnership that while they know the cases where tough action is called for (because the company is acting in bad faith) they also recognize where genuine good faith efforts are being made and where tough enforcement, or a reversion to Track One, would be counter-productive.[100]

It must be acknowledged that the potential for directly involving workers in SMSs, in the ways proposed above, varies substantially between countries and cultures. In Scandinavia, a high degree of participation is accepted by all parties as part of historic understandings which emerged out of the industrial democracy programme,[101] and principles of participatory management have been embodied in legislation such as the Norwegian Working Environment Act 1977.[102]

Although less extensive or deeply entrenched, worker participation is also accepted under Australian legislation[103] and is required under a Framework Directive within the EU. However, in the USA worker participation has never played a significant role, and is strongly resisted by employers. As trade unionism continues to decline even further from an initial small base, the chances of incorporating many of the recommendations made above, in that jurisdiction, seem remote in all but a small minority of enterprises. Since such participation is an important factor in the overall success of a management systems approach, it is likely that the USA will experience greater problems than other countries should it embark on this path.

However, even in countries where trade unionism has historically been stronger and more deeply embedded in the negotiation of OHS issues (and where rights to participation are entrenched in legislation), there are increasing problems in maintaining this role. For example, it is well documented that the system of worker appointed safety representatives and worker participation in safety committees works well largely to the extent to which it is supported by strong and safety conscious trade unionism at any particular workplace.[104] Yet trade unionism is in decline internationally.[105] We have also referred to the increase in outsourcing, and the increasing proportion of SMEs in most economies. Both these developments are antithetical to strong trade union participation in workplace safety. Whether, or to what extent, new participatory models, such as territorial representation,[106] or the capacity for large employers (within whose organizations trade unions may be stronger) to impose supply-chain pressure on their suppliers and contractors in respect of OHS system matters,[107] will be exercised, remains an open question. In the meantime, there must remain serious doubt about the viability of extending a systems-based approach to all or most enterprises, serving to reinforce the merits of a two-track approach which encourages this approach for enterprises most capable of adopting it, but does not mandate it.

Conclusion

We have argued that there are considerable opportunities for improving the use of administrative discretion and the functioning of existing approaches to inspection and enforcement of OHS regulation. Even under traditional approaches, there is much that could be achieved through the reallocation of resources, through the targeting of inspections, and through innovative strategies for inspection and enforcement which utilize *and integrate* an effective mixture of carrots and sticks. We indicated a number of specific initiatives that have already achieved significant success (for example, the Maine 200 and Minnesota voluntary compliance schemes) and a number of other opportunities that have not so far been implemented. We also argued the case for a lateral approach to regulation, taking advantage of the considerable leverage that large firms have over smaller trading partners. This would involve using the former as surrogate regulators, who through their management systems would exercize control over the OHS practices of suppliers and contractors.

However, it is in the use of SMSs that we see the greatest potential for achieving improvement in OHS performance. Certainly such an approach has dangers: of implementation failure; of the token adoption of 'paper systems'; of degenerating into a behaviourist approach which results in blaming workers; of top-downism which disempowers rather than directly involves workers; and of a mechanistic, box-ticking mentality which, far from achieving cultural change and continuous improvement, produces merely symbolic, rather than real benefits. However, these difficulties, while significant, are not insurmountable and we have pointed to the considerable evidence which shows that many organizations have been able to overcome these potential problems and to use systems to achieve substantial OHS improvements.

When effectively implemented, a systems approach might contribute to OHS performance at two levels. First, it might ensure that enterprises achieve the legal standard *and are prevented from slipping back below it*. This is essentially the contribution of QA. Many enterprises which, as a result of advice, persuasion, or threats from the regulatory inspectorate, bring themselves up to the legal standard, subsequently fall back below it. Others never achieve the legal standard in the first place. Accordingly, an approach which locks enterprises into achieving and maintaining the legal standard would be a very real achievement. This indeed is the focus of the inspectorate in Sweden, where the inspectorate regards its statutory role as to achieve compliance, and takes the view that if employers achieve more than this, this is largely irrelevant to their mandate.

Yet to stop at the legal standard, without aspiring to more, is to ignore the second, and most potent aspect of systems-based regulation, namely its capacity to lock in a commitment to continuous improvement and cultural change within the organization, to encourage top management commitment and its implementation by middle management, to avoid the 'sidecar model' in such a manner as to take enterprises 'beyond compliance' in an upward spiral of continuous improvement. The Du Pont aspiration of 'zero accidents' encapsulates this mentality. This is the essential contribution of TQM-based approaches.

The challenge for regulatory design is to provide encouragement and incentives for enterprises to embrace the 'continuous improvement' mentality and to find ways to take advantage of the considerable benefits of a systems-based approach to regulation. In our view, this implies a two-track system under which incentives are provided to those enterprises who are proven OHS good performers, who are ready, willing, and able to embrace a systems-based approach (Track Two), while leaving those who are not, the subject of a traditional (albeit improved) regulatory system (Track One).

We have set out an approach for Track Two firms which involves the use of third party auditors as surrogate regulators, a modified role for the inspectorate focusing on audits rather than inspection, and involving considerable resource implications. The latter should result in the redeployment of inspectors away from best practice enterprises (which will be largely self-regulating under an OHS management system) towards those with the worst OHS record which are most in need of scrutiny, advice, and, ultimately, enforcement. However, mindful of the risks of abuse of any self-regulatory approach, we propose the building in of a range of safeguards and oversight mechanisms including third party auditors (and means of maintaining their integrity and avoiding capture), the use of trigger mechanisms prompting government inspections and enforcement action, and a continuing and critical role for workers and trade unions, which seeks to embed their contribution at every stage of the process of designing, implementing, and monitoring management systems.

OHS systems are of recent origin; there is only limited empirical evidence of how they function in practice, and even less of how they might be integrated with regulation. In these circumstances, there is much to be said for an incremental approach, and for experiments involving limited sectors or numbers of enterprises, before embarking (as Norway and Sweden have done perhaps prematurely) on grand-scale regulatory reform. Nevertheless, there is considerable evidence, stated earlier, that the integration of OHS management systems into the regulatory framework offers one of the most potent opportunities for regulatory reform in many years. The model we have proposed is a means of grasping that opportunity.

NOTES

1 See our earlier discussion at p. 45.
2 It may also require greater control and job satisfaction.
3 However, this role is not necessarily exclusive. In particular, there is an expanding role of audits rather than traditional inspection, even for firms which do not adopt SMSs.
4 See A. Hopkins, 'Patterns of Prosecution', in R. Johnstone (ed.), *Occupational Health and Safety Prosecutions in Australia: Overview and Issues* (Centre for Employment and Labour Relations Law, The University of Melbourne, 1994), p. 177. For strong empirical support see C. Mayhew, R. Ferris, & C. Harnett, *An Evaluation of the Impact of Targeted Interventions On the OHS Behaviours of Small Business Building Industry Owners/Managers/Contractors* (National Health and Safety Commission, Australia, 1997).
5 I. Glendon & R. Booth, 'Risk Management in the 1990s: Measuring management performance in occupational health and safety' (1995) 11(6) *Journal of Occupational Health & Safety—Australia and New Zealand* 559.
6 Self-regulation by individual firms must be distinguished from industry association based initiatives. For example, I. Ayres & J. Braithwaite, *Responsive Regulation* (Oxford University Press, Oxford, 1992), distinguish between co-regulation (industry wide voluntary standards that are ratified by government and enforced by industry associations and public interest groups), enforced self-regulation (based on individual firms and government negotiating site-specific regulations, self-policed by the firm itself), and voluntary self-regulation (industry initiatives with little government involvement.
7 See further I. Ayres & J. Braithwaite, *Responsive Regulation* (Oxford University Press, Oxford, 1992), Ch. 4.
8 See N. Gunningham, 'Environment, Self-Regulation, and the Chemical Industry: Assessing Responsible Care' (1995) 17(1) *Law and Policy* 57.

9 For example, Ayres and Braithwaite argue the case for 'enforced self-regulation': an extension and individualization of the more widely-recognized strategy of co-regulation. Specifically, Ayres and Braithwaite are in favour not only of retaining public enforcement of privately promulgated standards, but also of invoking the enforcement pyramid, described above, as a means of discouraging foot-dragging by firms that claim to be regulating themselves, but who use it as a guise to avoid their legal obligations. In effect, then, Ayres and Braithwaite are endorsing a 'halfway house' of publicly blessed, but partly internal, modes of regulation as a means of encouraging more effective compliance. By so doing, they seek to avoid both the stultification of innovation, delay, and excessive costs commonly associated with direct government regulation, and the naïvety of trusting companies to regulate themselves. While their model differs in detail from our own (for example, we argue below the need for third party accreditation of SMSs rather than government approval, and the need for a number of specifics which they do not consider), its conceptual underpinning is similar (I. Ayres & J. Braithwaite, *Responsive Regulation* (Oxford University Press, Oxford, 1992), p. 103).

10 W. E. Orts, 'Reflexive Environmental Law' (1995) 89(4) *Northwestern University Law Review* 1227 at 1232. The term 'reflexive law' derives from G. Teubner, 'Substantive and Reflexive Elements in Modern Law' (1983) 17 *Law & Society Review* 239.

11 This list is adapted from the EMAS regulation of the EU. Significantly, worker participation is absent in EMAS.

12 On strategies to achieve this effectively, see J. A. Sigler & J. E. Murphy (eds.), *Corporate Lawbreaking & Interactive Compliance: Resolving the regulation-deregulation dichotomy* (Quorum Books, US, 1991), p. 82.

13 See United States Sentencing Commission Guidelines Manual, proposed 9D1.1 (J. C. Coffee, 'Environmental Crime and Punishment', Thursday 3 Feb. 1994, *New York Law Journal* 5, 10, and 29 at 29).

14 Ibid.

15 See also the EC's Health and Safety Framework and daughter Directives, which make it obligatory for a 'competent person' to be appointed, responsible for carrying out all OHS responsibilities.

16 See further p. 158–62 below.

17 This point is made by J. C. Coffee, 'Environmental Crime and Punishment', Thursday 3 Feb. 1994, *New York Law Journal* 5, 10, and 29 at 29.

18 See United States Sentencing Commission Guidelines Manual, proposed 9D1.1(a) (J. C. Coffee, 'Environmental Crime and Punishment', Thursday 3 Feb. 1994, *New York Law Journal* 5, 10, and 29 at 29).

19 D. K. Verma, 'Occupational Health and Safety Trends in Canada, in Particular in Ontario' (1996) 40(4) *Annals of Occupational Hygiene* 477–84.

20 J. L. Watson, 'The "New" OSHA: Reinventing Worker Safety and Health' (1998) 12(3) *Natural Resources and Environment* 183; K. A. Kovach, N. G. Hamilton, T. M. Alston, & J. A. Sullivan, 'OSHA and the Politics of Reform: An analysis of OSHA reform initiatives before the 104th Congress' (1997) 34 *Harvard Journal of Legislation* 169–90; and A. Foster, 'Safety—

Voluntary Program Exposes Industry Divide' (1998) 160(17) *Chemical Week* 34.

21 Note reference to legislative reform framed in Bills in the US Senate in 1995 which included initiatives such as Labour-Management Committees; Random Inspection Exemptions; Voluntary Compliance by Certification; Self-Disclosure Privilege; and Program Codifications (see K. A. Kovach, N. G. Hamilton, T. M. Alston, & J. A. Sullivan, 'OSHA and the Politics of Reform: An analysis of OSHA reform initiatives before the 104th Congress' (1997) 34 *Harvard Journal of Legislation* 169–90 at 171).

22 For an earlier 'progress-based' model see US EPA's Risk Management Program Rule 1996, 42 USC s. 7412(r)(7)(B)(ii). See also M. Collins, 'OSHA's safety and Health Program Standard' (1998) Apr. *The Synergist* 24–6.

23 Workers Compensation Board of British Columbia, Annual Report 1994, p. 9. See also N. Keith, *Canadian Health and Safety Law: A comprehensive guide to the statutes, policies and case law* (Canada Law Book, Aurora, Ontario, 1997).

24 K. Rest & N. Ashford, *Occupational Health and Safety in British Columbia: An administrative inventory of the prevention activities of the Workers' Compensation Board* (Ashford Associates, Cambridge, Mass., 1997). See also D. K. Verma, 'Occupational Health and Safety Trends in Canada, in Particular in Ontario' (1996) 40(4) *Annals of Occupational Hygiene* 477–84.

25 Acknowledging the significant role played by small businesses in the nation's economy, the US Congress passed the Regulatory Flexibility Act 1980 requiring Federal agencies to analyse the effects of proposed rules on small entities. In 1997 subsequent legislation, The Small Business Regulatory Enforcement Fairness Act, was introduced which requires that, before publishing a notice of proposed rule making that may have a significant economic impact on many small entities, the OSHA convene a small business advocacy review panel for the draft rule (United States General Accounting Office, *Regulatory Reform: Implementation of the small business advocacy review panel requirements*, Report to Congressional Requestors, GGD-98-36, The Office, Washington DC, 1998). See also United States, Congress, House of Representatives, Committee on Small Business, *The SAFE Act: How third party consultations have worked where OSHA has failed: Hearing before the Committee on Small Business*, US GPO, Washington DC, 1998; and Occupational Health and Safety Small Business Forum, *Occupational Health and Safety Small Business Forum: A forum to assist small business occupational health and safety performance improve* (Australian Chamber of Commerce and Industry, Unley, SA, 1996).

26 See below (question of trust pyramid) at pp. 206.

27 R. Stace, 'TQM and the Role of Internal Audit' (1994) 64(6) *Australian Accountant* 26 at 27.

28 A. Foster, 'Voluntary OSHA Program Exposes Industry Divide' (1998) 160(17) *Chemical Week* 34.

29 N. Gunningham, 'Environment, Self-Regulation, and the Chemical Industry: Assessing Responsible Care' (1995) 17(1) *Law and Policy* 57.

30 See p. 89–90 above.

31 See, further, summary provided in Industry Commission, Australia, *Work, Health & Safety: Inquiry into Occupational Health & Safety*, Volume II Appendices, Report No. 47 (AGPS, Conberra, 1 Sept. 1995), p. 368.

32 However, even in Norway, it is estimated that premises are inspected on average only once every 12 years (O. Saksvik & K. Nytro, *Implementation of Internal Control of Health, Environment and Safety: An Evaluation and a Model for Implementation* (SINTEF IFIM Institute for Social Research In Industry, Trondheim, Norway, 1995)), raising the possibility that Norway, itself, might benefit considerably from the resource savings which third party audits would provide.

33 In the section 'Track Two: From inspection to audit'.

34 South Australian Government, *Occupational Health & Safety* Submission Number DR275 to the Industry Commission, 11 May 1995.

35 An innovative alternative that has been trialled in Alberta, Canada, is a peer evaluation system whereby each participating company agrees to receive the services of a certified independent auditor from a participating company in the same industry group. Whether such a system would work at least as well as one utilizing auditors from outside the industry itself, whether it would result in collusion or the converse (auditors from rival firms exploiting opportunities to disadvantage their rivals), it is too soon to say. This is indeed one area where further empirical evidence is needed and where much may depend on the characteristics of the individual industry.

36 N. Gunningham & J. Prest, 'Environmental audit as a regulatory strategy' (1993) 15(4) *Sydney Law Review* 492–526 and references therein.

37 Health & Safety Executive (UK), *Successful Health and Safety Management* HS(G)65 (HMSO, London, 1991).

38 See, further, F. D. Lindsay, 'Successful Health and Safety Management: The contribution of management audit' (1992) 15 *Safety Science* 387.

39 See, further, the section below, 'Track Two: From inspection to audit'.

40 I. Glendon, 'Risk Management in the 1990s: Safety Auditing' (1995) 11(6) *Journal of Occupational Health & Safety—Australia and New Zealand* 569 at 570.

41 V. Blewett & A. Shaw, '*Quality Occupational Health and Safety?*' (1996) 12(4) *Journal of Occupational Health and Safety—Australian and New Zealand* 481–7.

42 Ibid.

43 See N. Gunningham & P. Grabosky, *Smart Regulation: Designing Environmental Policy* (Clarendon Press, Oxford, 1998).

44 A. Grafe-Buckens, 'Old and New EMAS: Challenges for the European Eco-Management and Audit Scheme' (1997) 6(11) *European Environmental Law Review* 300; R. Hillary, 'Environmental Reporting Requirements Under the EU: Eco-management and audit scheme (EMAS)' (1995) 15(4) *The Environmentalist* 293; M. Palomares-Soler & P. M. Thimme, 'Environmental Standards: EMAS and ISO 14001 Compared' (1996) 5(8/9) *European Environmental Law Review* 24; and M. Deturbide, 'Corporate Protector or Environmental Safeguard? The emerging role of the environmental audit' (1995) 5 *Journal of Environmental Law and Policy* 1.

45 Note that labour had serious objections to EMAS in this respect (see references in n. 44 above).

46 For a solution to this problem, see J. B. Busey, 'US Federal Administration', an announcement before the Aero Club of Washington DC (27 Mar. 1990) (1991) 12 *Cardozo* LR 1327. For a variation on this theme, see generally P. I. Ennis, 'Environmental Audits: protective shield, or smoking guns?' (1992) 42 *Washington University Journal of Urban and Contemporary Law*, 389. See also

TUTB Newsletter July 1996 No. 3, p. 8, European Trade Union Technical Bureau for Health and Safety, Brussels.

47 While self insurers and those who require a SMS to tender for government contracts might remain within Track Two, most others might find the remaining incentives insufficient to do so.

48 The one circumstance in which privilege should not be granted is where the duty holder seeks to invoke the audit in defence to a prosecution, in which case the prosecution should have a right to produce other evidence from the audit which counters this.

49 There is, after all, little incentive to conduct an audit if the information it generates serves to provide a basis for prosecution or other enforcement action.

50 It might be necessary to provide statutory guarantees that information gathered in such an audit cannot be used in any subsequent prosecution action. Such a strategy would work most effectively if the relevant inspectorate adopted a diagnostic role—at least in respect of voluntary audits. That is, it would see its primary means of obtaining compliance as the provision of technical assistance to companies in breach of regulatory standards, keeping advice and policing as quite separate functions. See further J. Braithwaite & P. Grabosky, *Of Manners Gentle: Enforcement Strategies of Australian Business Regulatory Agencies* (Oxford University Press and the Australian Institute of Criminology, Melbourne, 1986).

51 J. B. Busey, 'US Federal Administration', An announcement before the Aero Club of Washington DC (27 Mar. 1990) (1991) 12 *Cardozo* LR 1327. For a variation on this theme, see generally P. I. Ennis, 'Environmental Audits: Protective shield, or smoking guns?' (1992) 42 *Washington University Journal of Urban and Contemporary Law* 389. See also discussion of sentencing guidelines at pp. 282.

52 Australian Commonwealth Freedom of Information Act 1982 sl.1, but noting the exemptions in Part IV of that Act. The scope of some exemptions of particular relevance to matter supplied by business to government has been considered in *Searle Australia* v. *Public Interest Advocacy Centre and Dept. of Community Services and Health* (1992) 36 FCR 111, in particular the meaning of 'trade secrets' in s.43(I)(a), and the scope of s.45 (breach of confidence). New South Wales does not have Freedom of Information legislation.

53 Except where the regulated entity chooses to rely on some or all of those results to establish in court that it exercized 'due diligence', in which case the issue of confidentiality would be at the discretion of the court.

54 The internal auditor must be 'sufficiently independent of the activities they audit to make an independent and impartial judgment' (Council Regulation 1836/93, annex 2C).

55 The practical problems of ensuring the consistency and quality of third party audit would be considerably alleviated to the extent that there is an independent standard according to which such audits must be conducted. This is indeed the model that has been adopted in relation to environmental management systems, and has been contemplated but not, at the time of writing, endorsed, for OHS management systems.

56 M. S. Lopez, 'Application of the Audit Privilege to Occupational Safety and

Health Audits: Lessons learned from environmental audits' (1996) 12(2) *Journal of Natural Resources and Environmental Law* 21; S. C. Yohay, 'OSHA Compels Disclosure of Safety and Health Audits: Smart enforcement or misguided policy?' (1993) 18 *Employee Relations Law Journal* 663–8; and K. A. Kovach, N. G. Hamilton, T. M. Alston, & J. A. Sullivan, 'OSHA and the Politics of Reform: An analysis of OSHA reform initiatives before the 104th Congress' (1997) 34 *Harvard Journal of Legislation* 169–90 at 177.

57 See N. Gunningham, 'Environmental Auditing: who audits the auditors?' (1993) 10(4) *Environmental and Planning Law Journal* 229. See also R. Nadarajah, 'Ontario's Policy on Access to Environmental Evaluations: The creation of audit privilege' (1998) 7(3) *Journal of Environmental Law and Practice* 311.

58 J. C. Coffee, 'Environmental Crime and Punishment', Thursday 3 Feb. 1994, *New York Law Journal* 5, 10, and 29 at 10.

59 See, further, discussion of sentencing guidelines in Ch. 7.

60 J. C. Coffee, 'Environmental Crime and Punishment', Thursday 3 Feb. 1994, *New York Law Journal* 5, 10, and 29 at 10.

61 However, authorities are reluctant to insist on conformance with a company's own system, especially if the company is performing above minimum standards.

62 However, note that an OHS audit must address: conformance with the standard/company system; and suitability of that system to meet company OHS needs; and the implementation of the system. Arguably, outcomes are assured because it is necessary to survey compliance with accepted/company systems to obtain evidence of implementation and to confirm the key system requirement (in OHS systems) of compliance with OHS legislation. Personal communication with G. Mansell, Mar. 1996.

63 See the incentives described in Ch. 1.

64 It should be noted that in Sweden, under the IC approach, inspectors both inspect against the system *and* continue to perform conventional inspections. However, such an approach is extremely resource intensive and is not practicable in most jurisdictions.

65 Personal communication, P. Davies, Head of Chemical and Hazardous Information Division (CHID), Health and Safety Executive, UK, June 1997.

66 C. Mayhew, M. Quinlan, & R. Ferris, 'The Effects of Subconbtracting/Outsourcing on Occupational Health and Safety—Survey evidence from four Australian industries' (1997) 25(1–3) *Safety Science* 163–78.

67 Benchmarking is a process of auditing a function against a similar function in another organization. Ideally, this gives a standard of industry best practice. See further at n. 73.

68 These examples are taken from the US OSHA's Voluntary Protection Program, referred to in Ch. 2.

69 For example, TQM systems rely in part on statistical procedures applied to appropriate and accurate measurements, yet at the programme level the nature of the system is such that most of the OHS manager's professional practice tools are almost exclusively qualitative rather than quantitative. It is only recently that attention has been given to a 'softer' process which

stresses the importance of qualitative variables and of developing an approach to TQM that makes employees more involved, responsible, and accountable. Early forms of TQM used detailed statistical data to measure the success of a management system within an enterprise. Emphasis was placed on gaining access to 'hard' information about processes of production and service delivery from the measurement and documentation of procedures and outcomes through the use of instruments such as flow charts, scatter diagrams, and control charts (A. Wilkinson & H. Willmott (eds.), *Making Quality Critical: New perspectives on organizational change* (Routledge, London, 1995), p. 8).

70 See I. Glendon & R. Booth, 'Risk Management in the 1990s: Measuring management performance in occupational health and safety' (1995) 11(6) *Journal of Occupational Health & Safety—Australia and New Zealand* 559; and E. Emmett & C. Hickling, 'Integrating Management Systems & Risk Management Approaches' (1995) 11(6) *Journal of Occupational Health & Safety—Australia and New Zealand* 617–24.

71 A number of positive developments have taken place with regard to this aspect of TQM and OHS. As indicated above, initiatives include a possible expansion of the International Standards Organisation's ISO 9000 (AS 3900) series of voluntary quality management standards to OHS and the development of the ISO 14000 series on environmental management, which is also potentially relevant to OHS management systems. Standards Australia is in the process of developing an interim quality management standard to fill the gap pending completion of an ISO Standard (Industry Commission, Australia, *Work, Health & Safety: Inquiry into Occupational Health & Safety* Volume Report I, Report No. 47 (AGPS, Conberra, 1 Sept. 1995), p. 84).

A number of consistent trends are emerging in the development of appropriate measurement systems for the evaluation of TQM and risk assessment. These trends highlight the need for:

• data to be reported over time to track improvement;
• measurement programmes to be implemented to encourage ownership and provide useful feedback at levels where improvement can be implemented;
• results to be widely distributed to be effective;
• the right measures to be drawn from mainstream business data systems rather than exclusively from specialized, isolated environmental (OHS) applications; and
• measures which are as 'lean and clean' as possible for general interpretation and understanding.

Others have noted the importance of key indicators, including: assessment of the degree of compliance with performance standards; identification of areas where performance standards are absent or inadequate (i.e., areas where further action is necessary to develop the total health and safety management system); assessment of the achievement of specific objectives particularly those where implementation is secured over time; and injury, ill health, and incident data accompanied by analyses of both the immediate and underlying causes, trends, and common features (F. D. Lindsay, 'Successful Health and Safety Management: The contribution of management audit' (1992) 15 *Safety Science* 387).

72 See p. 158 above.

73 Benchmarking is the process whereby firms systematically compare themselves to firms that are recognized as the best in a given field, learning from those practices to improve their own performance. We are indebted to Darren Sinclair for development of this point. See also V. Blewett, 'OHS Best Practice Column: Benchmarking OHS: A tool for best practice' (1995) 11(3) *Journal of Occupational Health & Safety—Australia and New Zealand* 237.

74 Australian Manufacturing Council/Manufacturing Advisory Group (NZ), *Leading the Way: A study of best manufacturing practice in Australia and New Zealand* (1994).

75 Worksafe Australia, *OHS Good For Business* (AGPS, 1995).

76 See, further, K. Rest & N. Ashford, *Occupational Health and Safety in British Columbia: An administrative inventory of the prevention activities of the Workers' Compensation Board* (Ashford Associates, Cambridge, Mass., 1997), pp. 37–8, and 103.

77 Health & Safety Commission (UK), *Plan of Work for 1994/95*, Volume 1 Main Report and Volume 2 Summary (HMSO, Sheffield, UK, 1994).

78 Internal assessment of the Planned Investigation Program suggests that it is providing considerable benefits to audited agencies in terms of improved morale, productivity and lower costs. See Comcare Australia, *Analysis and Data from the 1994–95 Planned Investigation Program* AGPS, Conberra, (1996). See also the Queensland experience under the Compliance Audit Program (Submission of Queensland Department of Employment, Vocational Education, Training and Industrial Relations (1994) to the Industry Commission (1995) at p. 14).

79 T. Wilthagen, *External Regulation of Internal Control: Outside looking in, but how and by whom?*, Hugo Sinzheimer Institute, University of Amsterdam, Workshop on Integrated Control/Systems Control, Dublin, 29–30 Aug. 1996.

80 La Trobe/Melbourne Occupational Health and Safety Project (W. G. Carson, W. B. Creighton, C. Henenberg, & R. Johnstone), *Victorian Occupational Health and Safety: An assessment of law in transition* (Department of Legal Studies, La Trobe University, Victoria, 1989).

81 Specifically, Johnstone's research suggests that, for most of the 1980s, the inspectorate was uncomfortable with creative and imaginative investigation of accidents and with investigations of systems of work under the performance standard general duty provisions (R. Johnstone, *The Court and the Factory: The Legal Construction of Occupational Health and Safety Offences in Victoria*, unpublished Ph.D. thesis, The University of Melbourne, 1994). He rightly emphasizes that the effective enforcement of performance standards depends in large part on inspectors being given practical guidance on how to gather evidence to prove the 'practicability' or 'reasonable practicability' of measures to remove hazards. It also depends on the inspectorate having sufficient skills and training to be confident in identifying breaches of performance standards, particularly where employers challenge the inspector's diagnosis. If these prerequisites are absent, then inspectors will

characteristically take no action unless there is also a clear breach of a specification standard, and much of the value of a performance-oriented approach will be lost.

82 T. Wilthagen, *External Regulation of Internal Control: Outside looking in, but how and by whom?*, Hugo Sinzheimer Institute, University of Amsterdam, Workshop on Integrated Control/Systems Control, Dublin, 29–30 Aug. 1996. Those countries that have adopted the main recommendations of the Robens Report, including the UK and Australia, similarly make direct provision for safety representatives and committees, albeit with varying degrees of power and responsibilities (see N. Gunningham, *Safeguarding the Worker* (Law Book Company, Sydney, 1984)).

83 B. R. Gordon, 'Employee Involvement in the Enforcement of the Occupational Health and Safety Laws of Canada and the United States' (1994) 15(4) *Comparative Labor Law Journal* 527.

84 Shapiro & Rabinowitz note that: 'although the policy literature stresses that cooperative enforcement works better when regulatory beneficiaries are involved, OSHA has generally not found ways to integrate workers into its cooperative programs' (see S. S. Shapiro & R. S. Rabinowitz, 'Punishment Versus Cooperation in Regulatory Enforcement: A case study of OSHA' (1997) 49(4) *Administrative Law Review* 731–62 at 762).

85 *TUTB Newsletter* July 1996, No. 3, p. 8, European Trade Union Technical Bureau for Health and Safety, Brussels.

86 T. Larsson, Samhallsvetenskapliga forsknings-institutet I Uppsala, Sweden, personal communication, June 1997. See also J. A. Gross, 'The Broken Promises of the National Labor Relations Act and the Occupational Safety and Health Act: Conflicting values and conceptions of rights and justice' (1998) 73(1) *Chicago-Kent Law Review* 351.

87 P. Lindoe, *Internal Control* (Stavanger College, Rogaland Research, Stavanger, 1995).

88 General Confederation of Spanish Workers' Unions, UGT-E, Federal Office for Occupational Health, Comisiones Obreras, CC.OO, Federal Department of Occupational Health, quoted in *TUTB Newsletter* June 1997, No. 6.

89 S. Tombs, 'Law, Resistance and Reform: "Regulating" safety crimes in the UK' (1995) 4 *Social and Legal Studies* 343–65 at 346; and E. Tucker, *Worker Health and Safety Struggles: Democratic possibilities and constraints*, Paper presented at the International Symposium on the Social and Economic Aspects of Democratisation of Contemporary Society, Moscow, 12–17 Oct. 1992.

90 S. Tombs, 'Law, Resistance and Reform: "Regulating" Safety Crimes in the UK' (1995) 4 *Social and Legal Studies* 343–65 at 346.

91 P. Lindoe, *Internal Control* (Stavanger College, Rogaland Research, Stavanger, 1995), p. 99 (emphasis added).

92 See Ch. 1 above.

93 It might also be possible for enterprises to engage worker representatives more directly in the implementation and monitoring of SMSs. For example, one best practice enterprise suggested that worker representatives could be involved in all key steps of at least some aspects of system implementation: the standard setting process, the establishment of guidelines, training, audit, and evaluation, and hazard assessment.

94 K. Frick, *Enforced Voluntarism—purpose, means and goals of systems control*, National Institute for Working Life, Solna, Workshop on Integrated Control/Systems Control, Dublin, 29–30 Aug. 1996.

95 See, for example, K. Rest & N. Ashford, *Occupational Health and Safety in British Columbia: An Administrative Inventory of the Prevention Activities of the Workers' Compensation Board* (Ashford Associates, Cambridge, Mass., 1997), p. 213, noting labour concerns about Workers Compensation Board reluctance to address tough issues, including the right to refuse hazardous work. Note also US resistance to mandate worker participation.

96 N. Gunningham, *Safeguarding the Worker* (Law Book Company, Sydney 1984), Ch. 10.

97 See pp. 168 below.

98 See, for example, W. E. Orts, 'Reflexive Environmental Law' (1995) 89(4) *Northwestern University Law Review* 1227 at 1334.

99 Industry Commission, Australia, *Work, Health & Safety: Inquiry into Occupational Health & Safety*, Volume I Report and Volume II Appendices, Report No. 47 (AGPS, Conberra, 1 Sept. 1995). Accordingly, employees need to be able to obtain information about hazards and their control held by their employer. It is recommended that employees should have a right to know in relation to such information.

Given that it is the supplier who is in the best position to know the risks with their products and how they might be safely used, the Commission recommends that all suppliers of plant, equipment, materials, and services shall, as far as reasonably practical, inform purchasers and users about the identification, assessment, and management of risks to health and safety associated with their products.

100 J. Braithwaite, 'Foreword' in J. A. Sigler & J. E. Murphy (eds.), *Corporate Lawbreaking & Interactive Compliance: Resolving the regulation-deregulation dichotomy* (Quorum Books, US, 1991), p. x.

101 F. E. Emery and E. Thorsud, *Democracy at Work* (Nijhof, Leiden, 1976).

102 In Scandinavian terms, the requirement for a safety organization to be established in most workplaces can be regarded as the first and perhaps most important reflexive and systems-based element embodied in law. For example, the Danish Working Environment Act 1975 states that responsibility for ensuring safe and healthy working conditions rests with the employer but that this must be fulfilled in co-operation with those employed in the firm. Firms with more than 10 employees must form safety groups comprised of the local supervisor and a worker elected safety representative. In firms of more than 20 employees a safety committee has to be established, composed of two supervisors elected by the supervisors in the safety groups, two safety representatives elected by the safety representatives in the safety groups, and one person appointed by top management. Norway and Sweden similarly emphasize that management has the general legal responsibility for OHS but that this must be achieved through participation with elected representatives. See also B. Gustavsen & G. Hunnius, *New Patterns of Work Reform: The Case of Norway* (University Press, Oslo, 1981).

103 See, generally, R. Johnstone, *Occupational Health and Safety Law and Policy: Text and materials* (LBC Information Services, Sydney, 1997).

104 See D. Dawson, P. Willman, A. Clinton, & M. Bamford, *Safety at Work: The limits of self-regulation* (Cambridge University Press, Cambridge, 1988).

105 For example, in the UK, by 1992 there was a fall in trade union membership of 25% from its peak in 1979 (D. Walters (ed.), *The Identification and Assessment of Occupational Health and Safety Strategies in Europe*, Volume 1, The National Situations (European Foundation for the Improvement of Living and Working Conditions, Dublin, 1996)); and see A. Martin, A. Linehan, & I. Whitehouse, *The Regulation of Health and Safety in Five European Countries: Denmark, France, Germany, Spain, and Italy with a supplement on recent developments in the Netherlands*, HSE Contract Research Report No. 84/1996, (UK Health and Safety Executive, London, 1996).

106 Under this approach, a small number of worker representatives would cover all workplaces in a particular small business sector (e.g. restaurants) within a particular geographical area, talking to employees in that sector and representing their interests in negotiations with employers.

107 See above p. 130.

6 The Top of the Enforcement Pyramid: Rethinking the Place of Criminal Sanctions in OHS Regulation

Introduction

The roles of enforcement under two-track regulation

In Chapters 2 and 3 we examined issues pertaining to the design of OHS regulatory standards, particularly the development of systems-based standards. We suggested that enterprises might be encouraged to adopt systems-based approaches to improve their OHS performance through a combination of codes of practice setting out the fundamental principles of such an approach, and a range of administrative, public relations, and other incentives. Some of these incentives touched upon matters relating to the types and severity of sanctions which might be imposed upon offenders who are found to have contravened the obligations set out in OHS statutes.

In Chapters 4 and 5 we developed this model further, and outlined a 'two-track' approach to regulation. Track One, discussed in Chapter 4, offered a reformed model of the traditional form of regulation, a central part of which is a modified version of Braithwaite's enforcement pyramid. The pyramid aims to stimulate voluntary action to conform with, and go beyond, current OHS statutory standards, and at the same time offers increasingly forceful measures to induce compliance from those who are not inclined to comply voluntarily with OHS standards. We outlined inspection and enforcement strategies which are likely to be more effective in inducing employers and other duty holders to comply with their OHS obligations.

In Chapter 5 we set out a second, self-regulatory, approach (Track Two). Track Two put primary responsibility on employers and their workers to develop SMSs to reduce work-related injury and illness, subject to third party and government auditing and oversight. Despite its self-regulatory nature, firms taking the Track Two option might, in very limited circumstances, face state enforcement action. Where a Track Two enterprise's attempts at a SMS were unsuccessful, the enforcement agency's response would depend on whether the enterprise was identified as incompetent or recalcitrant. Incompetent firms would first be given advice and encouraged to improve. Where satisfactory results were still not forthcoming, they would be given a period of grace to rectify contraventions of the law, and then returned to Track One. Where it was clear that a firm was abusing Track Two, this betrayal of the trust implicit in the Track Two approach would give rise to a rapid escalation up the enforcement pyramid. A third possibility for pyramid-based enforcement action might occur where there were circumstances at the firm's workplace which resulted in serious injury, death, or the potential for serious disease. In sum, even Track Two enterprises might, in rare cases, be subjected to enforcement measures, which might, in the most serious cases, comprise prosecution. However, we envisage that most OHS prosecutions would arise under Track One.

Recasting the role of prosecution

One of the objectives of this book is to arrest the trend in most jurisdictions away from the use of criminal sanctions to enforce OHS obligations where more constructive responses fail.[1] More and more, OHS agencies appear to be moving away from 'punishment', and focusing on 'prevention',[2] on the assumption that retrospective punishment does not play a role in proactive 'prevention'. In this book we seek to recast the role of prosecution, and argue that the enforcement pyramid provides a framework for punishment (principally for the purposes of general deterrence) *to facilitate an agency's efforts to promote prevention*.

In this and the following Chapter we take a closer look at the top of the pyramid, where, in our model of OHS enforcement, agencies have the option of prosecuting OHS offences in the courts. We consider the measures which are available to OHS enforcement agencies to induce compliance from duty holders where enforcement methods based on persuasion and voluntary compliance have had little or no effect. In other words, we explore the big sticks referred to in Chapter 4. As we have emphasized throughout this book, we are conscious of the resource constraints placed on OHS enforcement agencies. While there is no doubt a case to be made in many jurisdictions for a greater *number* of OHS prosecutions, we recognize that, in jurisdictions where prosecution already plays a significant part in enforcement strategies, the issue will be not so much an increase in the number of prosecutions as a more effective use of criminal sanctions. We argue that prosecution need not always comprise 'after the injury' punishment by financial penalty, but can play a greater role in stimulating compliance efforts. Our overall aim is to argue for a more vigorous use of criminal sanctions in OHS regulation, and to rethink the instrumental and symbolic place of criminal sanctions in a modern, reflexive, model of OHS regulation, so that the law provides significant incentives for duty holders not only to comply with the minimum OHS statutory standards, but also to improve their OHS performance beyond those minima. In particular, we examine the varied roles of prosecution, and then discuss ways of extending its application such that potentially competing aims may be met. We also outline suggestions for improving the range of sanctions applied to offenders convicted of OHS offences.

Ideally, our methodology in exploring the issues raised in this Chapter would be to conduct an empirical analysis of the effectiveness of OHS enforcement strategies which extensively employ prosecution. However, as we suggest in Chapter 1, and illustrate in the Appendix to this book, no OHS agency has placed a heavy emphasis on criminal prosecutions to enforce OHS statutes, nor considered extending their application along the lines that we propose.[3] Accordingly, in this Chapter we build on our previous empirical research into OHS prosecutions in Australia,[4] and survey the literature on corporate criminal sanctions, which is most developed in the USA, in order to develop and outline our own model of OHS prosecutions. We emphasize that this hypothetical model is based on research emanating from countries with adversarial criminal justice systems (most notably Australia, the UK, Canada, and the USA), and that readers will have to reflect upon substantive, procedural, historical, and philosophical differences in assessing the applicability to their legal systems of our model.

In discussing the 'big sticks', we envisage the top of the enforcement pyramid to be comprised principally of prosecutions in the criminal courts, at both the lower (for example, magistrates' courts) and higher levels (Supreme or High Courts). We argue that the majority of prosecutions should be launched under the OHS statutes. In order to ensure the integrity, and hence effectiveness, of the pyramid, we will argue that more prosecutions need to be taken where there are defective systems of work, even though no injury or fatality has occurred (what we refer to as 'pure risk' prosecutions). We also suggest that the OHS enforcement agency will need to bring a number of prosecutions for industrial manslaughter and related offences where these are appropriate to demonstrate the severity of measures available to ensure compliance at the top of the pyramid (that is, to enhance the general deterrence effects of OHS prosecutions), and to satisfy public demands for retribution where workplace fatalities and injuries are attributable to gross negligence.

A further ingredient of a deterrent-based pyramid is a prosecution strategy that broadens the range of duty holders who are prosecuted. In particular we will argue in this Chapter that more prosecutions should be brought against designers, manufacturers, and suppliers of workplace plant, equipment, and substances, against franchisors, principal contractors, and subcontractors, and against directors and top management who are responsible for OHS contraventions.

In Chapter 7 we address issues to do with the range of sanctions for OHS offences. First, we suggest that there needs to be a tougher, and wider, choice of sanctions which will deter OHS duty holders from contravening their obligations under the OHS legislation, and which will also enable courts to make orders which require OHS offenders to reform defective OHS work systems. Accordingly, we discuss ways of improving the use of fines, and outline other sanctions which can be utilized by courts hearing OHS prosecutions. Second, we investigate mechanisms (such as sentencing guidelines) which will improve and rationalize the sentencing of OHS offenders.

But first we discuss the primary underlying purpose of sanctions for convicted OHS offenders at the top of the OHS enforcement pyramid.[5] Criminal enforcement can be motivated by a variety of concerns, and the design of criminal sanctions, and the development and implementation of prosecution strategies, will be strongly influenced by the rationale for using the criminal law in the first place.

The purpose of criminal sanctions in the OHS arena

In our introduction in Chapter 1 we noted that one of the consequences of the early history of the enforcement of the British factory legislation was that the OHS offences came to be seen as 'quasi-criminal' only, and quite different to 'real crime'. This view predominates in most countries at the end of the twentieth century. One of the aims of this Chapter is to reintegrate OHS crime with the mainstream criminal law, in order the maximize the effect of criminal sanctions in OHS enforcement.

The criminal law can serve functions which are either instrumentalist[6] (that is, to achieve an outcome such as reducing harm or maximizing happiness) or symbolic or ideological (that is, to assert moral (or other) values, and to demonstrate society's disapproval of conduct).[7] The model that we develop in this Chapter assumes that the primary purpose of

criminal prosecution at the top of the enforcement pyramid is instrumentalist—mainly general deterrence, but with a measure of specific deterrence, rehabilitation (organizational reform), and even incapacitation.
We acknowledge, however, that prosecutions perform very important
moral, symbolic, and retributive functions, and envisage that some forms
of prosecution, most notably under the general criminal law for industrial
homicide, will have strong symbolic and retributive motivations.

Deterrence

The primary purpose of OHS prosecutions is deterrence, which is prem-
ised on the notion that punishment will discourage the individual or
corporate wrongdoer (specific deterrence) and/or others (general deterrence) from engaging (or re-engaging, in the case of the former) in
proscribed conduct. Deterrence is based on the idea that criminal punishments can be used to outweigh the calculated benefits accruing to
those individuals or business organizations who consider committing
a crime.[8] The argument assumes that individuals and organizations
behave rationally, and choose paths of action leading to the greatest
benefit.[9] Accordingly, effective deterrence requires the costs of unsafe
work practices to exceed the corporate and/or personal benefits to be
derived from the business activity.[10]

According to deterrence theory, compliance is a function of the probability of an offender being punished and the severity of the subsequent
penalty.[11] As the Australian Industry Commission[12] noted, the expected
penalty (or deterrent) facing an individual OHS offender is determined
by the likelihood of being inspected (which it took as the number of
inspections as a proportion of workplaces), the likelihood of a breach
being detected, prosecuted, and convicted (the number of fines and on-
the-spot fines), and the average penalty imposed. It calculated that, at
best, the estimated expected penalty facing an OHS offender in Australia
was $33, which it conceded was likely to be an overestimate, because
most OHS breaches go undetected, and most detected contraventions
face zero probability of enforcement. There is no evidence to suggest that
the position is any different in any of the countries in Europe and North
America.[13]

Writers on deterrence theory have argued that policy makers should
avoid over-deterrence, and ensure that sanctions do not discourage
corporate conduct that is socially beneficial.[14] This is a difficult balance
to achieve. Even if legislators and judges were able to calculate with any

accuracy the likelihood that OHS contraventions will be detected, and the costs resulting from such conduct, theorists debate the optimal penalty to be imposed on the offender to provide the most efficient deterrent effect. Should the expected value of the penalty exactly equal the costs to society of the organization's conduct, or should a more stringent penalty be imposed to provide incentives for corporations and their managers and directors to improve their OHS management? Like Bonner and Forman,[15] we argue that:

> . . . criminal sanctions should . . . impose punishments upon corporations and/ or their executives which result in negative returns for [OHS duty holders]. Such a system would . . . counteract the willingness of many criminals to commit illegal acts if the costs of being convicted do not appear to vastly outweigh the benefits to be derived from illicit behaviour.

The limited liability company is usually only liable to the extent of its assets, which may be less than the possible benefits of high risk behaviour which threatens the health and safety of employees.[16] Consequently, monetary penalties are likely to leave a 'deterrence gap' because they are incapable of making corporations internalize all of the social costs resulting from inappropriate OHS policies.[17] This suggests that effective corporate sanctions require more than financial penalties to implement general deterrence. In Chapter 7 we discuss other sanctions which bolster the deterrent effect of financial penalties.

There may also be a tension in trying to reconcile specific and general deterrence. The latter might only be achieved by punishing the individual offender to a greater extent than was warranted by the crime that individual had committed.[18] Given the expense involved in launching and conducting prosecutions, it may be argued that prosecutions can only be fully justified if they have the effect of deterring more than the organization against which the action is taken, and if there is also a retributive element.[19] Although our enforcement pyramid is predicated upon the threat of strong specific deterrence through rapid escalation up the pyramid, this threat will be all the more credible if the sanctions are sufficiently punitive to provide a general deterrent.

We also note that in examining deterrence a distinction must be made between the actual chances of detection and punishment, and the perceptions thereof. What is important is the belief that business actors have of the likelihood and degree of punishment, even if, in actual fact, that belief is overstated. For example, in Australia, in 1989, the New South Wales government announced a policy of 'getting tough on polluters', and introduced new legislation involving $1 million fines for corporations and imprisonment of up to seven years for individual officers found guilty of serious offences.[20] Eight years later, no one had been imprisoned, nor had any senior officer of a major corporation been personally convicted of a serious offence. Yet senior officers continue to be deeply concerned about their personal liability under the legislation, and have been zealous in conducting environmental audits to discharge their 'due diligence' obligations.[21]

Some commentators argue that the underlying assumptions of deterrence theory are often ill-founded, particularly in crimes where expressive or emotional factors are likely to dominate.[22] However, deterrence theory would seem to be more applicable to corporate crime.[23] Braithwaite and Geis argue that corporate crimes 'are almost never crimes of passion . . . but calculated risks taken by rational actors. As such they should be more amenable to control by policies based upon utilitarian assumptions of the deterrence doctrine.'[24] Pearce and Tombs[25] suggest that OHS crimes are not 'one off acts of commission, but are rather ongoing states or conditions',[26] meaning that, unlike the case in street crime, the identification of the 'criminal' is not difficult, so that there need not be a low rate of crime detection.[27] They also point out that it is much easier to convey the 'message' of deterrence to business organizations than it is to traditional street criminals.[28]

This notion of the corporation as a rational actor has itself not gone unchallenged. Many commentators[29] have questioned the anthropomorphic assumption of the neoclassical economic model of the corporation, so essential to deterrence theory, that the corporation behaves like a unitary, rationally acting, and profit maximizing entrepreneur. For example, in many organizations the assumption of organizational unity is undermined by the differences in goals and perceptions held by organizational subunits, and individual managers.[30] Field research tends to suggest substantial limitations in the practical application of deterrence theory. In one of the most sophisticated empirical studies, Braithwaite and Makkai found, in the case of nursing home regulation, that it was only in certain contexts, and for certain minorities of actors, that there was an association between compliance and perceived severity and certainty of punishment. There was 'little support for the additive or multiplicative effects of the certainty of detection, the certainty of punishment, and the severity of punishment in the simple corporate context'.[31]

Pearce and Tombs, however, make the important point that there will always be individuals within organizations who are committed to reducing workplace hazards, and will be willing to act upon this commitment. Deterrence empowers these individuals, 'through creating conditions where their voices receive a hearing in the interests of the corporation as a whole'.[32] For these and other reasons,[33] as with conventional crime, we cannot treat individuals and corporations as homogeneous entities, abstracted from their operational environments.[34] We need to acknowledge that there are different types of organizations, with different organizational structures and decision-making processes, with the result that criminal sanctions will differ in their deterrent impact upon the organizations and the individuals within them.

How does the principle of deterrence translate to OHS crime? There is a considerable literature on the role of penalties in achieving both general and specific deterrence.[35] 'If both general deterrence and specific deterrence are strong over time, compliance may be incorporated by firms as an organizational norm.'[36] Brown, after reviewing USA and Canadian empirical data in particular, notes that:

There is good reason to believe that penalties enhance compliance in the short term by threatening would be offenders with punishment, and in the long term by changing attitudes about what is morally acceptable behaviour.

Over the long term, stigmatising health and safety offenders by subjecting them to legal punishment may generate a stronger moral commitment to protecting the well-being of employees.[37]

Empirical research in the USA, particularly in studies by Lewis-Beck and Alford[38] and Perry[39] in relation to coal mining legislation, and work carried out in the State of Oregon,[40] suggests that increased penalties do have an impact in reducing injury rates, although the latter is an incomplete indicator of compliance.[41] While even here the evidence is not consistent, there is certainly considerable data to suggest that both the severity and certainty of punishment do influence injury rates, with the latter exerting a much stronger influence than the former.[42] British evaluations of enforcement measures confirm these results.[43] These data appear to be contradicted by Gray and Scholz's study of the enforcement of the USA OSHA[44] which reported significant specific deterrent effects, but argued that larger penalties, at least those within the range imposed by the OSHA, did not induce greater general or specific deterrence.[45] They conceded, however, that significantly higher 'megapenalties' (involving hundreds of thousands of dollars or more) might have significant deterrent effects.[46] Gray and Scholz found that penalties prevented more injuries by influencing employers who had not themselves been penalized (that is general deterrence), rather than altering the behaviour of those upon whom penalties had been imposed (specific deterrence).[47]

Haines has cautioned against the possible negative side effects of court-based adversarial processes[48] on the development of OHS programmes within organizations.[49] She suggests that while prosecution needs to be used against those who flagrantly breach the law in order to maintain the credibility of the regulatory regime, when 'escalation of penalty occurs, motivation for corporate compliance shifts from co-operation and trust, to deterrence and mistrust'.[50] The threat of prosecution may induce duty holders to minimal compliance, but this may be at a formal level only, with no substantive change to OHS processes, procedures, or overall culture.[51] Haines's study also emphasizes that the impact of deterrent measures depends on the position of the target corporation or individual within the broader social structure. While small businesses may in fact be 'more responsive' than larger organizations to punitive sanctions, the impact of such sanctions will be confined to the short term if the competitive context within which the small business operates, and which constrains its ability to introduce OHS programmes, remains unaltered.[52]

This rather confusing empirical evidence suggests that deterrence apparently does not work across the board, even in the case of corporate crime. This is not an argument against the use of criminal sanctions,[53] but rather (particularly in the light of Braithwaite and Makkai's findings)[54] an argument for the use of the enforcement pyramid,[55] and for targeting those sanctions to circumstances and actors where persuasive measures have been ineffective, and where sanctions are most likely to be justified.[56] The evidence (to build on the point made by Gray and Scholz) suggests that the penalties imposed should be significant enough (in terms of the size of the fine, the degree of adverse publicity, or the other demands made upon the organization) to act as a general deterrent to others. Consistent with the notion that criminal prosecution is the big stick at the top of the enforcement pyramid, OHS agencies should ensure that it is used against the most egregious offenders, and should avoid using criminal sanctions before they are warranted.[57] Widespread publicity of penalties imposed plays an important role in duty holders' perceptions of possible punishment, and is an important aspect of general deterrence.

The enforcement pyramid, as Braithwaite and Ayres observe, does not rely on 'passive deterrence, that is deterrent credibility shaped by the potency of the sanctions waiting to be used'.[58] Rather it relies on a 'more dynamic modelling of deterrence as an unfolding process',[59] hence the importance of active escalation. Shapiro and Rabinowitz point to evidence from a number of jurisdictions showing that 'cooperative approaches can decrease compliance if agencies permit law breakers to go unpunished'.[60] Prosecutions should be targeted against those duty holders who have failed to comply with their OHS obligations despite attempts to induce such compliance with advisory and administrative enforcement measures, or who have contravened their OHS obligations in circumstances such that enforcement measures should be rapidly escalated to, or even begin at, the top of the pyramid. And, as we argue later in this Chapter, the OHS agency should ensure that, in addition to (or perhaps instead of) proceeding against the corporate employer at the workplace where the contravention occurred, prosecutions are brought against the organizational subunit, or individuals within the organization, most responsible for the commission of the offence. If the sanctions imposed on offenders are large enough and are widely publicized then even if the individual offender is not specifically deterred by the prosecution, the prosecution will serve as a general warning to other potential offenders that non-compliance will result in the imposition of costly criminal sanctions.

Finally, we note an important point made by Tombs:[61]

At the very least, deterrence poses both an ideological and a material challenge to employers. Their claims that corporations are rational systems that can be controlled effectively are central to their controls for thoroughgoing forms of what they term self-regulation . . . If such claims are taken at face value, deterrence should be effective, as employers would rationally respond and make calculations regarding the certainty of detection and the likelihood and nature of punishment. Conversely, if corporations are not rational systems of control, then the failure of deterrence only legitimates intervention as the external imposition of 'rationality'—and this would take safety organization out of the hands of 'managers alone'. This latter option approaches the endorsement of rehabilitation in the context of corporate crime.

In other words, deterrence is not the only motivation for criminal sanctions for OHS offenders.

Moving beyond deterrence in the use of criminal sanctions

In our view, the design of the top of the pyramid should be flexible, and go further than simply emphasizing the deterrent effect of prosecution.[62]

Just as the purposes of OHS enforcement may change as enforcement measures are escalated up the enforcement pyramid, so too might the purposes of criminal prosecution be varied depending on the nature of the contravention and the record, attitude to compliance, and other characteristics of the offender. A deterrence model relying mainly on financial penalties is a fairly blunt instrument, and can send only general messages to employers about the costs of non-compliance.

The instrumentalist aims of criminal punishment can be furthered by measures targeted at what is traditionally known as *rehabilitation*, which aims to ensure that wrongdoers acknowledge the wrongfulness of their deeds, acquire a sense of social responsibility, learn how to abide by society's laws, and are reintegrated back into society. Some commentators shortsightedly suggest that corporations cannot be rehabilitated because they are legal fictions, and because theories of corporate rehabilitation anthropomorphize the corporation.[63] However, in relation to OHS crimes committed by business organizations, the notion of rehabilitation might involve an organization taking measures to review and reform its policies and organizational structures and to develop policies and disciplinary and standard operating procedures to ensure that contraventions do not reoccur, or that modern SMSs are implemented.[64] We believe that prosecution must play a role that is consistent (and certainly not inconsistent) with systems approaches to OHS management.

A regulatory model which aims to provide threats to employers to adopt safe work systems, and incentives to develop OHS programmes and SMSs, will need sanctions which are more focused on corporate reform and restructuring. The emphasis on the encouragement of OHS programmes and SMSs suggests that a court imposing sanctions on convicted OHS offenders should be able to reward offenders for their previous efforts at developing OHS programmes and SMSs, penalize them for failing to take such initiatives, and, where appropriate, order them to introduce, develop, and monitor such systems. These issues will be explored in great detail in Chapter 7.

If measures aimed at deterrence and corporate restructuring fail, courts should be able to terminate the operation of the worst OHS offenders (*incapacitation*).

We argue further that a focus solely on the instrumentalist aspects of criminal punishment for OHS offenders ignores important symbolic and ideological dimensions of prosecution. The *symbolic* or *moral* aims of criminal sanctions seek to apportion moral blame for criminal acts, and officially demonstrate society's intolerance of harmful behaviour. As Glasbeek argues, '[w]e use the criminal law when our sensibilities are assaulted—when, in addition to redressing the particular problem, we want both to condemn the wrongdoers' conduct, and to stigmatise them'.[65] The criminal law both reflects existing public sentiments about the heinousness of certain activities, but can also be used to shape such perceptions, particularly if used in conjunction with media campaigns showing the reprehensible aspects of the behaviour, while simultaneously emphasizing society's condemnation of that behaviour. For example, prosecutions can challenge employers' claims that workplace hazards are natural and inevitable, or that workers consented to the risks.[66]

In regulatory offences (such as OHS contraventions), however, this moral force of law argument might be eroded in two ways. Labelling OHS offences as criminal but then applying meagre penalties[67] might erode the symbolic value of criminality. Alternatively, the use of strict liability, and the elision of the requirement of traditional *mens rea*, operates to direct emphasis away from the notions of individual guilt and moral culpability.[68] We suggest that these arguments bolster the case for larger monetary fines for OHS offenders (while being careful to avoid over-deterrence), for conducting OHS prosecutions in the mainstream criminal courts (see below), and for the broader range of sanctions for OHS offenders which we advocate in Chapter 7. The moral and symbolic dimensions of criminal punishment also suggest that components of the mainstream criminal justice system, particularly crimes punishing various forms of criminal homicide (discussed later in this Chapter), be integrated into an OHS prosecution strategy.

A significant motivation for many forms of criminal punishment is *retribution*, where society is said to exact vindication by punishing acts considered egregious, so as to express moral outrage and reaffirm a commitment to the maintenance of legal and moral standards.[69] Retribution is not concerned with the effect of punishment on future levels of crime. Rather it reflects public demands that offenders causing injury or death be made to suffer for their wrongful actions, 'irrespective of whether this will reform [the offender's] character, deter [its] misconduct, or set an example for others'.[70] Because moral and symbolic rationales underly retribution as a justification for criminal punishment, the state of mind (*mens rea*) of the corporation is important.[71] Despite past ambiguities, there is little doubt that today's business organizations are considered by the general public, and indeed the law, to be capable of being held morally blameworthy and responsible for unwanted practices and policies leading to worker ill health. Consistent with retribution is *'just desserts'*, or the satisfaction of justice. Here the criminal justice system assumes that offenders are responsible for their actions, and, accordingly, for their punishment, and therefore that there should be a direct connection between the crime committed and the punishment which ensues. Once again, public demands for retribution and just deserts reinforce the need for criminal prosecutions for serious contraventions of OHS standards, and, in addition, for homicide-related offences under the mainstream criminal law.

Civil or criminal penalties?

There remains the issue of whether the punishment of OHS offenders can be achieved using civil penalties, rather than resorting to criminal sanctions. Given that civil penalties can be imposed more quickly and cheaply than can criminal sanctions,[72] and that it is easier in most jurisdictions to satisfy the civil burden of proof (on the balance of probabilities) than the criminal standard (beyond reasonable doubt), why not simply punish OHS offenders by imposing civil penalties? Do criminal sanctions achieve regulatory objectives which cannot be met by the use of civil penalties? We emphasize at this point that these questions should be seen in the context of the enforcement pyramid, so that the issue is what does the prospect of criminal sanctions add to the pyramid that civil penalties (for example, on-the-spot fines or USA OSHA-type citations) at a lower level do not contribute?

Fisse[73] points out that criminal sanctions have two advantages over civil penalties in achieving the utilitarian objectives of deterrence. First, civil penalties do not have the 'deterrent value resulting from the stigma of criminal conviction and punishment'.[74] The offender's perception of punishment includes not only the size of the financial penalty, but also 'views of the symbolic and stigmatising function of the penal sanction'.[75] Empirical studies confirm that the possible loss of corporate image, prestige, and employee morale caused by adverse publicity, and the efforts required to defend the corporation from public attack, have a larger impact on the corporate wrongdoer than large fines.[76] The stigma of conviction cannot be written off as a business expense, or passed off to shareholders or consumers.[77] Of course, the stigmatic impact of conviction depends on the conviction being publicized, a point which we will return to in Chapter 7.[78] While the 'criminality' of OHS offences, as we pointed out in Chapter 1 and earlier in this Chapter, is shrouded with ambiguity, it appears that community attitudes to OHS and environmental crime are becoming tougher in many parts of the Western world, so that corporations are more likely to be stigmatized by OHS convictions, thus increasing their deterrent effect.

Second, a criticism of both civil and criminal monetary penalties is that they ignore the important non-financial values in corporate decision making, such as the urge for power, the desire for prestige, the creative urge, the desire for security, and to serve others, and the need to identify with a peer group.[79] In other words, corporate behaviour is motivated by more than the financial calculus assumed by deterrence theory. Criminal sanctions can be used to affect all of the values and variables, financial and non-financial, that influence corporate decision making. While, no doubt, civil sanctions can be developed to have an impact on the non-financial values of business organizations, these effects are more likely to be achieved by designing a wider range of criminal sanctions. In Chapter 7 we examine criminal sanctions that can affect these non-financial values within business organizations.

Criminal sanctions have an advantage over civil penalties in achieving objectives other than deterrence. The moral, symbolic, retributive, rehabilitative, and incapacitative aspects of criminal sanctions provide a rationale for criminal sanctions which civil penalties are unable to match.[80] Indeed, the various rationales for criminal sanctions should not be seen as mutually exclusive—in fact the most practical forms of rehabilitation and incapacitation are 'subgoals of deterrence'.[81] The punishment of OHS crime does not just focus on trying to inhibit dangerous work practices, but, as we stress in Chapter 2, aims for a change in management approaches, and in particular requires the organization to take the necessary steps to avoid a repetition of the offence. These steps cannot simply be taken by individuals within the firm, but require the organization itself to conduct internal investigation of the offence, discipline those members of the organization responsible for the organization's non-compliance, and to undergo a revision of policies and operating procedures.[82] Civil penalties will (may) not be sufficient to induce these measures. In such cases, criminal sanctions demanding action by the organization are required. We discuss these in Chapter 7.

Making the most of prosecution in the enforcement pyramid

The place of prosecution in the enforcement pyramid: Two levels of courts

When we mapped out the fundamental structure of the enforcement pyramid in Chapter 4, we argued that OHS prosecutions provide the 'big sticks' at the top of the enforcement pyramid to induce voluntary compliance by duty holders with their OHS obligations. The primary purpose of court prosecutions will be general deterrence, backed up by specific deterrence and some degree of rehabilitation, retribution, and denunciation. As a last resort, the courts may also have to terminate the operation of the worst, most recalcitrant OHS offenders. In our model, prosecutions will be brought where OHS duty holders have failed to comply with their OHS obligations despite the full use of enforcement measures at the lower levels of the pyramid, or where there are other reasons, noted in Chapter 4, for entering the pyramid at its upper reaches, or for 'leapfrogging' up the pyramid. We also observed in Chapter 4 that the taller the pyramid (that is, the more levels there are in the hierarchy), the greater the scope that the OHS enforcement agency has to escalate its enforcement response to ensure compliance with the OHS statute. The pyramid will be taller, and more flexible in operation, if prosecutions are able to be conducted at two levels of courts—a lower level court (such as a magistrates' or local court) and a higher level court (such as a District Court or even a Supreme or High Court).[83]

The two levels of court should be located within the mainstream criminal justice system. We believe that regulators should resist calls to have OHS prosecutions conducted before 'specialist' courts, whether they be courts specifically created for OHS prosecutions, or industrial relations tribunals. The argument usually used to justify the call for specialist tribunals is that the mainstream criminal courts demonstrate a lack of understanding of the principles of OHS and the nature of OHS offences. As we argued earlier in this Chapter, one of the aims of this book is to reassert the criminal nature of OHS prosecutions. The removal of OHS prosecutions from the mainstream criminal justice system simply reinforces the perception that OHS offences are 'regulatory' or 'quasi-criminal' matters.[84] If the concern is that magistrates and judges do not have enough of an understanding of OHS and OHS regulation, then specially trained panels of magistrates or judges can be created within the different levels of courts in the mainstream criminal justice system. In this way, OHS prosecutions can be adjudicated by properly trained judges, but without undermining the criminality of OHS offences.[85]

In order to achieve the flexible purposes of prosecution,[86] we envisage that the courts will be able to impose sanctions beyond fines.[87] Sanctions vested in the lower level courts might include community service orders and milder forms of corporate probation, such as the requirement that the firm report to the court on the results of the firm's internal investigations into why the offence was committed, and the OHS programmes that will be implemented to avoid further offending. The higher level court will be given a wider range of sanctions, designed to enhance general deterrence (through sanctions such as publicity orders) and organizational reform and restructuring (including stronger forms of corporate probation which might require firms to introduce SMSs). At the extreme tip of the enforcement pyramid lies the possibility of the higher level court ordering the dissolution of an organization which has demonstrably failed to comply with its OHS obligations and which poses an unacceptable threat to worker health and safety. The top of the pyramid, with its broader array of sanctions, is illustrated in Figure 5 below.

Because OHS prosecutions seek to achieve a number of flexible purposes,[88] they will be initiated in a variety of circumstances,[89] each intended to bolster the effectiveness of enforcement measures at the lower levels of the pyramid by demonstrating that lower level compliance measures take place in the shadow of the threat of serious sanctions. In our model, a greater number of prosecutions will come about as a result of an escalated enforcement response to non-compliance by duty holders, even though no injury or illness has actually resulted from non-compliance. Prosecutions might also be initiated where the OHS agency has reason to believe that enforcement action should begin near the top of the pyramid, for example where a serious incident, injury or 'near miss' reveals a history of wilful neglect of OHS issues, or where a Track Two firm has breached the trust placed in it by the enforcement agency, and has failed to a significant degree to take the promised measures to implement a SMS. We also accept that OHS agencies will always have to resort to prosecutions to respond to public pressure for retributive action where fatalities and serious injuries have occurred, even when these occur at workplaces under the management of Track Two firms. Empirical data suggests that at present the vast majority of prosecutions are initiated in response to serious injuries and fatalities. Accordingly, we

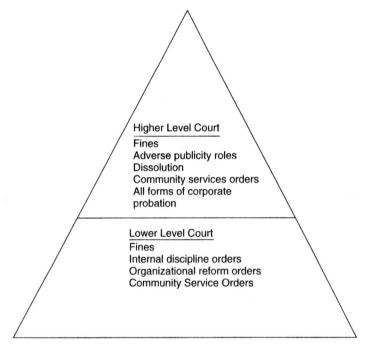

Higher Level Court

Fines
Adverse publicity roles
Dissolution
Community services orders
All forms of corporate
probation

Lower Level Court

Fines
Internal discipline orders
Organizational reform orders
Community Service Orders

Figure 5 The Sanctions at the Top of the Enforcement Pyramid

recognize that some prosecutions will always originate from breaches of the OHS legislation even where enforcement measures at the lower levels of the enforcement pyramid have not been fully utilized. As Figure 6 shows, we envisage a mix of prosecutions in the courts, some arising out of serious injuries and fatalities, some arising from serious breaches where no actual injury occurred, but where very serious consequences might have resulted and there is no evidence of good faith in developing OHS programmes, and others arising out of an escalation of enforcement measures through the enforcement pyramid.

We conclude this section with a brief justification for the two levels of court prosecutions for OHS contraventions. There is an argument that it is inappropriate to place lower level court prosecutions near the top of the enforcement hierarchy. Commentators have noted that there is a tendency for issues decided in the lowest level of the court hierarchy to be 'trivialized'. For example, McBarnet[90] argues that the essence of summary justice is that it be quick and cheap, with the minimization of 'legal

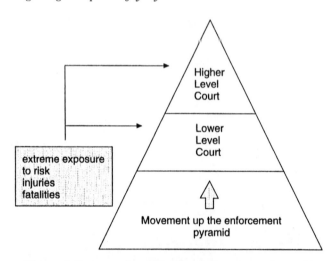

Figure 6 Entry to the top of the Enforcement Pyramid

niceties'. She observes that there is an ideology of the 'triviality' of the matters coming before the magistrates' courts, which are seen to deal with everyday matters, with low penalties, and little public scrutiny.

The magistrates' courts in the UK and Australia, for example, are at the lowest level of the judicial hierarchy, and traditionally adjudicated 'petty crime'. Johnstone, in his study of OHS prosecutions in Victoria, Australia, reports that one inspector noted in an interview 'the magistrates' courts are a bit Mickey Mouse',[91] suggesting that if prosecutions are to be taken seriously by prosecutors and defendants, they should take place at a high level within the court system. While the jurisdiction of the lower courts will vary from country to country, crime coming before these courts consists primarily in street crime in both its dimensions, that is traditional crime and traffic crime, where the defendants are individuals. Cases are decided quickly and routinely, with the minimum of fuss. This fast summary justice and the high proportion of cases which proceed on guilty pleas institutionalizes an analysis of culpability that is primarily based on lower court judges' common sense opinions of OHS, and the briefly constructed facts put to the court by the parties. The result is usually a lowish penalty. Most important of all, lower court judges are not accustomed to dealing with corporate offenders, and not used to imposing large penalties. Most offences prosecuted in the lower courts have maxima well below those in the OHS statutes, and most of their penalties are imposed on individuals with low capacity to pay.[92]

Whilst acknowledging these characteristics of the lower court, we nevertheless argue that lower court prosecutions have their place in the hierarchy. They can be initiated where the OHS offences are not extremely serious (in the sense that either the actual or potential risks arising from contraventions were not great, the defendant had a history of co-operation with the OHS agency, and so on). These prosecutions will act as a specific deterrent (particularly because of the fairly controversial evidence noted in Chapter 4 and earlier in this Chapter that it is the fact of the penalty rather than its size which spurs duty holders into action) and serve as a warning to others that the criminal law can be invoked quickly and effectively to punish non-compliance with the OHS legislation. A good Australian example of the present use of the lower courts for specific and general deterrent purposes is the recent New South Wales WorkCover Authority practice of bringing quick magistrates' court prosecutions with minimal evidence where there has been a contravention of the OHS legislation which has not resulted in an injury.

At the top level of the enforcement pyramid are prosecutions in the higher level courts, and manslaughter prosecutions (which we discuss later in this Chapter).[93] Provided these prosecutions are successful, they signal very clearly that the agency has the capacity and willingness to move up the enforcement pyramid to ensure compliance. Thus the more prosecutions brought before the higher level courts the greater the public perception that OHS offences are 'serious', and the greater the general deterrent effect of those prosecutions, particularly if they are widely publicized. The deterrent (and as we noted earlier, the rehabilitative) role of the courts will be enhanced if the court is given a wider range of sanctions to impose on OHS offenders.[94]

Using prosecution to maintain the integrity of the enforcement pyramid: Broad principles

If prosecution loses its credibility as a general deterrent the lower levels of the enforcement pyramid will be more difficult to implement. In this section we outline some simple principles which will ensure that prosecution plays an effective role in bolstering the enforcement pyramid.

The first basic principle is that the OHS agency must ensure that the enforcement pyramid is *applied uniformly and consistently across the board*. The

public perception of OHS enforcement must be that each OHS inspector uses the same criteria and decision-making process to work her or his
way up the enforcement pyramid.[95] The criteria for bringing prosecutions must be transparent; that is, clear, publicly known, and consistently
applied.[96]

The second basic principle is that prosecutions should not just focus on
employers, but on *other duty holders*, such as designers, manufacturers and
suppliers, principal contractors, contractors and subcontractors, self-
employed persons in relation to non-employees, franchisors, and managers and directors. We have already observed that we cannot assume
that all business organizations behave like unitary, rationally acting, and
profit maximizing entrepreneurs, and need to acknowledge that there
are different types of organizations, with different decision making structures, and with variable responses to criminal sanctions.[97] Later in this
Chapter we put the argument for more prosecutions of individual
corporate officers, to ensure that the deterrent effects of prosecution are
maximized.[98] Here we argue for a prosecution strategy that does not
confine its reach to employers in relation to offences pertaining to their
employees.

We note that most OHS statutes impose general duties of care on
persons involved in supplying, manufacturing, or importing equipment,
plant, and substances for use at the workplace, or designing, erecting, or
installing plant and equipment at the workplace.[99] There are two simple
reasons for an increased enforcement focus on these provisions.[100]

First, by doing all it can to ensure that manufacturers and suppliers of
plant and equipment comply with their legal obligations to manufacture
and supply plant and substances that are safe and without risks to health,
the OHS enforcement agency will be addressing hazards *at their source*.
This would appear to be a resource-efficient way of ensuring that plant
and substances in the workplace are hazard free. Second, it is inappropriate to focus all enforcement activity on employers, with little focus on
the enforcement of duties imposed upon other parties. Because of the
nature of the duty imposed on manufacturers, suppliers, and others,
prosecutions for contraventions of these duties should be mainly 'pure
risk' prosecutions, as discussed later in this Chapter. While, in many
cases, a prosecution under these provisions will be motivated by the
incidence of an injury or disease arising from unsafe work processes or
substances, there is a strong case for proactive prosecutions where manufacturers, suppliers, and others have made inadequate efforts to comply
with their statutory duties.

As business organizations create new commercial legal forms through which to conduct their business, OHS agencies will need to broaden the targets of their enforcement action. We discussed in earlier Chapters the increased use of outsourcing arrangements. Regulators need to ensure that their OHS statutes are flexible enough to extend their reach to principal contractors, head contractors, and subcontractors. We note the trend in many countries, and illustrated in the 1992 EC Directive To Implement Minimum Health and Safety Requirements at Temporary or Mobile Construction Sites, for OHS obligations to be extended to designers of buildings, planning supervisors, and clients. Franchising is becoming an increasingly popular form of distribution of goods and services. In 'business format' franchises, for example, the franchisor develops a system of doing business, and permits the franchisee to use that system in the operation of the franchisee's independently owned business, but closely following methods developed by the franchisor. The franchisor provides the franchisee with a blueprint for running the business, the initial training in the operations of the franchised business, and ongoing marketing, business, or technical assistance during the operation of the franchise.[101] The level of control that the franchisor exercises over the operations of the franchisee suggests that the franchisor should be held accountable for the OHS programme within the overall business system.

Enforcement measures, including prosecution, should be carefully targeted at the parties within these outsourcing and franchising relationships who had control over the OHS management systems, and who breached their OHS obligations. In some cases this might involve action against those in immediate control of workers (for example, franchisees or subcontractors), and in other cases it might include organizations which design and specify the overall system of work (franchisors and principal contractors).

Our third principle for ensuring that prosecution maintains the integrity of the enforcement pyramid is to ensure that the enforcement pyramid is *properly integrated*. In particular, it must be made clear to all OHS duty holders that the agency is willing and able to work its way up to the top of the pyramid in all cases where compliance with OHS standards is unsatisfactory.[102] As we argued in Chapter 4, the more benign approaches to enforcement at the bottom of the pyramid are negotiated in the context of the big sticks at the top of the pyramid, and thus there needs to be a demonstrated capacity to move up the pyramid where compliance is tardy or unforthcoming, or where the duty holder is not making good faith attempts to comply. In particular, this means that a prosecution profile must be redirected away from a major focus on injury-based prosecutions to encompass 'pure risk' prosecutions. We discuss 'pure risk' prosecutions in detail later in this Chapter.

The fourth principle is that there must be *tough sanctions at the top* of the pyramid so that compliance at lower levels of the pyramid is negotiated within the shadow of 'big sticks'.[103] This means that the fines imposed by the courts must be higher,[104] a broader range of sanctions must be utilized,[105] and the sanctions must be applied consistently from case to case.

The fifth principle is that *sanctions imposed must be consistent with, and reflect, the emphasis on systems-based approaches* to OHS, and the principles of sound OHS management. In other words, the sanctions imposed must provide incentives to OHS duty holders to develop SMSs, and if need be must require the development of such systems. Furthermore, the sanctions should give employers incentives to engage in a programme of continuous revision and improvement of their OHS management systems.[106] We deal with this point later in Chapter 7.

The sixth principle is that offences under the *mainstream criminal law must be integrated into the pyramid*, to maximize the deterrent and retributive effect of the top of the pyramid. For example, the pyramid must ensure that where prosecutions are taken as the result of injuries and fatalities, at least some of these prosecutions not only focus on the provisions in the OHS statutes, but also utilize manslaughter and related crimes, such as recklessly or negligently causing injury or disease.[107]

The seventh principle is that OHS prosecutions might be *initiated by parties affected by poor OHS management*. In some jurisdictions trade unions can initiate prosecutions for OHS offences.[108] OHS statutes could be amended to give the power to initiate prosecutions to other public interest groups, such as workers' health organizations. Difficult issues may arise as to how union and public interest group investigations and prosecutions (which should take place in lower level courts) can be dovetailed with those conducted by the OHS inspectorate. These issues will not be easily resolved, but should be addressed in the OHS agency's prosecution guidelines.[109] The point here is that worker and trade union involvement needs to be built into the enforcement pyramid, and most particularly in every step in our proposed Track Two.[110] This includes the possibility of trade union or public interest group initiated prosecutions as a 'back-up' mechanism where the OHS inspectorate is unable (or even unwilling) to initiate prosecutions in appropriate cases, particularly in cases where prosecutions should be taken under the agency's prosecution guidelines. This is particularly important given the limited resources of inspectorates in most countries to bring prosecutions for OHS offences. Union-initiated prosecutions will also increase the prospects of the prosecution profile in an industry reflecting workers' experiences of the hazards in the industry.[111] In Figure 7 we summarize the different types of prosecutions that can be initiated by OHS inspectorates and by unions and public interest groups.

The eighth principle is that, because the purpose of prosecutions is general deterrence, the outcomes of all successful prosecutions must receive the maximum possible *publicity*. Publicity will also serve to enhance the moral and symbolic purposes of prosecution.[112] OHS agencies should publish details of all OHS prosecutions, and the courts should be able to punish OHS contraventions with publicity orders.[113]

The final and most important principle, in terms of the operation of the pyramid, is that the role of prosecutions in it, and the likely penalty if a prosecution is successful, must be *transparent*. To the extent that we

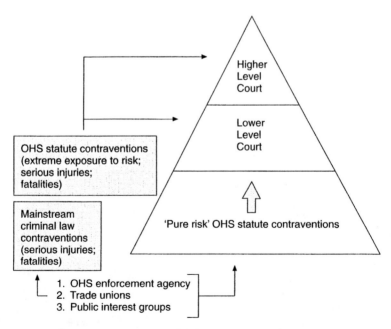

Figure 7 Summary of Different Types of Prosecution

can assume that OHS duty holders are rational actors,[114] duty holders should be given as much advance information as possible about the likelihood, and consequences, of a successful prosecution. All OHS duty holders must know when escalation will take place up the pyramid, and what will happen when a prosecution is initiated. This principle means that the enforcement guidelines, particularly prosecution guidelines, must be publicly available and must clearly set out the rationale of the enforcement pyramid and the purposes and criteria for prosecution.[115] Transparency and acceptance of the guidelines will be enhanced if they are developed in consultation with all parties (including employers, workers, and their representatives) involved in the enforcement of OHS legislation. Detailed prosecution guidelines serve at least three other important functions in an OHS enforcement pyramid. They ensure that OHS inspectors are accountable for the consistent implementation of the enforcement pyramid, and in particular the initiation of prosecutions at the top of the pyramid (consistency across the board). They also establish an ongoing review process by ensuring that, if circumstances require it, inspectors are able to step in to take over the escalation of enforcement measures initiated by another inspector (consistency in the particular case). Finally, they make the OHS enforcement agency's enforcement strategy clear to other parties (such as trade unions or workers' health organizations) who have an interest in prosecution, and who may be able to initiate prosecutions.[116] In this way the OHS agency can ensure adequate dialogue with these groups to promote a consistent but democratic approach to OHS prosecution.

More flexible sanctioning: Broadening the range of OHS prosecutions

In the remainder of this Chapter we consider in more detail some of the OHS enforcement strategies outlined in the previous section. These strategies will broaden the range of matters prosecuted, so that prosecution plays a more assertive role in encouraging OHS duty holders to comply with the broad range and scope of the duties in the OHS statutes, and provides an incentive for duty holders to be more proactive in developing OHS management programmes and effective SMSs. We argue that there should be a clearer OHS agency strategy on both the range of matters which it wishes to prosecute (including 'pure risk' and homicide prosecutions); and the appropriate defendants (including individual corporate officers) in these prosecutions.

Moving up the Pyramid: 'Pure risk' prosecutions

In Chapter 2 we discussed the general duty provisions which have been a feature of most OHS statutes in Europe, North America, and Australia since the British Robens Report in the early 1970s. The general duties are broad, flexible, and open-textured.[117] We noted in Chapter 2 the preventative nature of the general duties and, in particular, the dynamic concept of the employer's duty. It is clear that in most European, North American, and Australian jurisdictions there is no need for an employee to have suffered an injury for an employer to be in breach of the duty.[118]

As our analysis of the 'split pyramid' in Chapter 4 illustrated, prosecutions are rarely integrated into an escalated enforcement response against duty holders, but rather are as a reaction to a workplace injury or fatality. Even though OHS agencies in Europe, North America, and Australia have moved to a more proactive approach to inspections and enforcement, much of the agencies' enforcement activity is largely reactive in nature. A large proportion of investigations carried out by inspectorates and almost all prosecutions are focused on specific events, usually incidents which result in death or injury to employees. This approach tends to focus both the investigation and the prosecution on the particular event, rather than the surrounding system of work and other duty of care issues, which we believe should be the focus of OHS enforcement activity.[119] Important hazards such as manual handling, workplace stress, and substance exposure are rarely the subject of prosecution activity. Further, this approach to enforcement reduces the significance of OHS prosecutions in promoting good OHS practices, by not dealing with crucial issues such as management systems and risk assessment and control.[120]

As we explained, the rationale of the enforcement pyramid is to induce voluntary compliance by duty holders in the shadow of the possibility of serious penal sanctions for non-compliance. If OHS duty holders are to believe that persuasive and administrative enforcement mechanisms at the lower levels of the pyramid are being backed up by big sticks at the top of the pyramid, then there needs to be at least some well-publicized prosecutions showing that there is an integrated pyramid.

At least some of these prosecutions must be prosecutions *which are not event focused*, and which are the result of an agency escalating its enforcement activities up the pyramid. This may occur in cases where substantial hazards exist and a duty holder has repeatedly failed to respond to the inspectorate's requests for compliance and subsequent administrative sanctions, or where there is great potential for a fatality or a series of very serious injuries or diseases, in a workplace where the employer has a bad injury and claims record, and where there is a clear failure to follow widespread industry approaches to OHS management.[121] Such prosecutions will act as a specific deterrent to the non-compliant duty holder, but more importantly, if properly publicized, will provide an example or threat to other duty holders (general deterrence) that action will be taken against them if they choose not to comply voluntarily with their obligations, and that such action will be taken even if no injury or disease has yet resulted.

Prosecutions may also be initiated in relation to Track Two firms who abuse the OHS agency's trust by failing to take adequate steps to implement and monitor SMSs, even though no injury has yet resulted from such failures. These prosecutions reinforce the seriousness of the trust relationship between the agency, the Track Two firm, and affected third parties, particularly workers.

The prosecutions will also have symbolic effects, by officially demonstrating society's intolerance of duty holders who do not adopt proactive health and safety strategies, but rather respond tardily to issues as and when they arise; and by shaping public perceptions of the need to prevent injuries and disease through attention to systems of work. Such prosecutions would help to emphasize the importance of the risk assessment and control approach, the need for sophisticated SMSs and other preventative strategies, and the vital role of information provision under modern OHS regulations. The messages of these prosecutions would be that unless duty holders manage OHS and respond to the requirements of the inspectorate, they will be prosecuted. This is in contrast with the more traditional message which appears to be that as long as duty holders avoid industrial accidents they will not be prosecuted.

In Track Two prosecutions, the message would be that self-regulation through the implementation of SMSs requires responsibility and commitment. While temporary incompetence or misunderstanding of Track Two requirements will be met with advisory and educational measures, breach of trust or lack of commitment will be punished.

We accept that a major constraint on an agency's enforcement role is the limited resources available for investigation and prosecution. Again, we emphasize that we are not necessarily arguing for a greater number of prosecutions, but rather suggesting that more effective use be made of the prosecutions that the agency has the resources to initiate. We are arguing for a shift in the allocation of those resources to more cases alleging defective systems of work, even though injuries or illness have not yet resulted from the defective system. This requires OHS agencies to embrace, at least on small scale, a more labour intensive 'test-case' prosecution strategy. A minor shift in resources allocated to the investigation and prosecution of a few 'pure risk'[122] cases each year is better than no shift at all, and is crucial to ensuring that the agency's focus on advice, education, and persuasion at the bottom of the pyramid will yield results in terms of safer workplaces.

In Track One 'pure risk' cases the evidence of having proceeded through the lower stages of the pyramid becomes central to the case against the duty holder, because it demonstrates the ongoing contravention of the duty holder's obligation. Advisory and persuasive measures in relation to a duty holder would be recorded in appropriate documentation (stage one). In due course the agency would increase compliance activity in relation to those employers and other duty holders who were not complying within a reasonable period of time, and would invoke administrative sanctions. This activity would also be recorded (stage two). Where duty holders failed to respond to these measures, and the workplace hazards were significant, the agency would initiate a 'pure risk' prosecution. We recognize that the proper investigation of such cases may be quite challenging.[123] An injury is very compelling evidence of the manifestation of risk. The absence of an injury means that a broader range of evidence would need to be gathered, including the records of stages one and two, industry practice, epidemiological data collected domestically and in other countries, and the opinions of experts. Examples of such cases could include prosecutions of the manufacturers and suppliers of particular equipment which has the potential to cause injury or of a substance which may cause disease; and prosecutions of employers for failing to undertake risk assessment and to introduce appropriate control measures for the performance of manual handling tasks, the use of a hazardous substance, or to reduce stress at the workplace.[124]

Some Track One pure risk prosecutions will be brought after 'leapfrogging' up the pyramid in response to extremely dangerous systems of work leading to high risks to workers. Here the offence is not so much an 'ongoing' contravention as exposure to extreme hazards. The prosecutor will be able to rely on employee and expert evidence about the hazardous conditions at the workplace over a relatively short period of time.

For Track Two prosecutions for breach of trust, the evidence will demonstrate the extent to which the management systems introduced by the Track Two firm fell short of the requirements of the accredited SMS.

Prosecuting fatalities and serious injuries: Manslaughter and related crimes

We argued in the previous section that OHS prosecutions need to be diversified to include more 'pure risk' prosecutions, where unsafe work systems create serious risks of injury to workers, but injuries have not yet resulted. At the same time, we have argued earlier in this Chapter that a prosecution strategy seeking to maximize the incentives on duty holders to comply with their OHS obligations should also include prosecutions for manslaughter and related crimes. This apparently contradictory strategy recognizes the complexities in the use of criminal sanctions in OHS regulation. There will always be a number of egregious situations where, even though there is an emphasis purely on general and specific deterrence, and on enforcement measures which work their way up the pyramid, public demands for retributive justice for serious injuries, disease, and death at work will justify prosecutions under the OHS statutes and for manslaughter under the mainstream criminal law.

Statistics suggest that work-related fatalities are still high in most of Europe, North America, and Australia.[125] Until recently, industrial manslaughter prosecutions were rare in these jurisdictions. Manslaughter and related criminal prosecutions for workplace death have infrequently been attempted in Canada, with little success.[126] In the last decade or so such prosecutions have been brought successfully in the USA.[127] Most manslaughter prosecutions have been launched in California, New York, and Illinois. There have also been signs that the position is changing in the UK,[128] and in Australia. In Australia, OHS agencies have launched manslaughter prosecutions in Victoria,[129] in New South Wales,[130] and in Queensland.[131] In the UK the first corporate manslaughter case was in 1927, but only two further cases were prosecuted over the next fifty years. None of these cases were successful. The first conviction took place in 1994, in relation to a small company. Both the company and its managing director were found guilty of manslaughter, and the managing director was given a three-year custodial sentence, subsequently reduced to two years by the Court of Appeal.[132]

Corporate criminal liability

Apart from an ideologically-based reluctance of enforcement authorities to consider using general criminal law prosecutions in the workplace, there are also legal technical reasons for the paucity of prosecutions for industrial homicide and serious injury. The issue is whether corporations can commit crimes such as manslaughter. In contrast to the absolute liability general duty offences under OHS statutes, the successful prosecution of crimes such as manslaughter in most jurisdictions requires the proof of *mens rea*, or a guilty mind on the part of the defendant.[133] Corporations, as artificial legal entities, can only act through their human employees and agents. The courts have developed rules to determine *which* employees and agents may have their individual knowledge or intention attributed to a corporation for the purpose of determining the corporation's state of mind. These legal rules have developed unevenly in Europe, North America, and Australia.[134] Anglo-Australian jurisdictions, and some European countries, make it very difficult for OHS enforcement agencies to initiate such prosecutions.

The Anglo-Australian position[135] is that a corporation can be held criminally liable for an offence (such as manslaughter) which requires proof of knowledge or intention, but only where that knowledge or intention is attributable to an officer of the corporation (generally directors and superior officers of the company) who was senior enough to be considered a 'directing mind' of the corporation.[136] The courts have refused to allow the aggregation of the separate faults of individuals, where none of them has sufficient fault which can be attributed to the corporation. Critics[137] argue that the 'directing mind' principle 'in practice means that large corporations are virtually insulated from criminal liability for serious offences'.[138] Consequently prosecutions for industrial manslaughter tend to focus on smaller corporations.[139] Proposed reforms focus on failings of the corporation itself rather than adapting forms of law designed for determining individual liability to complex organizations such as corporations. Late in 1997 the British government announced that it would implement the recommendation of the English Law Commission to enact a new crime of 'corporate killing', which would not require the attribution test to be satisfied. A corporation would be guilty of corporate killing if at least one of the causes of a person's death could be attributed to management failure, and that failure constituted conduct falling far below what could reasonably be expected of a corporation in the circumstances. Management failure would occur if the way in which an organization's activities were managed or organized failed to ensure the health and safety of persons affected by those activities.[140] In Australia section 12.3 of the Criminal Code Act 1995 (Cth.)[141] establishes new methods for establishing the *mens rea*[142] of corporations including the concept of a 'corporate culture', defined in section 12.3(6) as:

> . . . an attitude, policy, rule, course of conduct or practice existing within the body corporate generally or in the part of the body corporate in which the relevant activities takes place.

In continental Europe, some countries have adopted principles of corporate criminal liability, while others have yet to embrace the concept.[143] Where countries do not recognize corporate criminal liability, only individuals can be held criminally liable. Some countries have adopted substitute (ersatz) systems to impose civil liability on the corporation for criminal fines imposed on natural persons acting on behalf of the corporation.[144] This suggests that fundamental reforms to the rules governing corporate criminal liability need to be made in countries with narrow corporate liability provisions before there can be any prospect of corporate manslaughter and related prosecutions playing an effective part in an OHS enforcement strategy. Such reforms might be based on the provisions in the model Australian Model Criminal Code described above,[145] or might follow the broader approach taken in the USA.[146] Under the basic USA common law principle of *respondent superior*, corporations are liable for acts of agents[147] as long as the agent is acting within the scope of her authority or apparent authority, for the benefit of the company, even if the acts were against corporate policy or express instructions.[148] In recent times courts have been more and more willing to find that state legislatures intend to extend liability for homicide to corporations. The Council of Europe in 1988 accepted a recommendation that member states consider the promotion of corporate criminal liability. It recommended that corporations be liable for offences committed in the exercize of their activities even if the offence was extraneous to the purposes of the organization, and even if a natural person could not be identified. The organization should be exonerated where its management had taken all necessary steps to avoid the offence.[149]

The importance of prosecuting for manslaughter/related offences

The principal motivations for industrial manslaughter prosecutions are moral, symbolic, and retributive, and show society's intolerance for organizational behaviour causing workplace deaths.[150] The argument for greater prosecution of manslaughter and related offences is that an increased focus on manslaughter prosecutions would address public disquiet, expressed more and more in jurisdictions all over the world, about the leniency afforded to workplace deaths in comparison to other forms of homicide occurring outside the workplace.[151] Such cases demonstrate that corporations and those who manage their operations are not exempt from the law which punishes criminal negligence that results in death or injury.

We also contend that manslaughter and similar prosecutions are an appropriate way of focusing on the *result* of negligent conduct by duty holders as well as their failure to meet OHS standards.

Further, we argue that a credible deterrence model of OHS regulation requires the threat of manslaughter and similar prosecutions. In a sense, manslaughter prosecutions play an important role in 'pyramid maintenance', by ensuring that tough action can be taken if things actually do go wrong. Manslaughter prosecutions attract considerable publicity and thus contribute to the general deterrent effect of OHS prosecutions. This increased awareness of the likely consequences of failing to implement proper OHS systems assists the inspectorate in its day-to-day enforcement of the legislation at the lower levels of the enforcement pyramid. We believe that a significant general deterrent effect will be achieved if an OHS agency only manages to launch one or two manslaughter prosecutions a year.[152]

The policy issues are, however, more complex than they appear at first glance. The complexity has to do with the differences which have developed historically between OHS and other forms of regulatory crime, and more traditional, mainstream crime.

OHS offences have traditionally been classified into a nebulous 'public safety' category, distinct from 'conventional' criminal offences such as robbery, rape, and murder.[153] OHS offences are a form of inchoate offence, such as attempt or incitement, which are not defined in relation to a specific result or harm (such as causing death or grievous bodily injury), but rather penalize the failure to take measures to prevent harm.[154] Under this type of offence, 'there is no offence which reflects the seriousness of the harm which actually ensues as opposed to the risk of that harm'.[155] This is one reason for the 'social construction of health and safety violations as something separate and distinct from manslaughter',[156] and 'quasi-criminal'.[157] Carson and Johnstone argue that the prosecution of OHS offences under the general criminal law may be counter-productive: by focusing on a few particularly serious cases and singling them out for 'special treatment', one risks undermining the 'normal' OHS offence under OHS legislation. They argue for the inclusion in the OHS legislation of an offence of causing death through breach of the Act or, alternatively, 'industrial manslaughter'.[158] Their point is not that there should not be prosecutions for manslaughter, but rather that OHS agencies should ensure that whatever use is made of traditional crimes, such as manslaughter, should not further entrench the historically created, ambiguous, 'quasi-criminal' status of OHS crime.

Consequently, incorporating mainstream criminal offences, such as industrial manslaughter, with the prosecution of OHS offences is not as straightforward as it may appear. Not only do we need to dovetail mainstream criminal offences with inchoate, regulatory offences, but we have to reconcile two approaches to criminal regulation with inconsistent rationales. As we have argued earlier in this book, in particular in Chapter 2, OHS statutes seek to influence organizational behaviour by taking a pragmatic and activist approach to regulation which involves setting down standards to signal from the outset the required level or type of performance, and making non-compliance with the standard a criminal offence. Compliance is dependent upon an ongoing relationship between the organization and the OHS agency, and is regarded as a long-term objective. It is achieved largely by conciliatory and informal methods (negotiation, education, and persuasion) with deterrence-based prosecution a last resort.

In contrast, resort to industrial manslaughter and related offences involves the state utilizing the traditional criminal law. The state takes a reactive (rather than an activist) role, for mainly moralistic, symbolic, and retributive (rather than pragmatic deterrence-based) purposes, although there is certainly scope for utilitarian aims such as deterrence to be incorporated into this approach. Rather than taking a standards-based approach, this traditional criminal model is passive at the outset, and the agency response is prompted not by a breach of standards, but by a serious consequence—a fatality or injury. Unlike the OHS legislation's focus on the risk of harm, of necessity the traditional criminal law approach is event focused. By resorting to the traditional criminal law, the agency is seeking to establish and reassert moral boundaries by publicly denouncing workplace behaviour resulting in injury or death. This traditional criminal law approach is less flexible than the OHS regulatory model, and prosecution is regarded as the most important tool in the regulatory armoury. A breach of the law is dealt with immediately, hence the focus on prosecution.

There is no satisfactory resolution to this dilemma. We suggest that

OHS agencies should incorporate manslaughter prosecutions into their OHS enforcement strategies in a way that recognizes the subtle tensions and disconnections between traditional crimes such as manslaughter, and regulatory crimes such as those found in the OHS statutes. One possible response is for the OHS statutes to include an offence of industrial manslaughter, or causing death through a breach of the OHS statute.[159] This offence would retain the significance of the term 'manslaughter', but would keep the offence within the OHS statute, thus preventing any suggestion that the OHS offences themselves were any less 'criminal' than the mainstream criminal law. Alternatively, the OHS offences themselves might be included in mainstream criminal statute itself, to emphasize that OHS offences are crimes properly so called. Either way, OHS agencies should accept that the bringing of manslaughter charges does raise a dilemma, and, where prosecutions result from fatalities, they should prosecute manslaughter charges in addition to prosecutions for breaches of the OHS statutes themselves. In so doing they should explain to the court how the emphasis of the OHS offence (on the defective system of work) differs from the manslaughter charge (on the result, death), and try to ensure that manslaughter prosecutions are used to bolster the deterrent value of all OHS enforcement activities.[160]

Who to prosecute: Looking beyond 'the corporate employer' to corporate officers

The empirical data suggests that OHS prosecutions have tended to focus overwhelmingly on corporate employers under the various OHS statutes. Earlier in this Chapter we argued that there should be more prosecutions under provisions imposing duties on designers of plant, and buildings, and manufacturers and suppliers of plant and substances, and, where appropriate, prosecutions of subcontractors, franchisors, and clients, principal contractors, and planning supervisors of building projects. We also suggested that there should be more prosecutions under provisions imposing duties on directors and managers of the corporate employer which has committed an OHS offence. Our concern here is to ensure that all avenues of specific and general deterrence, and of organizational reform, are utilized at the top of the pyramid to reduce unsafe work systems, plant, and substances. The message from the top of the pyramid must be clear: each dimension of the problem is covered by a specification standard, performance standard, or process/systems-based duty, and these standards will be enforced all the way through to the top of the pyramid, against organizations and against the culpable higher level managers and directors in those organizations. Here we examine in more detail the issue of enforcement measures against company directors and top level managers (corporate officers).

When should the OHS agency go behind the corporate veil and prosecute employees, managers, or directors of the company under the relevant provisions of the OHS statute, or for manslaughter or other traditional criminal offences for the negligent or reckless infliction of injury, particularly where the corporation is a large organization?

Again, the international data seems to suggest that OHS agencies bring relatively few prosecutions under the provisions in OHS statutes imposing individual liability upon corporate officers.[161] Most prosecutions are of corporate employers under the provisions of the OHS statute. Moreover, an examination of the prosecutions that have been taken against managers and directors suggests that there is an inappropriate skewing of these types of prosecutions to directors and managers in small organizations.[162]

Prosecutions of individual corporate officers can also take place under the mainstream criminal law. Until recently there have been very few homicide prosecutions against managers and directors. There are, however, some examples of executives in both the UK and the USA[163] being prosecuted for various forms of homicide after their gross negligence led to the deaths of workers. By the end of 1996, four British prosecutions against directors of companies for manslaughter had been successfully conducted.[164]

In the USA, since the mid-1980s authorities have increasingly turned to the mainstream criminal law to punish individual managers and directors responsible for unsafe work practices leading to employee deaths. In 1985, for example, in *People* v. *Film Recovery Systems*,[165] an Illinios circuit court found three individual corporate executives (the president of the corporation, and the plant manager and foreman) guilty of murder, after a worker died of acute cyanide toxicity after being exposed to critically dangerous levels of hydrogen cyanide gas.[166]

Internal discipline

Some commentators[167] argue that it is not merely a question of deciding when to go beyond a prosecution of the corporate employer to prosecute an individual director or manager who is particularly culpable for an OHS offence. As most OHS prosecutors will concede, this is a time-consuming, costly, and complex exercize, particularly when conducted in relation to large corporations.[168] Fisse and Braithwaite comment that:[169]

> . . . where corporations are sanctioned for offences, in theory they are supposed to react by using their internal disciplinary systems to sheet home accountability, but the law now makes little or no attempt to ensure that such a reaction occurs. The impact of enforcement can easily stop with a corporate pay-out of a fine or monetary penalty . . .

The challenge for OHS policy makers is to encourage, or even compel, organizations themselves to:

- initiate and carry out the process of conducting an internal investigation to determine who within the management of the organization was particularly responsible for the offence;
- take responsibility for disciplining the managers thus identified; and
- take measures to review and restructure their internal OHS management policies, procedures, and practices.[170]

In Chapter 3 we looked at ways in which we might provide incentives to organizations to improve their OHS management systems generally. In this section we look in greater detail at two issues: namely deciding when to bring proceedings against individuals within the organizational structure and when to prosecute the corporation; and developing ways of ensuring that internal disciplinary procedures are invoked by the company once it has been convicted of an OHS offence.

Why prosecute individuals?

Why prosecute individuals within the organizational structure?[171] As with corporate offenders, the main purpose of sanctions against corporate officers is deterrence, both general and specific.[172] But as with corporate offenders, sanctions should also be concerned with the subgoals of rehabilitation (or organizational reform), retribution, denunciation, and, if the occasion warrants it, incapacitation.

A primary reason for imposing criminal liability for OHS contraventions on individual corporate officers is that the imposition of civil or criminal penalties on corporations for these offences can simply be seen by corporations as a cost of doing business, and passed on to consumers, shareholders, or employees.[173] One solution is to impose criminal liability on corporate officers.[174] While it is true that individual officers can externalize civil penalties by demanding that the corporation pay the penalty, or increase their remuneration to meet the penalty, the stigma of criminal conviction, and the prospect of a gaol sentence, cannot be externalized in this way.[175] We are not here arguing that OHS agencies should focus on either corporations or individual corporate officers at the exclusion of the other. Corporate sanctions are necessary because 'it is often the complex interplay between managers, standard operating procedures, corporate priorities, market demands and various other forces at work within corporations, rather than simply the influence of a particular individual, which is the ultimate cause of corporate criminal behaviour'.[176] Rather, our argument is that an effective OHS prosecution strategy should ensure a mix of prosecutions.[177] Edelman argues that 'by directing the force of the law at both the corporate firm and its directors, the deterrence threat will force the corporate entity and its individual actors to monitor each other's activities . . . The threat of prosecution will encourage employees to seek out and remedy violations before they occur and it will encourage the corporate entity to implement policies aimed at preventing violations.'[178]

As we noted earlier, the evidence which casts doubt on the effectiveness of deterrence is primarily concerned with the effects of sanctions on corporations rather than individuals within those corporations. Deterrence theorists commonly make the mistake of assuming that top management will act as a fiduciary, protecting the interests of the organization, and that therefore sanctions against the corporation will impact directly on management decision making. This assumption is not borne out in practice. Managers may be willing to ignore OHS issues, or require dangerous work practices, if they can advance their careers by doing so,[179] without any fear of personal liability. Braithwaite and Makkai[180] argue that managers might also:

. . . act in accordance with group loyalties other than to the firm (for example, loyalty to their profession, their co-workers, the government). And top management often will not be 'rational'. Perhaps most critically, top management will often not know about decisions to break the law in the interests of the corporation, as effective control over decisions to comply with or break the law is in the hands of middle managers over whom top management has limited control.

This suggests that OHS prosecutions should be targeted at individual corporate decision makers, not just the organization itself, because individuals who are vulnerable to personal sanctions have both a much greater incentive and a greater capacity to avoid these penalties than do fiduciaries.[181] Commentators argue that criminal deterrence is likely to be most effective where a person is able to weigh up the risk, or where, even though the risk is low, the punishment is greatly feared.[182] Penalties imposed upon corporate officers are not bounded by the limited liability status of the organization.[183] Sanctions such as fines, imprisonment, probation, community service orders, disqualification of directors,[184] and even adverse publicity, may indeed have a very considerable deterrent effect on individual managers or directors, imposing relatively severe penalties on responsible individuals, which are incapable of being treated as simply part of the cost of doing business and passed on to consumers.[185] As Geis notes, gaol terms in particular 'have a self-evident deterrent impact upon corporate officials, who belong to a social group that is exquisitely sensitive to status deprivation and censure'.[186]

Others argue that fear of criminal prosecution, even in the absence of conviction, may effectively deter corporate executives from engaging in illegal activity.[187] In a recent study of OHS in Australian companies Hopkins found that it was the fear of personal criminal liability, in particular prosecutions being brought against them as individuals and the subsequent publicity, which is one of the most important factors motivating senior managers to pay attention to OHS.[188] He argues that the 'due diligence' type of provision imposing liability on managers and directors[189] is the best way to ensure that managers are motivated to improve OHS. For example, the imprisonment of Donald George McMurty, the former managing director of West Zinc, by the Western Australian Court of Petty Sessions in 1995 for a serious offence of polluting a wetland by ordering the discharge of toxic waste into a stormwater drain, provided a general deterrent to other company directors.[190] A similar impact was achieved in the USA after the *People* v. *Film Recovery Systems* case,[191] and in the late 1980s in California, when the Los Angeles District Attorney's OHS Program resulted in prosecutions of directors and managers of corporations for workplace injuries and death.[192] In 1996 a director of a demolition company was given an immediate gaol sentence for breaching the UK HSW Act 1974, for failing to take precautions when demolishing a factory which contained asbestos.[193] In that same year two directors of a carpet manufacturing company were sentenced to four months' imprisonment after a factory worker lost an arm while cleaning machinery.[194]

Experience in USA environmental regulation suggests that bringing prosecutions against individual corporate officers, with the possibility of gaol sentences, plays an important role in inducing voluntary compliance measures at lower levels of the enforcement pyramid, and provides a strong incentive for offending organizations to co-operate with enforce ment officials when officials resort to administrative or civil sanctions.[195] Most significantly, prosecutions have a 'profound educative and prefer- ence-shaping effect' which creates a corporate and public ethic which promotes voluntary compliance.[196] Smith argues that criminal prosecu- tions achieve these functions in the USA simply because managers and directors who do not ensure that their organizations comply with their environmental obligations can be gaoled for two or more years. She argues that there 'is little point to criminal prosecution of corporations and other organisations except in conjunction with the prosecution of responsible corporate managers and directors'.[197]

A further reason for imposing liability on individual officers is that the offending corporation can go out of business, be bankrupted, or other- wise dissolve itself, leaving no corporate entity to shoulder the responsi- bility for the offence. Prosecution of culpable officers will ensure that not only is adequate punishment extracted for the offences which took place in relation to the dissolved corporation, but will deter, even prevent, officers from moving on to another corporation and continuing the dangerous work practices.

Focus on top management

In so far as penalties *are* sought against individuals, then it is superior rather than subordinate officers of corporations who should be the principal targets. It is unlikely that the OHS management systems and culture of an organization can be reformed without a change in attitude by top management.[198] Even if they are not 'direct actors', top managers have formal responsibility within the organization for the acts or omis- sions which led to the offence, and are in a position to alter the operating procedures of the company to prevent the commission of the offence, and to develop effective SMSs.[199] Subordinates often have little discre- tion and are 'constrained by orders, rules, standard operating procedures and goals or targets set by their superiors',[200] and can be dismissed or have their careers jeopardized for disobeying company directions.

It is important, therefore, in developing a prosecution policy to ensure that the officer who is to be prosecuted was in a position to do something about the OHS issue (this is required by many of the legal provisions in any event); that individuals (and particularly lower level employees) are not prosecuted for their actions when the real issue is policy or proce- dures of the company or that the behaviour of the lower level employee has been induced by an incentive (emanating from the top of the management structure) to behave in an unsafe manner;[201] that scapegoats are not put forward by the management to take the full force of such individual prosecutions; and that talented and hard-working individuals not be put off from seeking positions of responsibility in OHS management because of fear of criminal prosecution. We will return to these issues later in this Chapter.

We noted earlier that prosecutions against corporate officers are rare, both under OHS statutes, and in other areas of the criminal law. The reasons for this are relatively simple:[202] it is often extremely difficult to obtain proof of the officer's criminal behaviour. OHS offences are often the result of omissions, where there is only evidence (if any) of acquiescence or a failure to prevent illegal conduct by others, or are the result of communication breakdowns within the organization (particularly the failure to respond to warnings or complaints). Contraventions might arise because subordinates incorrectly anticipate what their superiors require, or may result from pressure to comply with the organization's standards which themselves are contrary to law.[203] Often the contravention is truly the result of group acts, rather than the result of any one particular individual's conduct. It is also difficult in most cases to determine who, in the complex chain of command in the organization hierarchy, had responsibility for formulating or implementing OHS policies. 'A corporate society finds it easier to hide its skeleton in closets, and in a big corporation the closets are more numerous and more obscure.'[204] There are many reasons for this: internal communications within the company are easily kept secret, particularly if they are oral communications; corporate officials and employees often close ranks in the face of an external investigation; and higher level management and directors are often able to isolate themselves from decisions, or to claim ignorance about a particular policy, act, or omission.[205]

Fault-based corporate officer liability

The provision in the OHS statute which imposes individual liability upon corporate officers might itself specify which officer is to be prosecuted. There are a range of models of individual liability to be found in the various OHS and related statutes. A model frequently used is to base corporate officer liability on a fault principle. One approach[206] is to make company officers liable for OHS offences by corporate employers if the offence occurred with the consent or connivance of, or was attributable to the wilful neglect of, the officer.

A second approach[207] has lower requirements for the proof of fault, and presumes that, in addition to the corporation being liable for the offence, officers in a position to influence the conduct of the corporation are also liable, unless they can show that they have exercised due diligence to prevent such contraventions. In other words, the defendant corporate officer has the burden of proving that due diligence was exercised. Due diligence could be proved, for example, by showing that the officer was responsible for introducing, implementing, and maintaining an effective compliance programme[208] or SMS. Health and safety policy makers can thereby ensure that individual corporate officer liability provisions play a part in providing top-level managers with incentives to introduce state of the art OHS management systems. Hopkins[209] argues that this second approach is most likely to impose the greatest incentives on managers and directors to ensure that the corporate employer complies with its OHS obligations.[210]

A third approach is to impose liability on a manager as an employee of the organization. All managers would technically be employees of the organization. Almost all OHS statutes impose a general duty upon employees to safeguard their own health and safety and the safety and health of others. The duty can be expressed in terms of a duty not to be negligent, to a duty not to wilfully expose others to danger.

This raises the question of which corporate officer should be prosecuted in the event that there is a contravention of the OHS statute, and more than one officer technically can be said to fall within the scope of the provision conferring individual liability for the contravention. In small organizations it may be a simple exercize to find the corporate officer who is most appropriately prosecuted under these fault-based models.[211] Larger organizations may be more complex, for the reasons outlined above.

One possible approach to determining which corporate officer to prosecute is for the OHS enforcement agency to devote more training and resources to assisting inspectors to investigate the management structure of an offending organization,[212] in order to determine just how the organization is structured and who within the organization is especially responsible for the offence. This would appear to be a tall order, too time-consuming, and relying too heavily on the inspectorate's scarce resources, but may be facilitated by the ongoing relationship between the inspectorate and business organizations fostered by the process of pyramidal enforcement described in Chapter 4. In the process of giving informal directions, and then issuing improvement, prohibition, or penalty notices, the inspector will build up knowledge of the responsible higher level manager. Where the organization itself is unclear about which manager is responsible, the inspector, over time, can make it clear which manager she or he sees as being responsible for OHS issues.[213] The point is not so much to satisfy the legal requirements of the OHS statute, but to ensure that, if a person is to be prosecuted within the organization, it is the right person, so that specific and general deterrence is maximized.

Alternatively the OHS agency could simply decide to conduct an investigation against the most senior corporate officer, and see whether such an investigation could 'flush out' a manager to whom responsibility had been delegated. Written evidence of such a delegation would be required, as would clear evidence that the manager was given appropriate resources, personnel, and time to implement appropriate OHS policies. We suspect, however, that this approach will in most cases consume a great deal of time and resources, and will in all likelihood be obstructed by the factors we have outlined above.

Nominated responsibility

A second possible approach for focusing liability for OHS breaches depends on modifying the basis of individual responsibility for OHS away from a purely fault-based model, to a combination of nominated[214] and fault-based individual responsibility.[215] Developments in OHS law in Europe,[216] North America, and Australia make it clear that the corporate employer cannot delegate OHS responsibilities so as to be exculpated for non-compliance. But at the same time, the employer needs to ensure that a senior officer oversees the implementation of the employer's responsibilities. This approach would require employers to appoint senior level responsible officers who must take reasonable steps[217] or who must take all reasonably practicable measures and exercize due diligence[218] to ensure that the employer complies with its OHS obligations. The responsible officer should be a director or senior manager—OHS is not advanced within an organization if, as is suggested is happening in some countries including The Netherlands,[219] chief executives delegate OHS issues to subordinates outside the senior management team.

There are a number of possible approaches to this sort of model.[220] For instance,[221] the OHS statute itself could define in advance who has such responsibility (for example, the OHS manager, the General Manager, or Managing Director, or a member of the Board who is given overall responsibility for OHS).[222] The statute could then provide that the responsible person (or people) will be held liable in the event of a contravention of the statute, but that the organization must ensure that these officers have the time, resources, and expertise to ensure that the organization complies with its OHS obligations. The OHS statute might specify the kind of authority and resources that the organization should give to the responsible officer, and might allow responsible managers to seek a court order for appropriate organizational authority or resources to carry out the responsibilities.

Where the organization wished to deviate from the responsible officer designated by statute, the organization must very clearly specify which directors and senior managers have prime responsibility for OHS within the organization, and provide notification in writing to the local OHS inspector of the new allocation of responsibilities. The notification must indicate that the designated manager has the time, resources, and training to ensure that the organization complies with its OHS obligations, and must be signed by the designated manager. The legislation should set out a procedure which ensures that managers can refuse to accept a delegation that is unreasonable, and might set out a dispute resolution procedure where delegations are hotly contested. The dispute resolution process would include an investigation into whether the proposed responsible officer was given adequate resources and authority to discharge her or his OHS compliance responsibilities.

Alternatively, the OHS statute might require employers to develop an OHS policy and set of procedures which might include a requirement that the policy identify senior managers who are responsible for developing OHS procedures and who are responsible for developing operating procedures which have an impact on OHS. These managers would then have an affirmative duty to correct or attempt to correct operating procedures which fell short of required OHS standards.[223] Where the policy did not identify a responsible officer, the responsibility would fall upon the CEO.

Whichever of these two basic approaches is taken, the OHS statute should provide the responsible officer with a due diligence defence, which would be satisfied by appropriate efforts to implement a compliance programme or SMS. The statute might bolster the authority of the responsible officer by specifying that it is an offence for any person to disobey a direction of the responsible officer which was given to ensure compliance with the OHS responsibilities. The responsible officer might also be given a statutory right to request written confirmation of instructions from more senior management which have the effect of undermining the responsible officer's efforts to ensure compliance with the organization's OHS obligations. The responsible officer should have the right to call in an OHS inspector to investigate the contested instruction, and should have the right not to implement the instruction until it is given in writing. Such written confirmation would serve to exculpate the responsible officer, and shift liability for non-compliance with the organization's OHS obligations onto the senior manager issuing the instruction.[224]

These procedures would go some way towards addressing the tendency of many organizations to blame workers for OHS risks, and will promote better systems of internal accountability. They might only be necessary in large companies, and their complexity and potential lack of flexibility, and the fact that they (particularly the first approach) impose a particular organizational structure on organizations, might render them unsuitable for certain industries.[225] In small companies the flatness and narrowness of the organizational structure will usually mean that it is quite clear which managers and directors have responsibility for decision making, including decisions affecting OHS.

Activate and monitor private justice systems

A third broad approach to the issue of corporate officer liability for OHS offences, put forward by Fisse and Braithwaite,[226] is to 'activate and monitor the private justice systems of corporate defendants'.[227] It may not be necessary for the OHS enforcement agency to prosecute individual managers who are responsible for OHS contraventions if, when convicted of OHS offences, corporations undertake steps to identify and discipline those managers who were responsible for the contravention.

The Fisse and Braithwaite 'accountability Model' is based on a number of clearly stated principles or rules of action (which they call 'desiderata'), the most important of which is to:

. . . seek to publicly identify all those who are responsible and hold them responsible, whether the responsible actors are individuals, corporations, corporate subunits, gatekeepers, industry associations or regulatory agencies themselves.[228]

The model[229] recognizes that the allocation of responsibility should be cost-efficient, and that managerial flexibility should be preserved, rather than straitjacketed by law. Those who are responsible for equal wrongs must be treated equally. The accountability model is built on an enforcement pyramid. The adoption of Fisse and Braithwaite's model[230] for OHS prosecutions might involve prosecution guidelines that specify that:

- the corporation should be prosecuted where the corporation has not activated its private justice systems sufficiently to *identify publicly* that it has diagnosed the cause of the offence, remedied the identified causes, disciplined those managers found to be responsible, and compensated injured workers and members of the public;
- the corporation should be prosecuted where there is evidence that the offence has been committed in accordance with a calculated policy or because of grossly defective OHS management systems or an entrenched ethos or culture of non-compliance of the company; and
- responsible individuals should be prosecuted where:
  - they are beyond the disciplinary reach of corporations (for example, consultant accounting firms who advise against OHS procedures on the grounds of cost);
  - the disciplinary action taken by the corporation against the individual is insufficient to reflect the level of responsibility that individual bears for the offence; or
  - there is no level of disciplinary action that the corporation can implement to reflect the severity of the individual's offence, for example, where a senior manager orders work to be done in the face of obvious dangers to health.[231]

Most importantly, the integrity of the internal disciplining model requires safeguards to be introduced to ensure that scapegoating and other unjust practices do not take place within the organization when it implements its internal disciplining procedures. Such safeguards might include judicial scrutiny of corporate action when the organization reports to the court on its internal disciplining processes; empowering employees with the right to complain to the court (or an internal accountability monitoring committee or tripartite OHS monitoring committee in the case of Track Two organizations)[232] if scapegoating occurs; legal recognition of private systems of justice so as to involve participatory self-determination of issues involved in allocating responsibility for the OHS offences committed by the corporation; and minimal procedural protections for individuals exposed to internal disciplinary proceedings. If scapegoating is identified, escalated sanctions might be applied. In particular, scapegoating offences would be prosecuted in a higher level court.

Fisse and Braithwaite's model goes further than any other model in ensuring that state resources are not frittered away in the difficult exercize of trying to pin criminal responsibility on individual managers and directors within organizations which have contravened their duties under the OHS statute. It also fits in well with a regulatory scheme that seeks to emphasize the development of OHS management systems. The identification of those managers responsible for OHS contraventions can be the first step in a process of identifying weaknesses in and redesigning the organization's OHS management systems. The OHS agency needs to ensure that its prosecution policies, and the OHS sentencing guide-lines,[233] provide sufficient inducement to organizations to undertake the tasks of disciplining managers responsible for OHS contraventions, and of improving OHS management systems.

For self-policing to work, organizations must not fear that information gained from internal investigations and self-reporting will be used in later prosecution proceedings against the organization.[234] In her study of business responses to workplace deaths, Haines, for example, shows how the threat of legal proceedings against organizations gave rise to defen-sive behaviour aimed at reducing the regulator's 'capacity to scrutinize activities internal to the company'.[235] Our emphasis in this section on organizational self-monitoring, and in Chapter 7 on OHS compliance programmes, will only be effective if there is a free flow of information within the organization. Provisions may need to be introduced for or-ganizations to be given some form of amnesty or privilege in relation to such information. The experience in the USA of corporate amnesty in relation to self-reporting in the antitrust area suggests that 'amnesty is more attractive than mitigation to corporations',[236] because even if a penalty is reduced as a result of mitigating circumstances, the offender still has to pay the minimum fine, the costs of defending the trial both in terms of money and time expended by employees of the organization, and suffer adverse publicity and the stigma of conviction.[237] Against this, policy makers may be unwilling to embrace selective amnesty if it ap-pears that corporate offenders are receiving benefits not available to other types of offender, and the administration of a selective amnesty scheme may be unwieldy and overly burdensome. These latter objec-tions may be overcome if details of self-reporting and internal discipli-nary proceedings by the corporation are made public, even though no prosecution results, and if the criteria for amnesty are clearly and publicly enunciated.[238] Alternatively, or in addition to selective amnesty, information gathered from internal investigations, internal audits, or self-reporting might be privileged (self-evaluative privilege), so that the information itself cannot be used in external investigations or for prosecution purposes.

Conclusion

In this Chapter we have illustrated how OHS prosecutions fit into the enforcement pyramid outlined in Chapter 4. Prosecution constitutes the 'big' sticks at the top of the pyramid, and provides general and specific deterrence as a shadow within which voluntary compliance measures are carried out. We have emphasized that an effective OHS prosecution process needs to have flexible goals, and, in particular, it needs to be able to achieve goals relating to organizational reform, retribution, and denunciation.

We have also argued that an effective prosecution policy would involve a more varied range of OHS prosecutions than has been the case to date in most countries. For example, the OHS agency should resort to prosecution when a deterrent was required to deal with a duty holder who has not complied with OHS standards despite the persuasive and mild specific deterrence measures used in the middle of the pyramid, even though an injury had not yet resulted. We also acknowledge that OHS agencies are likely always to be subject to public pressure to bring OHS prosecutions where serious injuries and fatalities have occurred in the workplace. Most of these prosecutions will focus on contraventions of the OHS legislation. We have argued that manslaughter prosecutions be integrated into the pyramid, not only to give recognition to public demands for retribution for workplace fatalities, but to bolster the appearance of severe deterrent and punitive measures at the top of the enforcement pyramid. We suggest that the OHS agency initiate a number of manslaughter prosecutions each year, in response to the worst workplace fatalities, in order to bolster the general deterrent effect of the top of the pyramid.

As we argued earlier in this Chapter, not only should different types of prosecutions be brought against employers, but a broader range of duty holders should be prosecuted, including culpable high-ranking managers, and directors, franchisors, subcontractors, and designers, manufacturers, and suppliers of unsafe plant and substances. All of these measures will ensure that the work of inspectors at the lower levels of the pyramid can be backed up with credible sanctions if compliance is not forthcoming.

In Chapter 7 we examine the sanctions that might be imposed on those who are prosecuted for OHS contraventions.

NOTES

1 Although we note that in the past few years there have been signs in a number of European and Australian jurisdictions that OHS regulators are looking to increase penalties for contraventions of OHS statutes (see European Foundation for the Improvement of Living and Working Conditions, 'Union proposals to enforce health and safety legislation at the workplace' (1998) Jan. *Evroline* (Greece); European Foundation for the Improvement of Living and Working Conditions, 'Penalties for labour law infringements under review' (1998) Feb. *Evroline* (Portugal); and Industry Commission, Australia, *Work, Health and Safety* (AGPS, Canberra, 1995), Vol. I, pp. 109 and 118). As we note later in this Chapter, it also appears that there is an increased focus on industrial manslaughter in the UK.

2 European Foundation for the Improvement of Living and Working Conditions, *Policies on Health and Safety in Thirteen Countries of the European Union: The European Situation*, Vol. II (Office of Official Publications of the European Communities, Luxembourg, 1996), p. 61.

3 In the USA, where discussion of corporate criminal sanctions generally has been most vigorous, the debate has to a large extent bypassed the OHS arena because the OSH Act is enforced mainly by civil penalties. Some jurisdictions, such as Nova Scotia in Canada, have introduced a wider range of sanctions, but have yet to use these provisions to an extent which would yield useful data about their effectiveness.

4 See R. Johnstone, *The Court and the Factory: The Legal Construction of Occupational Health and Safety Offences in Victoria*, unpublished Ph.D. thesis, The University of Melbourne, 1994; R. Johnstone, 'The Legal Construction of Occupational Health and Safety Offences in Victoria: 1983–1991', in R. Johnstone (ed.), *Occupational Health and Safety Prosecutions in Australia* (Centre for Employment and Labour Relations Law, The University of Melbourne, Melbourne, 1994); and our survey of prosecutions in New South Wales undertaken in late 1995.

5 See again our discussion of what constitutes the top of the pyramid at pp. 114 and 185 above.

6 Or utilitarian or consequentialist.

7 See, for example, C. Wells, *Corporations and Criminal Responsibility* (Clarendon Press, Oxford, 1993), pp. 15–16.

8 See K. C. Kennedy, 'A Critical Appraisal of Criminal Deterrence Theory' (1983) 88 *Dickinson Law Review* 1; and J. L. Miller & A. B. Anderson, 'Updating the Deterrence Doctrine' (1986) 77 *The Journal of Criminal Law and Criminology* 418.

9 J. P. Bonner & B. N. Forman, 'Bridging the Deterrence Gap: Imposing Criminal Penalties on Corporations and their Executives for Producing

Hazardous Products' (1993) 1 *San Diego Justice Journal* 1 at 7, quoting F. Lederman, 'Criminal Law, Perpetrator and Corporation: Rethinking a Complex Triangle' (1985) *76 Journal of Law and Criminology* 285 at 335. See also S. Shavell, 'Criminal Law and the Optimal Use of Nonmonetary Sanctions as a Deterrent' (1985) 85 *Columbia Law Review* 1232; and A. Norrie, *Crime, Reason and History: A critical introduction to criminal law* (Weidenfeld & Nicolson, London, 1993), p. 199. Norrie (at p. 199) states that: 'At the heart of the deterrence model is the notion of "economic man", capable of rational calculation of the consequences of his action.'

10 J. P. Bonner & B. N. Forman, 'Bridging the Deterrence Gap: Imposing Criminal Penalties on Corporations and their Executives for Producing Hazardous Products' (1993) 1 *San Diego Justice Journal* 1 at 10.

11 See the analysis of the impact of OHS enforcement in the USA and Canada by R. M. Brown, 'Administrative and Criminal Penalties in the Enforcement of Occupational Health and Safety Legislation' (1992) 30(3) *Osgoode Hall Law Journal* 691. See also J. P. Bonner & B. N. Forman, 'Bridging the Deterrence Gap: Imposing Criminal Penalties on Corporations and their Executives for Producing Hazardous Products' (1993) 1 *San Diego Justice Journal* 1 at 10; S. A. Shapiro & R. S. Rabinowitz, 'Punishment versus Cooperation in Regulatory Enforcement: A Case Study of OSHA' (1997) *Administrative Law Review* 713 at 716; and J. Braithwaite & T. Makkai, 'Testing an Expected Utility Model of Corporate Deviance' (1991) 25 *Law and Society Review* 7 at 35.

12 Industry Commission, Australia, *Work, Health and Safety* (AGPS, Canberra, 1995), Vol. II, pp. 393–7. See also J. L. Miller & A. B. Anderson, 'Updating the Deterrence Doctrine' (1986) 77 *The Journal of Criminal Law and Criminology* 418; and F. Zimring & G. Hawkins, *Deterrence: The Legal Threat of Crime Control* (The University of Chicago Press, Chicago, 1973).

13 For Europe, see T. Wilthagen, 'Reflexive Rationality in the Regulation of Occupational Health and Safety', in R. Rogowski & T. Wilthagen, *Reflexive Labour Law* (Kluwer, Deventer, 1994), at pp. 356–62; D. Walters (ed.), *The Identification and Assessment of Occupational Health and Safety Strategies in Europe* (European Foundation for the Improvement of Living and Working Conditions, Dublin, 1996). For the UK in particular, see B. Hutter, *Compliance: Regulation and Environment* (Clarendon Press, Oxford, 1997), especially pp. 220–32; P. James, 'Reforming British Health and Safety Law: A Framework for Discussion' (1992) 21 *Industrial Law Journal* 83 at 97–104; S. Tombs, 'Law, Resistance and Reform: "Regulating" Safety Crimes in the UK' (1995) 4 *Social and Legal Studies* 343 at 344; and D. Walters (ed.), *The Identification and Assessment of Occupational Health and Safety Strategies in Europe* (European Foundation for the Improvement of Living and Working Conditions, Dublin, 1996), pp. 191–2 (where the HSE is reported as calculating that the annual probability of a firm receiving an enforcement notice is 1 in 80 and of being prosecuted is 1 in 800) and p. 200. For discussion of the USA, see S. A. Shapiro & R. S. Rabinowitz, 'Punishment versus Cooperation in Regulatory Enforcement: A Case Study of OSHA' (1997) *Administrative Law Review* 713 at 716–17; T. H. McQuiston, R. C. Zakocs, & D. Loomis, 'The

Case for Stronger OSHA Enforcement—Evidence from Evaluation Research' (1998) 88 *American Journal of Public Health* 1022 (OSHA will inspect each workplace on average once every 144 years, and each workplace covered by state plans will be inspected once every 44 years; and the average citation for a serious violation was $763 in 1995); and R. M. Brown, 'Administrative and Criminal Penalties in the Enforcement of the Occupational Health and Safety Legislation' (1992) 30(3) *Osgoode Hall Law Journal* 691; and for Canada, see R. M. Brown, 'Administrative and Criminal Penalties in the Enforcement of Occupational Health and Safety Legislation' (1992) 30(3) *Osgoode Hall Law Journal*, 691; R. M. Brown, 'Theory and Practice of Regulatory Enforcement: Occupational Health and Safety Regulation in British Columbia' (1994) 16 *Law and Policy* 63; E. Tucker, 'And Defeat Goes On: An Assessment of Third Wave Health and Safety Legislation', in P. Stenning (ed.), *Accountability for Corporate Crime* (University of Toronto Press, Toronto, 1995).

14 'Policy makers must consequently focus upon designing penalties whose marginal deterrence value outweighs their societal cost', where 'costs in this sense are the excess of the social harm caused by a product over the benefits the product confers on society': J. P. Bonner & B. N. Forman, 'Bridging the Deterrence Gap: Imposing Criminal Penalties on Corporations and their Executives for Producing Hazardous Products' (1993) 1 *San Diego Justice Journal* 1 at 9.

15 J. P. Bonner & B. N. Forman, 'Bridging the Deterrence Gap: Imposing Criminal Penalties on Corporations and their Executives for Producing Hazardous Products' (1993) 1 *San Diego Justice Journal* 1 at 11.

16 Ibid. 17.

17 Ibid. 18, in the context of punitive damages for product liability actions.

18 See, generally, J. M. Morris, 'The Structure of Criminal Law and Deterrence' (1986) *Criminal Law Review* 524.

19 We discuss retribution above at p. 194.

20 See Environmental Offences and Penalties Act 1989 (NSW).

21 In this particular case, major law firms seem to have acted as inadvertent 'amplifiers' of this misconception, sensitizing corporations and their officers to the possible consequences of breaching their legal duties, as part of a marketing exercise to promote the benefits of environmental audits, conducted by themselves.

22 For example, see E. Sutherland, *White-Collar Crime: The Uncut Version* (Yale University Press, New Haven, 1983), pp. 236–8; S. Tombs, 'Law, Resistance and Reform: "Regulating" Safety Crimes in the UK' (1995) 4 *Social and Legal Studies* 343–65 at 355; and F. Pearce & S. Tombs, 'Hazards, Law and Class: Contextualizing the Regulation of Corporate Crime' (1997) 6 *Social and Legal Studies* 79 at 92–9.

23 W. J. Chambliss, 'Types of Deviance and the Effectiveness of Legal Sanctions' (1967) *Wisconsin Law Review* 703.

24 J. Braithwaite & G. Geis, (1982) 28 *Crime and Delinquency* 282 at 301–2. Croall remarks that 'Corporations are often assumed to be future orientated, concerned with their reputation and "quintessentially rational". Unfavourable publicity and the harrowing experience of investigation,

prosecution and trial, therefore, can and do act as deterrents': H. Croall, *White-Collar Crime* (Open University Press, Buckingham, 1992), p. 147.

25 F. Pearce & S. Tombs, 'Hazards, Law and Class: Contextualizing the Regulation of Corporate Crime' (1997) 6 *Social and Legal Studies* 79 at 86.

26 Ibid. 92.

27 Pearce and Tombs therefore conclude, quite reasonably, that if greater resources were given to OHS enforcement agencies, the rate of offence detection and enforcement, and hence deterrence, would increase.

28 Employers are influenced by publicity given to corporate prosecutions, and OHS regulators are more easily able to communicate information to employers than the police are to traditional offenders.

29 See M. B. Mentzger & C. R. Schwenk, 'Decision Making Models, Devil's Advocacy, and the Control of Corporate Crime' (1990) 28 *American Business Law Journal* 323 at 337. See also R. Ellickson, 'Bringing Culture and Human Frailty to Rational Actors: A. Critique of Classical Law and Economics' (1989) 65 *Chicago-Kent Law Review* 23.

30 Individual managers will not necessarily share the long-term profit maximization goals of the organization, but may have shorter term urges and desires for power, prestige, creativity, security, or career advancement. Organizations operating in concentrated markets with relatively high barriers to entry may not be motivated solely by profit maximization goals, but may be concerned with innovation, growth, organizational prestige, and other goals: M. B. Mentzger & C. R. Schwenk, 'Decision Making Models, Devil's Advocacy, and the Control of Corporate Crime' (1990) 28 *American Business Law Journal* 323 at 341. We address this issue later in this Chapter when we examine non-financial values of the organization.

31 J. Braithwaite & T. Makkai, 'Testing an Expected Utility Model of Corporate Deviance' (1991) 25 *Law and Society Review* 7 at 35.

32 F. Pearce & S. Tombs, 'Hazards, Law and Class: Contextualizing the Regulation of Corporate Crime' (1997) 6 *Social and Legal Studies* 79 at 95.

33 Social psychologists have shown that corporate decision making 'may at times produce decisions that are riskier than those preferred by any of the individual decision makers, and riskier than those the hypothetical rational actor would make': see M. B. Mentzger & C. R. Schwenk, 'Decision Making Models, Devil's Advocacy, and the Control of Corporate Crime' (1990) 28 *American Business Law Journal* 323 at 345. Individuals within organizations may be so concerned with maintaining a common view in the organization that they ignore a realistic analysis of the various courses of action. In large organizations it is necessary for decision making to be diffused, and authority to be shared. 'The resultant control loss by the organization's top managers enhances the likelihood of economically irrational behaviour by employees further down in the organizational hierarchy' (M. B. Mentzger & C. R. Schwenk, 'Decision Making Models, Devil's Advocacy, and the Control of Corporate Crime' (1990) 28 *American Business Law Journal* 323 at 346).

34 F. Pearce & S. Tombs, 'Hazards, Law and Class: Contextualizing the Regulation of Corporate Crime' (1997) 6 *Social and Legal Studies* 79 at 82, argue that while corporations are clearly motivated by profit maximization,

they may be less than successful in this quest. This is because they are 'sites of bounded, multiple and competing rationalities'. Their profit maximizing rationality can be influenced by external conditions and pressures.

35 For a review of this literature, see J. Braithwaite & T. Makkai, 'Testing an Expected Utility Model of Corporate Deviance' (1991) 25 *Law and Society Review* 7; and R. M. Brown, 'Administrative and Criminal Penalties in the Enforcement of the Occupational Health and Safety Legislation' (1992) 30(3) *Osgoode Hall Law Journal* 691.

36 T. H. McQuiston, R. C. Zakocs, and D. Loomis, 'The Case for Stronger OSHA Enforcement—Evidence from Evaluation Research' (1998) 88 *American Journal of Public Health* 1022 at 1022.

37 R. M. Brown, 'Administrative and Criminal Penalties in the Enforcement of the Occupational Health and Safety Legislation' (1992) 30(3) *Osgoode Hall Law Journal* 691.

38 Lewis-Beck and Alford (1980) conducted a rigorous analysis of American coal mining legislation between 1940 and 1970 and showed that a major factor in reducing the level of fatalities in the coal mining industry was the size of the Federal government's budget allocation to health and safety regulation and the seriousness of enforcement measures (M-S. Lewis-Beck & J. R. Alford, 'Can Government Regulate Safety: The Coal Mine Example' (1980) 76 *American Political Science Review* 745).

39 Perry's (1982) study over the period 1930–79 has confirmed Lewis-Beck and Alford's results. Perry concluded: 'The research indicates that strong safety laws reduce coal-mine fatalities and that, if laws are strong, coal-mine fatalities decrease with increases in Federal spending on mine health and safety.' Overall, a coal miner in 1976 was over four times less likely to be killed at work than a 1932 coal miner (C. Perry, 'Government Regulation of Coal Mine Safety: Effects of Spending under Strong and Weak Law' (1982) 10 *American Politics Quarterly* 303).

40 The State of Oregon increased its OHS penalties threefold from 1987 to 1992, together with other changes to its enforcement, prevention, and workers compensation programmes, and found that workers compensation claims fell by over 30% and fatalities by over 21% from 1988 to 1992, even though employment increased 10%. The incidence of lost workday cases fell by over 21% from 1988 to 1991. See the Study by the Oregon Department of Insurance and Finance in 1993, noted in Industry Commission, Australia, *Work, Health and Safety* (AGPS, Canberra, 1995), Vol. II, p. 402.

41 Compliance with regulations does not in itself ensure prevention of injury.

42 See also F. Zimring & G. Hawkins, *Deterrence: The Legal Threat of Crime Control* (The University of Chicago Press, Chicago, 1973), pp. 84–7 and 158–73; A. Bartal & and L. Thomas, 'Direct and Indirect Effects of Regulation' (1985) 28 *Journal of Law and Economics* 1; studies cited in J. Mendeloff, *Regulating Safety* (MIT Press, Cambridge, Mass., 1979); J. Mendeloff, 'The Role of OSHA Violations' (1984) 26 *Journal of Occupational Medicine* 353; W. B. Gray & J. T. Scholz, 'Does Regulatory Enforcement Work? A Panel Analysis of OHSA Enforcement' (1993) 27 *Law and Society Review* 177; and R. Brown, 'Administrative and Criminal Penalties in the Enforcement of the

Occupational Health and Safety Legislation' (1992) 30(3) *Osgoode Hall Law Journal* 691. See also the studies cited in the Industry Commission, Australia, *Work, Health and Safety* (AGPS, Canberra, 1995), Vol. II, pp. 401–2. J. L. Miller & A. B. Anderson, 'Updating the Deterrence Doctrine' (1986) 77 *The Journal of Criminal Law and Criminology* 418 at 421, confirm that empirical studies in deterrence theory generally show that it is the 'perception of the odds of getting convicted and punished for a crime that will have the greatest influence in determining [a] decision whether or not to commit a crime'.

43 Pearce and Tombs point out that, where the British Health and Safety Executive have evaluated their enforcement strategies, their conclusions have provided clear support for deterrence, both general and specific (F. Pearce & S. Tombs, 'Hazards, Law and Class: Contextualizing the Regulation of Corporate Crime' (1997) 6 *Social and Legal Studies* 79 at 94).

44 Gray and Scholz (W. B. Gray & J. T. Scholz, 'Analysing the equity and efficiency of OSHA enforcement' (1991) 3(3) *Law and Policy* 185; W. B. Gray & J. T. Scholz, 'Does Regulatory Enforcement Work? A Panel Analysis of OHSA Enforcement' (1993) 27 *Law and Society Review* 177; and W. B. Gray & J. T. Scholz, 'Can Government Facilitate Cooperation? An informational model of OSHA enforcement' (1997) 41(3) *American Journal of Political Science* 693–717) conducted a study of injuries and inspections under the USA Occupational Safety and Health Act 1970 (OSH Act) from 1979 to 1985. We discuss this study in Ch. 3.

45 W. B. Gray & J. T. Scholz, 'Analysing the equity and efficiency of OSHA enforcement' (1991) 3(3) *Law and Policy* 185 at 201. Their OHS enforcement data suggests that there 'is no evidence that higher penalties provide more specific or more general deterrence than lower penalties, at least not in the general range of penalties imposed by OSHA in our sample plants'.

46 W. B. Gray & J. T. Scholz, 'Analysing the equity and efficiency of OSHA enforcement' (1991) 3(3) *Law and Policy* 185 at 203, suggest that 'The ineffectiveness of penalties over $5,000 cannot be generalized to the extremely high "megapenalties" involving hundreds of thousands or millions of dollars that OSHA has imposed in the past several years. These more dramatic penalties may indeed have large effects. Large fines of this size get media attention that more normal fines do not, and send strong signals about enforcement priorities. Such symbolic fines may be particularly important for initial enforcement of a new safety standard that has met with considerable resistance. Furthermore, completely eliminating all higher and medium penalties would be likely to result in a decrease in deterrence, particularly if all penalties were trivialized to the point that they could be safely ignored.'

47 R. M. Brown, 'Administrative and Criminal Penalties in the Enforcement of the Occupational Health and Safety Legislation' (1992) 30(3) *Osgoode Hall Law Journal* 691 at 705.

48 Most of Haines's data on the impact of court processes was concerned with coronial inquests and common law compensation proceedings. She extrapolates the negative effects on organizational virtue of these processes to the prosecution process.

49 F. Haines, *Corporate Regulation: Beyond 'Punish or Pursuade'* (Clarendon Press, Oxford, 1997), especially Ch. 8.

50 F. Haines, *Corporate Regulation: Beyond 'Punish or Pursuade'* (Clarendon Press, Oxford, 1997), p. 219. See also E. Bardach & R. Kagan, *Going by the Book: The problem of regulatory unreasonableness* (Temple University Press, Philadelphia, 1982).

51 Likewise, Shapiro and Rabinowitz suggest that managers may resort to minimal compliance or refuse to co-operate with OHS agencies in identifying and solving new problems if they perceive that they have been prosecuted despite their good faith efforts to comply (S. A. Shapiro & R. S. Rabinowitz, 'Punishment versus Cooperation in Regulatory Enforcement: A. Case Study of OSHA' (1997) 49(4) *Administrative Law Review* 713 at 718–19).

52 F. Haines, *Corporate Regulation: Beyond 'Punish or Pursuade'* (Clarendon Press, Oxford, 1997), p. 223.

53 See S. Tombs, 'Law, Resistance and Reform: "Regulating" Safety Crimes in the UK' (1995) 4 *Social and Legal Studies* 343–65 at 356.

54 J. Braithwaite & T. Makkai, 'Testing an Expected Utility Model of Corporate Deviance' (1991) 25 *Law and Society Review* 7 at 35, discussed earlier in this section.

55 I. Ayres & J. Braithwaite, *Responsive Regulation: Transcending the Deregulation Debate* (Oxford University Press, Oxford, 1992), pp. 21–35.

56 D. Chappell & J. Norberry, 'Deterring Polluters: The Search for Effective Strategies' (1990) 13 *University of New South Wales Law Journal* 97 at 102–3, surveyed the literature on deterrence in relation to the related area of environmental regulation. They concluded that 'there has been weak but supporting evidence that under limited conditions the threat of punishment may be effective in preventing or reducing certain undesirable behaviours. There is evidence, too, that the certainty of punishment and, to some extent, its severity, contribute to a deterrent effect.' See, also, N. Hemmings, 'The New South Wales Experiment: The Relative Merits of Seeking to protect the Environment through the Criminal Law by Alternative Means' (1993) Oct. *Commonwealth Law Bulletin* 1987. Susan Smith argues that the use of criminal prosecution for environmental offences in the USA has proved effective to promote voluntary compliance: S. Smith, 'An Iron Fist in a Velvet Glove: Redefining the Role of Criminal Prosecution in Creating an Effective Environmental Enforcement System' (1995) 19 *Criminal Law Journal* 12; and S. Smith, 'Doing Time for Environmental Crimes: The United States Approach to Criminal Enforcement of Environmental Laws' (1995) *Environmental and Planning Law Journal* 168. Smith's arguments are more theoretical than empirically-based. Her work is discussed later in this Chapter.

57 See again our discussion in Ch. 4.

58 I. Ayres & J. Braithwaite, *Responsive Regulation: Transcending the Deregulation Debate* (Oxford University Press, Oxford, 1992), p. 39.

59 Ibid.

60 S. A. Shapiro & R. S. Rabinowitz, 'Punishment versus Cooperation in Regulatory Enforcement: A. Case Study of OSHA' (1997) 49(4) *Administrative Law Review* 713 at 722.

61 S. Tombs, 'Law, Resistance and Reform: "Regulating" Safety Crimes in the UK' (1995) 4 *Social and Legal Studies* 343–65 at 356. See also F. Pearce & S. Tombs, 'Hazards, Law and Class: Contextualizing the Regulation of Corporate Crime' (1997) 6 *Social and Legal Studies* 79 at 97; and F. Pearce, 'Corporations and Accountability', in P. Stenning (ed.), *Accountability for Corporate Crime* (University of Toronto Press, Toronto, 1995).

62 See S. Tombs, 'Law, Resistance and Reform: "Regulating" Safety Crimes in the UK' (1995) 4 *Social and Legal Studies* 343– 65 at 356.

63 See J. P. Bonner & B. N. Forman, 'Bridging the Deterrence Gap: Imposing Criminal Penalties on Corporations and their Executives for Producing Hazardous Products' (1993) 1 *San Diego Justice Journal* 1 at 5, n. 11; and M. Metzger, 'Corporate Criminal Liability for Defective Products: Policies, Problems, and Prospects' (1984) 73 *Geo. L J* 1 at 9.

64 See B. Fisse, 'Reconstructing Corporate Criminal Law: Deterrence, Retribution, Fault and Sanctions' (1983) 56 *Southern California Law Review* 1141 at 1163.

65 H. Glasbeek, 'A Role for Criminal Sanctions in Occupational Health and Safety', in *New Developments in Labour Law, Meredith Memorial Lectures*, 1988 (Yvon Blais, Montreal, 1989), p. 125. As Durkheim would argue, this putative process represents a concrete expression of the collective conscience which in turn serves to strengthen social solidarity. See further H. Glasbeek, 'Occupational Health and Safety Law: Criminal Law as a Political Tool' (1998) 11 *Australian Journal of Labour Law* 95.

66 See E. Tucker, 'The Westray Mine Disaster and its Aftermath: The Politics of Causation' (1995) 10 *Canadian Journal of Law and Society* 91 at 113.

67 See our discussion of this in Ch. 1.

68 See B. M. Hutter, *Compliance: Regulation and Environment* (Clarendon Press, Oxford, 1997). For a discussion of strict liability and OHS crimes, see W. G. Carson, 'Symbolic and Instrumental Dimensions of Early Factory Legislation', in R. G. Hood (ed.), *Crime, Criminology and Public Policy* (Heinemann, London, 1974); W. G. Carson, 'The Conventionalisation of Early Factory Crime' (1979) 7 *International Journal of the Sociology of Law* 37; W. G. Carson, 'The Institutionalization of Ambiguity: Early British Factory Acts', in G. Geis and E. Stotland (eds.), *White Collar Crime: Theory and Research* (Sage, London, 1980); W. G. Carson, 'White Collar Crime and the Enforcement of Factory Legislation' (1970) 10 *British Journal of Criminology* 383 (1970); and W. G. Carson, 'Some Sociological Aspects of Strict Liability and the Enforcement of Factory Legislation' (1970) 33 *Modern Law Review* 396.

69 A. Curcio, 'Painful Publicity: An Alternative Punitive Damages Sanction' (1996) 45 *De Paul Law Review* 341 at 349.

70 D. B. Dobbs, 'Ending Punishment in "Punitive" Damages: Deterrence Measured Remedies' (1989) 40 *Alabama Law Review* 831 at 844.

71 See our discussion of manslaughter below. See J. P. Bonner & B. N. Forman, 'Bridging the Deterrence Gap: Imposing Criminal Penalties on Corporations and their Executives for Producing Hazardous Products' (1993) 1 *San Diego Justice Journal* 1 at 6.

72 See R. M. Brown, 'Administrative and Criminal Penalties in the Enforcement of the Occupational Health and Safety Legislation' (1992) 30(3) *Osgoode Hall Law Journal* 691.

73 B. Fisse, 'Reconstructing Corporate Criminal Law: Deterrence, Retribution, Fault and Sanctions' (1983) 56 *Southern California Law Review* 1141 at
1147.

74 See B. Fisse, 'Reconstructing Corporate Criminal Law: Deterrence, Retribution, Fault and Sanctions' (1983) 56 *Southern California Law Review* 1141 at
1147–54. See also R. M. Brown, 'Administrative and Criminal Penalties in
the Enforcement of the Occupational Health and Safety Legislation' (1992)
30(3) *Osgoode Hall Law Journal* 691; and P. T. Edelman, 'Corporate Criminal
Liability for Homicide: The Need to Punish Both the Corporate Entity and
its Officers' (1987) 92 *Dickinson Law Review* 193 at 216–18.

75 J. L. Miller & A. B. Anderson, 'Updating the Deterrence Doctrine' (1986)
77 *The Journal of Criminal Law and Criminology* 418 at 425. See also J. P.
Bonner & B. N. Forman, 'Bridging the Deterrence Gap: Imposing Criminal
Penalties on Corporations and their Executives for Producing Hazardous
Products' (1993) 1 *San Diego Justice Journal* 1 at 9.

76 B. Fisse & J. Braithwaite, *The Impact of Publicity on Corporate Offenders* (1983);
R. M. Brown, 'Administrative and Criminal Penalties in the Enforcement
of the Occupational Health and Safety Legislation' (1992) 30(3) *Osgoode
Hall Law Journal* 691 at 707; and D. Chappell & J. Norberry, 'Deterring
Polluters: The Search for Effective Strategies' (1990) 13 *University of New
South Wales Law Journal* 97 at 108. This argument assumes that the organi-
zation does enjoy some prestige, and aspires to maintain that prestige.

77 B. Fisse, 'Sentencing Options against Corporations' (1990) 1 *Criminal Law
Forum* 211 at 229. Whereas social condemnation of individuals can make the
individual feel outlawed and drive them towards further deviant behaviour,
Fisse argues that 'corporations are more likely to react positively to criminal
stigma by attempting to repair their images and regain public confidence'
(B. Fisse, 'Reconstructing Corporate Criminal Law: Deterrence, Retribution, Fault and Sanctions' (1983) 56 *Southern California Law Review* 1141 at
1154).

78 See A. Cowan, 'Note: Scarlett Letters for Corporations? Punishment by
Publicity Under the New Sentencing Guidelines' (1992) 65 *University of
Southern California Law Review* 2387.

79 See B. Fisse, 'Reconstructing Corporate Criminal Law: Deterrence, Retribution, Fault and Sanctions' (1983) 56 *Southern California Law Review* 1141 at
1154–9.

80 Ibid. 1167.

81 Ibid. 1159.

82 Ibid. 1160.

83 As we noted in Ch. 4, the enforcement pyramid will operate side by side
with other forms of OHS regulation, which will include civil compensation
schemes. The different forms of regulation will operate independently and
contemporaneously, but with different purposes, and not in any particular
sequence. Our enforcement pyramid pertains to measures initiated by the
OHS enforcement agency. Other court processes, aimed for example at
compensation for workplace injuries or disease, may cut across the enforcement pyramid, and may distort the impact of the OHS agency's enforcement measures. The agency will, therefore, have to be sensitive to the
impact of other legal procedures on the alleged offender. See F. Haines,

Corporate Regulation: Beyond 'Punish or Persuade' (Clarendon Press, Oxford, 1997), p. 221.

84 See R. Johnstone, *The Court and the Factory: The Legal Construction of Occupational Health and Safety Offences in Victoria*, unpublished Ph.D. thesis, The University of Melbourne, 1994, Ch. 11.

85 K. Carson & R. Johnstone, 'The Dupes of Hazard: Occupational Health and Safety and the Victorian Sanctions Debate' (1990) 26 *Australian and New Zealand Journal of Sociology* 126; Industry Commission, Australia, *Work, Health and Safety* (AGPS, Canberra, 1995), Vol. I, pp. 113–14.

86 See our discussion of deterrence, organizational reform, retribution, and moral denunciation, above, and our discussion of manslaughter and pure risk prosecutions, below at pp. 206–15.

87 We discuss these other sanctions, particularly adverse publicity orders, community service orders, corporate probation and dissolution, in detail in Ch. 7.

88 See again our discussion of deterrence, organisational reform, retribution, and moral denunciation at the beginning of this Chapter.

89 See again our discussion of points of entry to the pyramid in Ch. 4.

90 D. J. McBarnet, *Conviction: Law, the State and the Construction of Justice* (Macmillan, London, 1981), pp. 138–46.

91 R. Johnstone, *The Court and the Factory: The Legal Construction of Occupational Health and Safety Offences in Victoria*, unpublished Ph.D. thesis, The University of Melbourne, 1994.

92 Nevertheless, it should be noted that there are some characteristics of OHS prosecutions that distinguish these offences from the 'trivia' described by McBarnet and Croall. Croall refers to the 'paradox' of the triviality of the setting and the conventionalized nature of the offences, on the one hand, and the seriousness of the level of danger or fraud involved in consumer offences (H. Croall, 'Mistakes, Accidents, and Someone Else's Fault: The Trading Offender in Court' (1988) 15 *Journal of Law and Society* 293 at 295). Most defendants in OHS prosecutions are represented by legal counsel, often at high cost, thus indicating the seriousness of the offence, at least in the eyes of the defendant. Second, the maximum fines were at the top end of the range of magistrates' courts' fines. Finally, the nature of the injuries suffered by workers was usually apparent to the court, and went some way to reducing the triviality of the matter.

93 In some Australian jurisdictions (most notably New South Wales and Victoria) the OHS enforcement agencies are initiating a greater proportion of prosecutions before the higher level courts.

94 We discuss these broader sanctions in Ch. 7.

95 For example, a similar principle is enunciated in the British Health and Safety Commission's *Enforcement Policy Statement*, 1995.

96 We discuss this further at the end of this section.

97 See our discussion of this point at p. 189–90 above.

98 We examine this point in detail later in this Chapter.

99 To date, the OHS legislation in British Columbia has not included such a provision. The recent report of the Royal Commission on Workers Compensation in British Columbia, *Report on Sections 2 and 3 of the Commission's*

Terms of Reference, 31 Oct. 1997, recommended that such a duty be imposed in the proposed new OHS statute in British Columbia.

100 In a number of jurisdictions, there are technical legal reasons for the under-enforcement of duties imposed upon manufacturers, and others. One reason is the difficulty of conducting the investigation and bringing the prosecution within the statutory limitation period, because the time of the offence will be the supply or manufacture of the plant or substances, rather than the time when the offence is discovered by the OHS inspectorate. In some jurisdictions this problem has been solved by extending the limitation period, or providing that the period commences at the time the contravention is discovered (and perhaps reducing the limitation period to six months or a year). See, for example, section 49 of the Occupational Health and Safety Act 1983 (NSW). A second difficulty is that some of the statutory duties imposed upon manufacturers, suppliers, and others have two major qualifications: that of reasonable practicability and 'when properly used' (a qualification which appears in most Australian OHS statutes and in the British HSW Act 1974). The latter qualification would appear to be unnecessarily restrictive because it may be quite foreseeable to the manufacturer or supplier that plant and substances will not be properly used, and quite practicable for the manufacturer or supplier to take steps to ensure that the plant or substances are safe and without risks to health even when the plant or substance is not properly used. This approach is taken in s.19(1)(c)(ii) of the Australian Occupational Health and Safety (Commonwealth Employment) Act 1991.

101 G. Hadfield, 'Problematic Relations: Franchising and the Law of Incomplete Contracts' (1990) 42 *Stanford Law Review* 927.

102 See our discussion of the 'split pyramid' in Ch. 4.

103 Ayres and Braithwaite use the expression 'height' of the pyramid to refer to the 'punitiveness of the most severe sanction'. See I. Ayres and J. Braithwaite, *Responsive Regulation: Transcending the Deregulation Debate* (Oxford University Press, Oxford, 1992), p. 40.

104 See Ch. 7.

105 See Ch. 7.

106 We discuss a wider range of sanctions in Ch. 7.

107 We discuss manslaughter prosecutions later in this Chapter.

108 For example, in Australia s. 48(1) of the New South Wales Occupational Health and Safety Act 1983 provides that a prosecution may be initiated by the secretary of a New South Wales registered trade union, any member or members of which are concerned in the matter to which the proceedings relate. OHS statutes might also give unions powers to investigate OHS offences. See ss. 31AF–31AP of the Occupational Health and Safety Act 1983 (NSW).

109 We discuss prosecution guidelines later in this section, where we suggest that unions and employers should be consulted in their development. This consultation should include detailed discussion about the coordination of the respective roles of trade union and OHS agency investigations and prosecutions. We suggest that a starting point in the discussions be:

- that all prosecutions, whether initiated by the agency or unions, only be initiated in accordance with the accepted prosecution guidelines;
- that agency investigations and prosecutions take precedence over trade union initiated investigations and prosecutions, but that employee representatives be consulted while investigations are carried out by the agency, and informed of the findings and proposed action resulting from the investigation;
- that where separate investigations are conducted, they are conducted with the maximum possible co-operation between the agency and the union;
- that the OHS statute contain a provision enabling the union to ask the agency three months after the investigation commences to confirm whether a prosecution will be taken; and
- that where the agency decides not to proceed with prosecution, information-sharing arrangements be developed between agency and the union.

110 See I. Ayres & J. Braithwaite, *Responsive Regulation: Transcending the Deregulation Debate* (Oxford University Press, Oxford, 1992), pp. 57–8, where they make the observation that a tripartite approach to regulation permits public interest groups 'the same standing to . . . prosecute under the regululatory statute as the regulator'.

111 Because unions will be accountable to their membership for the types of prosecutions they bring, they are likely to seek to ensure that prosecutions are conducted in relation to hazards that most threaten workers.

112 See again our discussion of the purposes of criminal sanctions at the beginning of this Chapter.

113 We discuss sanctions which require contraventions to be publicized in Ch. 7.

114 See the discussion of deterrence theory, above at p. 186 and following.

115 See Industry Commission, Australia, *Work, Health and Safety* (AGPS, Canberra, 1995), Vol. I, rec. 20, pp. 127–30.

116 See our brief discussion of this point earlier in this section.

117 See R. Johnstone, *Occupational Health and Safety Law and Policy: Text and materials* (LBC Information Services, Sydney, 1997), Ch. 5.

118 See R. Johnstone, *Occupational Health and Safety Law and Policy: Text and materials* (LBC Information Services, Sydney, 1997), Chs. 5 and 6.

119 The USA Federal environmental criminal statutes provide a further example of this event focus. The legislation is designed primarily to concentrate on environmental protection (preventative) rather than on the resultant outcome (punishment). Tucker reports that Canadian OHS prosecutions focus on the most immediate events (the actual breach leading to the injury), ignoring the most important systematic factors which are responsible for the harm producing activity. See E. Tucker, 'The Westray Mine Disaster and its Aftermath: The Politics of Causation' (1995) 10 *Canadian Journal of Law and Society* 91. See also Industry Commission, Australia, *Work, Health and Safety* (AGPS, Canberra, 1995), Vol. I, pp. 109–12 and rec. 13. The recent trend to broaden the range of hazards investigated to include such issues as taxi murders and the provision of personal

protective equipment and training to police faced with dangerous situations is noted.

120 See generally M. Quinlan & P. Bohle, 1991, *Managing Occupational Health and Safety in Australia* (Macmillan, Melbourne, 1991), Ch. 11; and Industry Commission, Australia, *Work, Health and Safety* (AGPS, Canberra, 1995), Vol. I, pp. 29–30.

121 See our discussion of different points of entry to the enforcement pyramid in Ch. 4.

122 By this we mean cases in particular areas such as hazardous substances or manual handling which are run because workers and others are exposed to unacceptable risks, even though there is not yet evidence of the manifestation of an injury or illness.

123 Inspectors may need greater investigative and forensic training, which might include a discussion of the literature and empirical research on the operation of OHS management systems; an overview of the key features of properly functioning OHS management systems; a discussion of organizational theory; techniques for investigating systems such as identification of key management personnel, following paper trails, and so on; how to distinguish between 'paper systems' and properly functioning OHS management systems—by involving employees and committees; techniques for determining hazard profiles, such as injury and illness statistical analysis, analysis of coronial databases, etc.; and the role of the expert witness.

124 Recently the Swedish inspectorate have begun to issue injunctions against public employers to improve the psychosocial aspects of employees' working conditions: see European Foundation for the Improvement of Living and Working Conditions, 'Public employers ordered to act against difficult working conditions' (1998) Feb. *Euroline*.

In some jurisdictions, most notably those in Britain, North America, and Australia where the adversarial model predominates, there may be rules of criminal procedure which specify that provisions such as the employers' general duty of care create a series of specific offences rather than the one ongoing general duty: *Boral Gas (NSW) Pty. Ltd.* v. *Magill* (1995) 58 IR 363; *R* v. *Australian Char Pty. Ltd.* (1996) 64 IR 387. The ruling may undermine the success of a 'pure risk' prosecution, because such prosecutions are heavily reliant on a court being prepared to take a broad view of the general duty as a dynamic duty requiring the implementation by an employer of a systemic approach to the management and control of risk. This approach is not assisted by the view that the general duty is comprised of many discrete offences such as a duty to train, a duty to instruct, etc. In order to enable pure risk prosecutions to be taken, the procedural rules would have to be reformed so that prosecutors could frame a single charge under a general duty to encompass the entire system of work alleged to be deficient.

125 A recent study for the British Health and Safety Executive prepared by the British Government Statistical Service (1997) showed that in 1994 there were 3.2 fatalities per 100,000 workers (employees and self-employed people) in the USA, 1.7 in Denmark in 1992, 2.6 in the Netherlands in 1991, 3.9 in France in 1993, 3.2 in Germany in 1994, 5.1 in Spain in 1993, 5.5 in Italy in 1991, and 0.9 in Britain in 1994–5. In 1996–7 there was a 17%

increase in workplace fatalities in the UK (D. Bergman, 'Government is Weak on Corporate Crime Says Law Campaigner' (1997) Dec. *Health and Safety* 3).

126 See E. Tucker, 'The Westray Mine Disaster and its Aftermath: The Politics of Causation' (1995) 10 *Canadian Journal of Law and Society* 91, at 113–16. Despite the strong evidence of management failure in the Westray mine disaster, no prosecutions for manslaughter have been initiated.

127 I. Reiner & J. Chatten-Brown, 'Deterring Death in the Workplace: The Prosecutor's Perspective' (1989) 17 *Law Medicine and Health Care* 23; I. Reiner & J. Chatten-Brown, 'When is it not an Accident but a Crime?' (1989) 17 *Northern Kentucky Law Review* 84. See also our discussion of manslaughter prosecutions against individual corporate officers, below at pp. 216. In the USA, a focus on manslaughter and related crimes has resulted from a perception that the enforcement of the OSH Act has focused too strongly on civil citations, and that the provisions for criminal prosecution under the OSH Act are inadequate, and rarely used. The trend towards the use of criminal prosecutions for corporate health and safety violations has not been confined to OHS. Since the mid-1980s there have been many criminal prosecutions for violation of environmental crimes, and particularly hazardous waste disposal. Criminal prosecutions for consumer protection violations have been common since the mid-1970s.

128 See S. Tombs, 'Law, Resistance and Reform: "Regulating" Safety Crimes in the UK' (1995) 4 *Social and Legal Studies* 343 at 351; and G. Slapper, 'Corporate Manslaughter: An Examination of the Determinants of Prosecutorial Policy' (1993) 2 *Social and Legal Studies* 423. Bergman notes that until 1993 the British Health and Safety Executive had no guidelines for its inspectors on corporate manslaughter. Such guidelines were introduced in 1993. From then until the end of 1997 there were 1,452 workplace deaths, and the HSE referred 48 of those (3%) to the police or Crown Prosecution Services (D. Bergman, 'Government is Weak on Corporate Crime Says Law Campaigner' (1997) Dec. *Health and Safety*). See also House of Commons Official Report, *Parliamentary Debates* (Hansard) 16 May 1995, Vol. 260, no. 105, 75. Tombs (S. Tombs, 'Law, Resistance and Reform: "Repulating" Safety Crimes in the UK' (1995) 4 *Social and Legal Studies* 343 at 351) attributes the changed attitude to corporate manslaughter to political struggles by unions, 'Hazards' groups, and victim groups, in response to a series of disasters. See D. Bergman, *Deaths at Work: Accidents or Corporate Crime. The Failure of Inquests and the Criminal Justice System* (Workers' Educational Association, London, 1991); and European Foundation for the Improvement of Living and Working Conditions, Walters, D. (ed.), *The Identification and Assessment of Occupational Health and Safety Strategies in Europe*, Vol. 1 (Office for the Official Publications of the European Communities, Luxembourg, 1996), pp. 201 and 205.

129 *R* v. *Nadenbusch*, Supreme Court of Victoria, Teague J (unreported), 14 June 1994; *R* v. *A C Hatrick Chemicals Pty. Ltd.*, Supreme Court of Victoria, Hampel J (unreported), 29 Nov. 1995. The latter was unsuccessful, as was a third manslaughter prosecution brought in 1998.

130 *Occupational Health and Safety Newsletter*, Issue No. 363, 15 Feb. 1996, p. 1.

131 The handful of manslaughter prosecutions initiated in Queensland have been unsuccessful.

132 *R* v. *OLL Ltd.* (Winchester Crown Court, 9 Dec. 1994). The managing director was held to have had actual knowledge of the risk that led to the death. See G. Slapper, 'A Corporate Killing' (1994) 144 *New Law Journal* 1735; S. Tombs, 'Law, Resistance and Reform: "Regulating" Safety Crimes in the UK' (1995) 4 *Social and Legal Studies* 343 at 351; A. Edgar, 'Directors' Liability' (1997) 141 *Solicitors' Journal* 328 at 329; M. Smith, 'The Company Behind Bars' (1995) Feb. *Health and Safety at Work* 8; F. Wright, *Law of Health and Safety at Work* (Sweet & Maxwell, London, 1997), p. 150; A. Ridley & L. Dunford, 'Corporate Killing--Legislating for Unlawful Death' (1997) 26 *Industrial Law Journal* 99; and C. Wells, 'Corporate manslaughter: A Cultural and Legal Form' (1995) 6 *Criminal Law Forum* 45.

133 For example, under Anglo-Australian law, the relevant *mens rea* for the crime of manslaughter is gross or criminal negligence. See *R* v. *Bateman* (1925) 19 Cr. App. R 8; *Andrews* v. *DPP* [1937] 2 All ER 552, [1937] AC 576; *R* v. *Adomako* [1994] 3 All ER 79. Manslaughter by criminal negligence involves 'a great falling short of the standard of care' which a reasonable person would have exercized involving 'such a high risk that death or grevious bodily harm would follow that the doing of the act merited criminal punishment': *Nydam* v. *R* [1977] VR 430 at 445. See also *R* v. *Wilson* (1992) 174 CLR 313. It may also be possible to prosecute industrial manslaughter as 'unlawful and dangerous act manslaughter'—see generally B. Fisse, *Howard's Criminal Law*, 5th edn. (Law Book Company, Sydney, 1990), pp. 123–31. In the USA, the legal rules concerning homicide differ from jurisdiction to jurisdiction, but in general, where workplace deaths occur, state prosecutors can lay charges for negligent homicide (a 'gross deviation from the standard of a reasonable person'), manslaughter (reckless actions), or murder (homicide committed intentionally or as a consequence of reckless indifference). Model Penal Code ss. 202 and 210 (1980). On *mens rea* see generally B. Fisse, *Howard's Criminal Law*, 5th edn. (Law Book Company, Sydney, 1990), pp. 478–536.

134 For a description and comparisons of these historical developments, see, for example, G. Stessens, 'Corporate Criminal liability: A Comparative Perspective' (1994) 43 *International and Comparative Law Quarterly* 493. See also M. Metzger, 'Orgnisations and the Law' (1987) 25 *American Business Law Journal* 257; D. J. Miester, 'Criminal Liability for Corporations That Kill' (1990) 64 *Tulane Law Review* 919; W. Laufer, 'Culpability and the Sentencing of Corporations' (1992) 71 *Nebraska Law Review* 1049; D. Stuart, 'Punishing Corporate Criminals with Constraint' (1995) *Criminal Law Forum* 219; P. T. Edelman, 'Corporate Criminal Liability for Homicide: The Need to Punish Both the Corporate Entity and its Officers' (1987) 92 *Dickinson Law Review* 193; E. Colvin, 'Corporate Personality and Criminal Liability' (1995) 6 *Criminal Law Forum* 1; and C. Wells, *Corporations and Criminal Responsibility* (Clarendon Press, Oxford, 1993).

135 Starting with *Lennard's Carrying Co. Ltd.* v. *Asiatic Petroleum Co. Ltd.* [1915] AC 705 and culminating in *Tesco Supermarkets Ltd.* v. *Natrass* [1972] AC 153. This reasoning has been accepted uncritically by Australian courts: see, for

example, *Universal Telecasters (Qld.) Ltd.* v. *Guthrie* 18 ALR 531 (FC); *Hamilton* v. *Whitehead* (1988) 82 ALR 626 (HC). See also *R* v. *A C Hatrick Chemicals Pty. Ltd.*, unreported, Supreme Court of Victoria, 29 Nov. 1995. In recent years a new trend has started to emerge in the law of the UK. In two cases involving the prosecution of corporations, the courts have refined the attribution principle by emphasizing that each case must be decided according to its own facts. In particular, it is necessary to focus on the particular statute under which the company was being prosecuted and to determine its purpose. See *Director General of Fair Trading* v. *Pioneer Concrete (UK) Ltd.* [1995] 1 AC 456 at 465 per Lord Templeman and *Meridian Global Funds Management Asia Ltd.* v. *Securities Commission* [1995] 3 All ER 918.

136 See, for example, *Tesco Supermarkets Ltd.* v. *Natrass* [1972] AC 153 at 170 per Lord Reid.

137 In England, see S. Field & N. Jorg, 'Liability and Manslaughter: Should We Be Going Dutch?', (1991) *Criminal Law Review* 156; D. Bergman, *Deaths at Work: Accidents or Corporate Crime. The Failure of Inquests and the Criminal Justice System* (Workers' Educational Association, London, 1991); C. Wells, *Corporations and Criminal Responsibility* (Clarendon Press, Oxford, 1993), pp. 131–40; and European Foundation for the Improvement of Living and Working Conditions, Walters, D. (ed.), *The Identification and Assessment of Occupational Health and Safety Strategies in Europe*, Vol. 1 (Office for the Official Publications of the European Communities, Luxembourg, 1996), p. 202. In Australia, see B. Fisse, *Howard's Criminal Law*, 5th edn. (Law Book Company, Sydney, 1990); and K. Polk, F. Haines, & S. Perrone, 'Homicide, Negligence and Work Death: The Need for Legal Change' in M. Quinlan (ed.), *Work and Health: The Origins, Management and Regulation of Occupational Illness* (Macmillan, Melbourne, 1993).

138 B. Fisse & J. Braithwaite, *Corporations, Crime and Accountability* (Cambridge University Press, Sydney, 1993), p. 8; see also B. Fisse, *Howard's Criminal Law*, 5th edn. (Law Book Company, Sydney, 1990), pp. 600–4 and C. Wells, 'Manslaughter and Corporate Crime' (1989) 139 *New Law Journal* 931 at 931.

139 F. Wright, *Law of Health and Safety at Work* (Sweet & Maxwell, London, 1997), p. 150.

140 Law Commission, *Legislating the Criminal Code: Involuntary Manslaughter* (HMSO, London, 1997); and *Health and Safety Bulletin* No. 252, Industrial Relations Services, Dec. 1996, p. 2. For comment on the proposals, see C. Wells, 'The Corporate Manslaughter Proposals: Pragmatism, Paradox and Peninsularity' [1996] *Criminal Law Review* 545; and A. Ridley and L. Dunford, 'Corporate Killing—Legislating for Unlawful Death' (1997) 26 *Industrial Law Journal* 99.

141 The Criminal Code Act 1995 (Cth.) comes into effect in the Commonwealth jurisdiction on 15 Mar. 2000, unless proclaimed earlier. It is envisaged that it will be adopted by all Australian states and territories and will in due course form the basis of all Australian Criminal law. The provisions of the Code clearly owe much to the work of Fisse (see, for example, B. Fisse, *Howard's Criminal Law*, 5th edn. (Law Book Company, Sydney, 1990), pp. 605–8), but see his criticism of the Interim Report in B. Fisse, 'Recent Developments in Corporate Criminal Law and Corporate Liability to

Monetary Penalties' (1991) 13 *University of New South Wales Law Review* 1; and B. Fisse, 'Corporate Criminal Responsibility' (1991) 15 *Criminal Law Journal* 166. For a discussion of the Criminal Code Act 1995, see T. Woolf, 'The Criminal Code Act 1995 (Cth)—Towards a Realist Vision of Corporate Criminal Liability' (1997) 21 *Criminal Law Journal* 257.

142 In relation to offences requiring proof of intention, knowledge, or recklessness. Offences such as manslaughter by gross negligence are dealt with by s. 12.4.

143 France for a long time resisted the idea of corporate criminal liability, but in 1992 enacted a new Penal Code which made provision for criminal sanctions on corporations, based on a very restrictive form of the attribution principle. In The Netherlands, s. 51 of the Criminal Code provides for a general system of corporate criminal liability based on vicarious liability. The Supreme Court has laid down the basic criteria of liability as being whether the defendant company had the power to determine whether the employee did or did not do the act in question, and whether the corporation usually 'accepted' such acts. See S. Field & N. Jorg, 'Liability and Manslaughter: Should we be going Dutch?' [1991] *Criminal Law Review* 156; G. Stessens, 'Corporate Criminal liability: A Comparative Perspective' (1994) 43 *International and Comparative Law Quarterly* 493 at 501, 507; C. Wells, *Corporations and Criminal Responsibility* (Clarendon Press, Oxford, 1993), pp. 121–2; and M. Mialon, 'Safety at Work in French Firms and the Effect of the European Directives of 1989' (1990) 6 *The International Journal of Comparative Labour Law and Industrial Relations* 129 at 144.

144 G. Stessens, 'Corporate Criminal liability: A Comparative Perspective' (1994) 43 *International and Comparative Law Quarterly* 493 at 501.

145 See the models, based on the proposed Australian reform, in D. Stuart, 'Punishing Corporate Criminals with Constraint' (1995) *Criminal Law Forum* 219; and E. Colvin, 'Corporate Personality and Criminal Liability' (1995) 6 *Criminal Law Forum* 1.

146 See K. B. Huff, 'The Role of Corporate Compliance Programs in Determining Corporate Criminal Liability: A Suggested Approach' (1996) *Columbia Law Review* 1252 at 1255–63; W. Laufer, 'Culpability and the Sentencing of Corporations' (1992) 71 *Nebraska Law Review* 1049; D. Stuart, 'Punishing Corporate Criminals with Constraint' (1995) *Criminal Law Forum* 219; E. Colvin, 'Corporate Personality and Criminal Liability' (1995) 6 *Criminal Law Forum* 7; and P. T. Edelman, 'Corporate Criminal Liability for Homicide: The Need to Punish Both the Corporate Entity and its Officers' (1987) 92 *Dickinson Law Review* 193.

147 Agents include officers, directors, managers, supervisors, subordinate employees, and independent contractors.

148 Where no single agent has the requisite *mens rea*, the court can use the 'collective knowledge' doctrine which deems the corporation's knowledge to be the aggregated knowledge of all its agents: *United States* v. *Bank of New England*, 821 F 2d 844 (1st Cir.), cert. denied, 484 US 943 (1987). Most state law in the US is based upon statutory adoption of the American Law Institute's Model Penal Code (see D. Stuart, 'Punishing Corporate Criminals with Constraint' (1995) *Criminal Law Forum* 219 at 240), which imposes vicarious liability for regulatory offences, subject to a due diligence defence,

and enacts an identification doctrine for penal violations. The identification doctrine specifies that the corporation may be convicted where 'the commission of the offence was authorised, requested, commanded, performed or recklessly tolerated by the board of directors or by a high managerial agent acting on behalf of the corporation within the scope of his office or employment'.

149 C. Wells, *Corporations and Criminal Responsibility* (Clarendon Press, Oxford, 1993), pp. 121–2.

150 See, for example, H. Glasbeek, 'A Role for Criminal Sanctions in Occupational Health and Safety', in *New Developments in Labour Law, Meredith Memorial Lectures*, 1988 (Yvon Blais, Montreal, 1989).

151 See S. Tombs, 'Law, Resistance and Reform: "Regulating" Safety Crimes in the UK' (1995) 4 *Social and Legal Studies* 343 at 343, who points out that the contemporary 'law and order' debate has been focused mainly on 'the relatively powerless "street" offender'. There is an understandable sense of public disquiet in the face of law enforcement policies which appear to differentiate between offences on the basis of where they occur: criminal negligence on the road will be prosecuted under the general criminal law but criminal negligence in the workplace will be prosecuted under the OHS statute.

152 See I. Reiner & J. Chatten-Brown, 'Deterring Death in the Workplace: The Prosecutor's Perspective' (1989) 17 *Law Medicine and Health Care* 23; and I. Reiner & J. Chatten-Brown, 'When is it not an Accident but a Crime?' (1989) 17 *Northern Kentucky Law Review* 84; and the McMurty case, discussed above at p. 219.

153 C. Wells, *Corporations and Criminal Responsibility* (Clarendon Press, Oxford, 1993), p. 5.

154 Ibid. p. 6; and P. Gillies, *Criminal Law*, 3rd edn. (Law Book Co., 1993), p. 649.

155 C. Wells, *Corporations and Criminal Responsibility* (Clarendon Press, Oxford, 1993), p. 6. But see, for example, s. 19(7) of the Occupational Safety and Health Act 1984 (WA), which provides for an increase in the maximum penalty for a breach of the employers' duty from $100,000 to $200,000 in a case where the breach 'causes the death of, or serious injury to, an employee'.

156 C. Wells, *Corporations and Criminal Responsibility* (Clarendon Press, Oxford, 1993), p. 6. See also W. G. Carson & R. Johnstone, 'The Dupes of Hazard: Occupational Health and Safety and the Victorian Sanctions Debate' (1990) 26 *Australian and New Zealand Journal of Sociology* 126 at 126–34.

157 See W. G. Carson, 'White Collar Crime and the Enforcement of Factory Legislation' (1970) 10 *British Journal of Criminology* 383; W. G. Carson, 'Some Sociological Aspects of Strict Liability and the Enforcement of Factory Legislation' (1970) 33 *Modern Law Review* 396; W. G. Carson, 'Symbolic and Instrumental Dimensions of Early Factory Legislation', in R. G. Hood (ed.), *Crime, Criminology and Public Policy* (Heinemann, London, 1974); W. G. Carson, 'The Conventionalisation of Early Factory Crime' (1979) 7 *International Journal of the Sociology of Law* 37; W. G. Carson, 'The Institutionalization of Ambiguity: Early British Factory Acts', in G. Geis and E. Stotland

(eds.), *White Collar Crime: Theory and Research* (Sage, London, 1980); and W. G. Carson, 'Hostages to History: Some Aspects of the Occupational Health and Safety Debate in Historical Perspective', in W. B. Creighton and N. Gunningham (eds.), *The Industrial Relations of Occupational Health and Safety* (Croom Helm, Sydney, 1985).

158 W. G. Carson & R. Johnstone, 'The Dupes of Hazard: Occupational Health and Safety and the Victorian Sanctions Debate' (1990) 26 *Australian and New Zealand Journal of Sociology* 126 at 140. See also the Industry Commission, Australia, Work, *Health and Safety* (AGPS, Canberra, 1995), Vol. I, p. 121.

159 See W. G. Carson & R. Johnstone, 'The Dupes of Hazard: Occupational Health and Safety and the Victorian Sanctions Debate' (1990) 26 *Australian and New Zealand Journal of Sociology* 126. Under the USA Occupational Safety and Health Act 1970 there is an offence of wilful violation of an OSH Act standard which results in death. For a discussion of the enforcement of this provision, see T. G. Gorbatoff, 'OSHA Criminal Penalty Reform Act: Workplace Safety May Finally Become a Reality' (1991) 39 *Cleveland State Law Review* 551.

160 R. Johnstone, *Occupational Health and Safety Law and Policy: Text and materials* (LBC Information Services, Sydney, 1997), pp. 431–2.

161 F. Wright, *Law of Health and Safety at Work* (Sweet & Maxwell, London, 1997), pp. 140–3; Industry Commission, Australia, *Work, Health and Safety* (AGPS, Canberra, 1995), Vol. I, p. 122. For the position in Denmark, see European Foundation for the Improvement of Living and Working Conditions, 'Danish building and construction sites are hazardous workplaces' (1997) April *Euroline*. See Sweden, where prosecution statistics suggest that prosecutions are brought only against individuals, from employees through to business proprietors: Swedish National Board of Occupational Safety and Health (1995) 2 *Newsletter*, p. 15.

162 A. Hopkins, *Making Safety Work: Getting Management Commitment to Occupational Health and Safety* (Allen & Unwin, Sydney, 1995), p. 107; R. Johnstone (ed.), *New Directions in Occupational Health and Safety Prosecutions: The Individual Liability of Corporate Officers, and Prosecutions for Manslaughter and Related Offences* (Centre for Employment and Labour Relations Law, Working Paper No. 9, The University of Melbourne, 1996); and A. Edgar, 'Directors' Liability' (1997) 141 *Solicitors' Journal* 328 at 329. We also note that bringing prosecutions against small corporations and against their individual managers and directors raises issues of double jeopardy if the director or manager is so closely identified with the corporation that they are effectively one and the same.

163 See I. Reiner & J. Chatten-Brown, 'Deterring Death in the Workplace: The Prosecutor's Perspective' (1989) 17 *Law Medicine and Health Care* 23; I. Reiner & J. Chatten-Brown, 'When is it not an Accident but a Crime?' (1989) 17 *Northern Kentucky Law Review* 84; M. Bixby, 'Workplace Homicide: Trends, Issues and Policy' (1991) 70 *Oregon Law Review* 333; and W. B. Creighton and P. Rozen, *Occupational Health and Safety Law in Victoria* (Federation Press, Sydney, 1997), p. 133.

164 For example, late in 1996 Alan Jackson and his company Jackson Transport

were both convicted of the manslaughter of an employee who died
after being splashed with a deadly chemical, parachloro-orthecresol. The
employee was only wearing a boiler suit and a baseball cap, and had no
personal protective equipment to protect him against the substance: *Safety
Management,* Oct. 1996; *Health and Safety Bulletin* No. 252, Industrial Relations
Services, Dec. 1996, p. 2. See also the case of David Holt Plastics Limited,
and the successful prosecution of its directors David and Norman Holt for
manslaughter, described in F. Wright, *Law of Health and Safety at Work* (Sweet
& Maxwell, London, 1997), p. 148.

165 *People* v. *Film Recovery Systems,* No. 83-11091, Cook County, Ill., Circ. Ct. 14
June 1985. The convictions were later quashed on appeal: *Illinois* v. *O'Neill,
Film Recovery Systems, Inc. and others* (1990) 55 NE 2d 1090.

166 The executives knew that contact with cyanide could be fatal, and that
workers became nauseous after contact with the chemical, but actively
sought to conceal the dangers from employees.

167 Most notably B. Fisse & J. Braithwaite, *Corporations, Crime and Accountability*
(Cambridge University Press, Sydney, 1993).

168 See, further, our discussion of this later in this chapter at pp. 220–21. These
points were made by prosecutors from New South Wales, Victoria, and
Queensland at the *Prosecution Developments—Directors' and Officers' OHS Liability
and the Use of Crimes Act/Manslaughter in the Industrial Context* Conference,
Regent Hotel, Melbourne, 18 Oct. 1995. See R. Johnstone (ed.), *New
Directions in Occupational Health and Safety Prosecutions: The Individual Liability of
Corporate Officers, and Prosecutions for Manslaughter and Related Offences* (Centre for
Employment and Labour Relations Law, the University of Melbourne,
Melbourne, 1996).

169 B. Fisse & J. Braithwaite, *Corporations, Crime and Accountability* (Cambridge
University Press, Sydney, 1993), pp. 1–2.

170 For the origins of this model, see I. Ayres & J. Braithwaite, *Responsive
Regulation: Transcending the Deregulation Debate* (OUP, Oxford, 1992), Ch. 4;
and B. Fisse & J. Braithwaite, *Corporations, Crime and Accountability* (Cambridge University Press, Sydney, 1993).

171 We note that in the recent report of the Westray Mine Public Inquiry in
Nova Scotia, the Commissioner, Justice Richard, recommended that both
the Canadian government and the Nova Scotia government review the
accountability of corporate executives for their (the corporate executives')
wrongful and negligent acts, and make sure that OHS statutes are reformed
to ensure proper accountability for failure by the corporation to secure and
maintain a safe workplace: Justice K. Peter Richard, *The Westray Story: A
Predictable Path to Disaster,* Report of the Westray Public Inquiry, Province of
Nova Scotia, 1997, recs. 73 and 74, p. 601.

172 A. Hopkins, *Making Safety Work: Getting Management Commitment to Occupational
Health and Safety* (Allen & Unwin, Sydney, 1995), p. 108.

173 See our discussion of overspills earlier in our examination of under-
deterrence at p. 187, and in our critique of the fine later in Ch. 7.

174 J. P. Bonner & B. N. Forman, 'Bridging the Deterrence Gap: Imposing
Criminal Penalties on Corporations and their Executives for Producing
Hazardous Products' (1993) 1 *San Diego Justice Journal* 1 at 19. As we noted

earlier, it could also, to some extent, be countered by the non-pecuniary aspects of criminal sanctions.

175 See M. Minister, 'Federal Facilities and the Deterrence Failure of Environmental Laws: The Case for Criminal Prosecution of Federal Employees' (1994) 18 *Harvard Environmental Law Review* 137 at 146.

176 J. P. Bonner & B. N. Forman, 'Bridging the Deterrence Gap: Imposing Criminal Penalties on Corporations and their Executives for Producing Hazardous Products' (1993) 1 *San Diego Justice Journal* 1 at 20.

177 See above pp. 202–04.

178 P. T. Edelman, 'Corporate Criminal Liability for Homicide: The Need to Punish Both the Corporate Entity and its Officers' (1987) 92 *Dickinson Law Review* 193 at 221.

179 See also J. C. Coffee, ' "No Soul to Damn: No Body to Kick": An Unscandalized Inquiry into the Problem of Corporate Punishment' (1981) 79 *Michigan Law Review* 386 at 393–9: for example (p. 393), a 'lower echelon executive with a lacklustre record may deem it desirable to resort to illegal means to increase profits (or forestall losses) in order to prevent his dismissal or demotion'.

180 J. Braithwaite & T. Makkai, 'Testing an Expected Utility Model of Corporate Deviance' (1991) 25 *Law and Society Review* 7 at 10. See also J. C. Coffee, ' "No Soul to Damn: No Body to Kick": An Unscandalized Inquiry into the Problem of Corporate Punishment' (1981) 79 *Michigan Law Review* 386 at 408.

181 See the evidence referred to by the Industry Commission, Australia, *Work, Health and Safety* (AGPS, Canberra, 1995), Vol. I, p. 122. See also J. P. Bonner & B. N. Forman, 'Bridging the Deterrence Gap: Imposing Criminal Penalties on Corporations and their Executives for Producing Hazardous Products' (1993) 1 *San Diego Justice Journal* 1 at 21–2.

182 H. L. Packer, *The Limits of the Criminal Process* (Stanford University Press, Stanford, 1968), p. 269; J. P. Bonner & B. N. Forman, 'Bridging the Deterrence Gap: Imposing Criminal Penalties on Corporations and their Executives for Producing Hazardous Products' (1993) 1 *San Diego Justice Journal* 1 at 22.

183 J. P. Bonner & B. N. Forman, 'Bridging the Deterrence Gap: Imposing Criminal Penalties on Corporations and their Executives for Producing Hazardous Products' (1993) 1 *San Diego Justice Journal* 1 at 22. Polinsky and Shavell argue that, from the perspective of the economic theory of deterrence, it is socially desirable to punish executives even where corporations themselves face liability. If corporations are made strictly liable for the harms they cause, as they are with OHS legislation, they will design rewards and punishments to induce employees to reduce the risks of harm. If the highest penalty the corporation can impose on the employee is less than the harm the employee can bring about, the employee's incentive to reduce the risk will be small. The extent to which corporations can discipline employees is likely to be restricted, because employees have limited assets, are protected by unfair dismissal legislation, and, in any event, the consequences of dismissal are reduced by the alternative employment opportunities available to employees. The state can impose bigger penalties (fines,

imprisonment) on employees than can the corporation. Consequently, state imposed sanctions will have a greater effect in inducing employees to take the appropriate degree of care. See A. M. Polinsky & S. Shavell, 'Should Employees be Subject to Fines and Imprisonment Given the Existence of Corporate Liability?' (1993) 13 *International Review of Law and Economics* 239.

184 See G. Slapper, 'Crime without conviction' (1992) Feb. 14 *New Law Journal* 192 at 193. Slapper refers to a British proposal that company directors whose commercial recklessness resulted in death be disqualified from business management for up to 10 years. In 1992 a company director was disqualified under s. 2(1) of the Company Directors Disqualification Act 1986, after contravening s. 37 of the British Health and Safety at Work Act 1974, the provision attributing prosecution of a director when a corporate OHS offence has been committed with their 'consent, connivance or . . . attributable to any neglect on the[ir] part' (see D. Bergman, *The Perfect Crime? How Companies Get Away with Manslaughter in the Workplace* (HASAC, Birmingham, West Midlands, 1994), p. 15; S. Tombs, 'Law, Resistance and Reform: "Regulating" Safety Crimes in the UK' (1995) 4 *Social and Legal Studies* 343 at 353; and F. Wright, *Law of Health and Safety at Work* (Sweet & Maxwell, London, 1997), p. 152).

185 T. L. Spiegelhoff, 'Limits on Individual Accountability for Corporate Crimes' (1984) 67 *Marquette Law Review* 604 at 624–30. But see R. Fox, 'Corporate Sanctions: Scope for a New Eclecticism' (1982) 24 *Malaya Law Review* 26 at 35–6 for a discussion of difficulties which may arise in imposing these sanctions on individuals within the corporate structure. For example, it is difficult to ensure that the individual, and not the corporation, pays the fine imposed on a corporate officer.

186 G. Geis (ed.), *On White Collar Crime: Offences in business, politics, and the professions* (Free Press, New York, 1982), p. 53.

187 See T. L. Spiegelhoff, 'Limits on Individual Accountability for Corporate Crimes' (1984) 67 *Marquette Law Review* 604 at 625.

188 See also the discussion of the deterrent effect of sanctions against executives in W. A. Mukatis & P. G. Brinkman, 'Managerial Liability for Health, Safety and Environmental Crime: A Review and Suggested Approach to the Problem' (1987) 25 *American Business Law Journal* 323 at 335–43.

189 See below pp. 221–22.

190 See N. Brunton, 'Directors, Companies and Pollution in Western Australia' (1995) 3 *Environmental and Planning Law Journal* 159 at 159–60.

191 The convictions were later quashed on appeal: *Illinos v. O'Neill, Film Recovery Systems, Inc. and others* (1990) 55 NE 2d 1090.

192 I. Reiner & J. Chatten-Brown, 'Deterring Death in the Workplace: The Prosecutor's Perspective' (1989) 17 *Law Medicine and Health Care* 23; and I. Reiner & J. Chatten-Brown, 'When is it not an Accident but a Crime?' (1989) 17 *Northern Kentucky Law Review* 84.

193 A. Edgar, 'Directors' Liability' (1997) 141 *Solicitors' Journal* 328 at 329. See also the UK manslaughter cases discussed earlier in this Chapter.

194 A. Edgar, 'Directors' Liability' (1997) 141 *Solicitors' Journal* 328 at 329.

195 S. Smith, 'An Iron Fist in a Velvet Glove: Redefining the Role of Criminal

Prosecution in Creating an Effective Environmental Enforcement System' (1995) 19 *Criminal Law Journal* 12; and S. Smith, 'Doing Time for Environmental Crimes: The United States Approach to Criminal Enforcement of Environmental Laws' (1995) *Environmental and Planning Law Journal* 168.

196 See also F. Pearce & S. Tombs, 'Hazards, Law and Class: Contextualizing the Regulation of Corporate Crime' (1997) 6 *Social and Legal Studies* 79 at 95, discussed at p. 187–92 above.

197 S. Smith, 'An Iron Fist in a Velvet Glove: Redefining the Role of Criminal Prosecution in Creating an Effective Environmental Enforcement System' (1995) 19 *Criminal Law Journal* 12 at 14. See also S. Smith, 'Doing Time for Environmental Crimes: The United States Approach to Criminal Enforcement of Environmental Laws' (1995) 12(3) *Environmental and Planning Law Journal* 168. Smith's arguments are more theoretical than empirically-based.

198 See Ch. 2; J. P. Bonner & B. N. Forman, 'Bridging the Deterrence Gap: Imposing Criminal Penalties on Corporations and their Executives for Producing Hazardous Products' (1993) 1 *San Diego Justice Journal* 1 at 23; and M. B. Clinard & P. C. Yeager, *Corporate Crime* (1980), pp. 43 and 58.

199 Lim Wen Ts'ai, 'Corporations and the Devil's Dictionary: the problem of individual responsibility for corporate crimes' (1990) 12 *Sydney Law Review* 311 at 331 and 335.

200 Ibid. 322–6.

201 See the spirited debate in Denmark as to whether employees should be penalized for OHS breaches (European Foundation for the Improvement of Living and Working Conditions, 'Danish building and construction sites are hazardous workplaces' (1997) Apr. *Euroline*).

202 This discussion is drawn heavily from N. Gunningham, *Safeguarding the Worker* (Law Book Company, Sydney, 1984), pp. 337–38.

203 See R. Fox, 'Corporate Sanctions: Scope for a New Eclecticism' (1982) 24 *Malaya Law Review* 26 at 36–7.

204 B. Fisse & J. Braithwaite, *Corporations, Crime and Accountability* (Cambridge University Press, Sydney, 1993), p. 14.

205 Lim Wen Ts'ai, 'Corporations and the Devil's Dictionary: the problem of individual responsibility for corporate crimes' (1990) 12 *Sydney Law Review* 311 at 324–5; and C. Stone, *Where the Law Ends: The Social Control of Corporate Behaviour* (Harper and Row, New York, 1975), p. 61.

206 See the UK HSW Act 1974, Consumer Protection Act 1987, Environmental Protection Act 1990, Water Resources Act 1991, and Water Industry Act 1991, and OHS legislation in the Australian jurisdictions of Victoria, Western Australia, and the Northern Territory.

207 United States Model Penal Code ss. 2.07(2) and (5) (1980); and New South Wales and Queensland (and Tasmania in relation to directors).

208 See our discussion of compliance programmes later in this Chapter, and R. S. Gruner & L. M. Brown, 'Organizational Justice: Recognizing and Rewarding the Good Corporate Citizen' (1996) *The Journal of Corporation Law* 731.

209 A. Hopkins, *Making Safety Work: Getting Management Commitment to Occupational Health and Safety* (Allen & Unwin, Sydney, 1995), p. 107. He also suggests that it will be easier to prosecute officers of large companies who in the past

too easily have been able to argue that (because of the size of the company) they had no knowledge of the circumstances of the actual contravention of the OHS statute. Nevertheless, we note that a similar provision in s. 10 of Environmental Offences and Penalties Act 1989 (NSW) appears not to have resulted in a major increase in prosecutions against managers and directors for environmental offences.

210 The Industry Commission, Australia, *Work, Health and Safety* (AGPS, Canberra, 1995), Vol. I, pp. 122–3 supported this due diligence approach. But note the strong criticism by Fisse of similar (even more benign) provisions in the Environmental Offences and Penalties Act 1989 (NSW), s. 10, in B. Fisse, 'Recent Developments in Corporate Criminal Law and Corporate Liability to Monetary Penalties' (1991) 13 *University of New South Wales Law Review* 1 at 5–7. See also Lim Wen Ts'ai 'Corporations and the Devil's Dictionary: the problem of individual responsibility for corporate crimes' (1990) 12 *Sydney Law Review* 311 at 326–7, and 341–2.

211 See A. Edgar, 'Directors' Liability' (1997) 141 *Solicitors' Journal* 328 at 329.

212 Various models of business organizations are described in S. M. Kriesberg, 'Decisionmaking Models and the Control of Corporate Crime' (1976) 85 *Yale Law Journal* 1091; H. Mintzberg, *The Structuring of Organizations* (Prentice-Hall, Englewood Cliffs, New Jersey, 1979); B. Fisse & J. Braithwaite, *Corporations, Crime and Accountability* (Cambridge University Press, Sydney, 1993), Ch. 4; and N. Gunningham, *Safeguarding the Worker* (Law Book Company, Sydney, 1984), pp. 325–7.

213 This will require the OHS agency to have a good system of record keeping so that, when inspectors leave the agency, new inspectors can be properly briefed about each organization's management structure and track record.

214 *Nominated accountability* occurs when responsibility is imposed on a nominated person, for example the person with special responsibility for an area of concern.

215 See J. Braithwaite & B. Fisse, 'Varieties of Responsibility and Organizational Crime' (1985) 7 *Law and Policy* 315.

216 The effect of the EEC Framework Directive 89/391 of 12 June 1989, On the Introduction of Measures to Encourage Improvements in the Safety and Health of Workers at Work (OJ [1989] L183/1), especially article 5, is to make it clear that the employer's *responsibilities* are not delegable: see M. Biagi, 'From Conflict to Participation in Safety: Industrial Relations and the Working Environment in Europe 1992' (1990) 6 *The International Journal of Comparative Labour Law and Industrial Relations* 67 at 70–1; and M. Mialon, 'Safety at Work in French Firms and the Effect of the European Directives of 1989' (1990) 6 *The International Journal of Comparative Labour Law and Industrial Relations* 129 at 134–5. In Britain and Australia see *R* v. *British Steel plc* [1995] IRLR 310; *R* v. *Associated Octel Co. Ltd.* [1996] 4 All ER 846, [1996] 1 WLR 1543.

217 See South Australia.

218 See Tasmania.

219 D. Walters (ed.), *The Identification and Assessment of Occupational Health and Safety Strategies in Europe* (European Foundation for the Improvement of Living and Working Conditions, Dublin, 1996), p. 151.

220 Note that this model does not involve exculpating the corporate employer. It is a means for determining which individual corporate officer is also to incur liability.

221 This model is based, at least in parts, on the provisions of French law, to be found in the Law of 1976, decisions of the courts, and the Circular of 2 May 1977 (see M. Mialon, 'Safety at Work in French Firms and the Effect of the European Directives of 1989' (1990) 6 *The International Journal of Comparative Labour Law and Industrial Relations* 129 at 135); and the New South Wales Coal Mines Regulation Act 1982 and the Queensland Coal Mining Act 1925. See B. Fisse & J. Braithwaite, *Corporations, Crime and Accountability* (Cambridge University Press, Sydney, 1993), pp. 126–7; Lim Wen Ts'ai, 'Corporations and the Devil's Dictionary: the problem of individual responsibility for corporate crimes' (1990) 12 *Sydney Law Review* 31 at 339–40; and J. Braithwaite & P. Grabosky, *Occupational Health and Safety Enforcement in Australia* (Australian Institute of Criminology, Canberra, 1985), pp. 44–5 and 90–2.

222 This was proposed in 1996 by the then Labour opposition in Britain.

223 See N. Gunningham, *Safeguarding the Worker* (Law Book Company, Sydney, 1984), pp. 340–1.

224 A similar defence, enabling a worker prosecuted for an OHS offence to prove that the offence was committed as a result of formal instructions given by her employer and despite the worker's objection, is to be found in Quebec's An Act Respecting Occupational Health and Safety, s. 240.

225 See B. Fisse & J. Braithwaite, *Corporations, Crime and Accountability* (Cambridge University Press, Sydney, 1993), pp. 126–7; Lim Wen Ts'ai, 'Corporations and the Devil's Dictionary: the problem of individual responsibility for corporate crimes' (1990) 12 *Sydney Law Review* 311 at 339–40; and J. Braithwaite & P. Grabosky, *Occupational Health and Safety Enforcement in Australia* (Australian Institute of Criminology, Canberra, 1985), pp. 44–5 and 90–2.

226 B. Fisse & J. Braithwaite, *Corporations, Crime and Accountability* (Cambridge University Press, Sydney, 1993); B. Fisse & J. Braithwaite, 'The Allocation of Responsibility for Corporate Crime: Individualism, Collectivism, and Accountability' (1988) 11 *Sydney Law Review* 469; B. Fisse 'Corporations, Crime and Accountability' (1995) 6 *Current Issues in Criminal Justice* 378; and B. Fisse, 'Individual and Corporate Criminal Responsibility and Sanctions Against Corporations', in R. Johnstone (ed.), *Occupational Health and Safety Prosecutions in Australia: Overview and Issues* (Centre for Employment and Labour Relations Law, The University of Melbourne, 1994).

227 B. Fisse, 'Individual and Corporate Criminal Responsibility and Sanctions Against Corporations', in R. Johnstone (ed.), *Occupational Health and Safety Prosecutions in Australia: Overview and Issues* (Centre for Employment and Labour Relations Law, The University of Melbourne, 1994), p. 107.

228 Ibid. p. 108.

229 B. Fisse & J. Braithwaite, *Corporations, Crime and Accountability* (Cambridge University Press, Sydney, 1993), Chs. 5 and 6.

230 Ibid. pp. 180–1.

231 The model also requires that the corporate defendant and relevant

personnel are to perform disciplinary and other duties under accountability agreements, orders, and assurances. A key feature is that the corporation is to specify in advance the individuals and units which are primarily responsible for ensuring compliance with the accountability agreements, orders, and assurances. The court or enforcement agency should ensure that, where necessary, provision is made for the supervision or monitoring of accountability agreements, orders, and assurances.

232 See Ch. 3.
233 See Ch. 7.
234 Note, 'Growing the Carrot: Encouraging Effective Corporate Compliance' (1996) 109 *Harvard Law Review* 1783 at 1795–8. See also J. W. Nunes, 'Comment: Organizational Sentencing Guidelines: The Conundrum of Compliance Programs and Self-reporting' (1995) 27 *Arizona State Law Journal* 1039.
235 F. Haines, *Corporate Regulation: Beyond 'Punish or Pursuade'* (Clarendon Press, Oxford, 1997), p. 220.
236 Note, 'Growing the Carrot: Encouraging Effective Corporate Compliance' (1996) 109 *Harvard Law Review* 1783 at 1796.
237 See P. H. Bucy, 'Organizational Sentencing Guidelines: The Cart Before the Horse' (1993) 71 *Washington University Law Quarterly* 329 at 352.
238 Note, 'Growing the Carrot: Encouraging Effective Corporate Compliance' (1996) 109 *Harvard Law Review* 1783 at 1796.

7 Bigger Sticks: Tougher and more Flexible Sanctions for OHS Offenders

Introduction

We began the previous Chapter with a discussion of the purposes of OHS prosecutions at the top of the enforcement pyramid. We concluded that the principal rationale is general deterrence, an instrumentalist, consequentialist, and utilitarian rationale which requires punishment for OHS crime to outweigh the calculated benefits of the crime to those who commit the crime. Deterrence is measured by the likelihood of an OHS duty holder being inspected, prosecuted, and convicted, and the average penalty imposed. We argued that deterrence would be most effective if at the top of the pyramid courts could impose sufficiently large penalties to demonstrate to individual duty holders that punishments outweigh the benefits of OHS crime. We observed that there were theoretical and empirical reasons for suggesting that deterrence would not work for all corporate OHS offenders, particularly when punishments were confined to financial penalties. We argued that deterrence would be more effective if OHS prosecutions sought to achieve various sub-goals related to deterrence, some of which, like rehabilitation (or what might better be called organizational reform), are instrumentalist, and some, like retribution, are symbolic. We also suggested that deterrence would be increased if penalties were directed to non-financial as well as financial values within organizations, and if prosecutions were taken against responsible directors and top-level managers.

In this Chapter we look more closely at the sanctions at the top of the pyramid, and raise and address the major criticisms of sanctions for OHS offences. One consistent criticism of OHS enforcement regimes[1] from the early nineteenth century until the present day is that the maximum penalties available under the OHS statutes have been inadequate, and the actual penalties imposed by the courts have been too small a proportion of these maxima. In particular, the level of fine imposed has fallen far short of the penalties required for specific and general deterrence. Further, 'there appears to be no rational method, indeed no method at all, by which courts come to determine the level of fine'.[2] Critics have also argued that sanctions have been restricted to fines, which have proved to be completely ineffective in ensuring that OHS offenders restructure their workplace to comply with OHS standards. On a broader level, under-punishment of corporations for OHS and other sorts of crime will mean that the price of the goods and services produced by an organization will not fully reflect the cost of production. Rather the costs of corporate crime will be borne by society in general.[3]

We argue for a tougher, and wider, range of sanctions which will deter OHS duty holders from contravening their obligations under the OHS legislation, and which will also enable courts to make orders which require OHS offenders to reform defective OHS work systems. Accordingly, we discuss ways of improving the use of fines, and outline other sanctions which can be utilized by courts hearing OHS prosecutions against corporations. This discussion is concerned principally with sanctions against business organizations. We assume that in most jurisdictions the sanctions available for individual corporate officers (natural persons) within the mainstream criminal justice system provide sufficient sentencing options for courts.[4] Finally, we investigate mechanisms (including pre-sentence reports, victim impact statements, and sentencing guidelines) which will improve and rationalize the sentencing of OHS offenders, and which will make it easier for OHS duty holders to anticipate the likely outcomes of prosecutions. A greater level of certainty, consistency, and transparency in the sentencing process will ensure that OHS duty holders have incentives to adopt sound OHS management practices.

The deficiencies of the fine as an OHS sanction

A major criticism of current OHS statutory regimes is that they provide the courts with very limited options when it comes to imposing sanctions for contraventions of OHS offences. The options are limited because most OHS offences are likely to be committed by employers, and most employers are corporations. As Stone has commented:[5]

To understand why we are not doing a better job of it, we must look to legal history. When the law was forming, it was individual, identifiable persons who trespassed, created nuisances, engaged in consumer frauds. The law responded with contemporary notions about individuals—what motivated them, terrified them, and constituted justice toward them. Later, as corporations became the dominant vehicle for social action, only rarely did the law [respond] with specifically tailored adaptations. Since a body of law addressed to 'persons' already existed, it was simply transferred to corporations without distinction.

When corporate criminal sanctions have been developed, it has been argued that, because corporations are not natural persons, many of the traditional sanctions for criminal activity (such as imprisonment) are not applicable, and other sanctions (such as community service orders) are not appropriate. Most OHS statutes therefore restrict penalties to a monetary fine (which in most jurisdictions has been increased over time, and in particular in the last decade or so), and the possibility of imprisonment in rare cases, usually for offences committed by individual employees, managers, or directors. The almost exclusive use of monetary penalties is further justified by the argument that fines are the most economically efficient (i.e. least costly) means of penalizing corporate offenders: they are easy to calculate; they are simple to administer; in keeping with a *laissez-faire* philosophy, they do not involve any intervention by state agencies in the internal affairs of corporations; and they provide the state with funds to offset the cost of enforcement.[6]

However, in most OHS prosecutions the fine levied is generally considered to be inadequate,[7] thus reducing whatever deterrent, rehabilitative, or retributive effect the fine may have. For example, while recently fines for OHS contraventions in the UK have increased from an average of £40 per offence in 1970, to £3,061 in 1993–4, but dropping to £2,567 in 1995–6, these levels of fine are extremely small, and unlikely to provide a serious deterrent to anything other than a very small organization.[8] Dutch evidence of prosecution outcomes from 1989 to 1991 suggests that the average fine was 300 Dutch guilders, or £1,500.[9] The average fine for OHS offences in Denmark in 1995 was DKK14,000.[10] In most Australian jurisdictions, average fines in 1996 ranged from about $2,500 (Western Australia) to just under $7,800 (Victoria). Penalties imposed under the USA OHS Act rarely reach the maximum penalties, apart from a few cases involving egregious violations.[11] The average Federal penalty for a 'serious violation' in 1995 was $763 (11 per cent of the maximum), and for 'other than serious' violations was $52 (7 per cent of the maximum).[12]

In addition, we argue that even if higher fines were imposed, high fines in themselves do not provide sufficient incentives for OHS offenders to adopt better OHS management approaches. Commentators have strongly criticized the usefulness of the fine as the principal sanction for corporate wrongdoing.[13]

One criticism is that fines signal to corporations that 'offences are purchasable commodities'[14] rather than acts or omissions considered by the state to be intolerable. Second, fines only have an impact upon the financial concerns of the corporation, rather than on managerial motivation and other non-financial concerns of the company or entities within the company which might shape the activities, policies, and procedures towards OHS. Third, the fine may punish innocent parties (in the sense of lacking direct responsibility for the offence) instead of or as well as the guilty.[15] Depending on whether and how the fine is 'passed on', fines may inflict heavy losses on more innocent parties, such as reduced dividends for shareholders[16] (in the case of privately owned corporations), reduced returns to taxpayers (in the case of state-owned enterprises), reduced wages or job opportunities for workers, or increased prices for consumers.[17]

Fourth, fines do not ensure that corporate offenders will take disciplinary action against those officers or managers within the company who are responsible for the actions or omissions which led to the offence (particularly since such action may be disruptive, embarrassing for the responsible company officers, encourage whistle-blowers within the company, or open up possibilities of civil litigation against the company). Fifth, fines do not require the corporation to review its management systems (including its systems of work) to ensure that there is not a repetition of the OHS offence or that there are no other such offences committed in the future. Fines treat offending organizations as ' "black boxes", attempting to shape organisational behaviour by applying external stimuli instead of directly attacking the organizational factor that produced the violation at issue'.[18] It may even be that fines actually diminish the resources available to the offender to reform the OHS systems within the organization.

Sixth, as we outlined in Chapter 6, the level of fine required to reflect the seriousness of the offence and to act as retribution or as a deterrent may far exceed a corporation's capacity to pay (the 'deterrence trap' or 'retribution trap', depending on the purpose of the penalty).[19] Seventh, because there are no mechanisms to recover fines from entities other than those upon which the fine is levied, fines can be avoided by organizations changing their corporate structures or identities, or moving their assets to other corporate entities. Finally, fines imposed on state instrumentalities simply mean that assets are moved from one state agency or instrumentality to general revenue or to the OHS agency without any necessary improvement in OHS practices, policies, or procedures in the offending agency.

These well-accepted criticisms of the fine do not mean that the fine should be abandoned as a sanction for organizations which have breached OHS legislation, but rather that measures should be taken to ensure that fines are severe enough to deter OHS offenders and potential offenders, and that other sanctions need to be introduced to complement the fine and to ensure that OHS penalties play a role not only in deterring further OHS contraventions, but also rehabilitating the offending company. The literature canvasses a wide range of alternative corporate sanctions which might be introduced in order to ensure that OHS penalties are effective. These sanctions include community service orders; publicity orders; equity fines; injunctions; corporate probation or supervisory orders (internal discipline orders, organizational reform orders, or punitive injunctions); and, in severe cases, incapacitation.

Rethinking corporate sanctions for OHS offences

Improving the fine

As we noted in Chapters 4 and 6, the theory underpinning the imposition of fines is that, in making OHS decisions, profit maximizers will take OHS measures into account if the perceived benefits exceed the costs of doing so. In practice, however, whatever deterrent effect there is in the imposition of a fine has been undermined by the reluctance of the courts to impose substantial penalties for serious breaches of the OHS legislation.[20] To a large extent this reflects the fact that the maximum fines in most OHS statutes are themselves inadequate to act as a serious deterrent to corporations. The data outlined earlier in this Chapter suggests that in practice, even where maximum fines are quite large, OHS prosecutions result in relatively low fines in absolute terms, and also viewed as a percentage of the maximum available fine.

Amongst the measures that can be introduced into OHS statutes to ensure that the monetary fine is a more effective deterrent are: (i) an increase in the maximum available fines; and (ii) tailoring fines to take into account the defendant's resources.[21]

Increase the maximum fines

A simple application of deterrence theory suggests that the maximum available fines for OHS offences should be increased, so as to present all OHS duty holders with the clear message that OHS contraventions lead to serious criminal action. The size of the maximum fine, as the courts repeatedly observe, indicates the seriousness that the community, through the legislature, attaches to OHS offences. It should also give the court sufficient scope to impose a penalty large enough to deter a well-resourced corporate offender. In the past decade numerous Canadian provincial inquiries into OHS have recommended increased maximum fine levels[22]—up to C$10,000 in Saskatchewan,[23] and C$250,000 in Nova Scotia.[24] The recent report of Royal Commission on Workers Compensation in British Columbia recommended maximum fines of $500,000 for corporations.[25] In the UK, maximum penalties for breaches of the HSW Act 1974 are unlimited in the higher courts, and up to £20,000 in the lower courts. The Australian Industry Commission observed that the maximum penalties it recommended for OHS offences (A$500,000 for corporations, A$100,000 for individuals)[26] 'may not be the most appropriate in the longer run. In the Commission's view the appropriate level is likely to be higher.'[27] This suggests that OHS agencies should monitor the penalties resulting from these increased maxima, with a view to further increases in the maxima.

In addition, the OHS statutes might include a continuing fine for each day that the offence continues beyond a certain period after the hazard is discovered, unless the convicted offender is able to prove that it took the appropriate steps to remedy the hazard as expeditiously as possible. This will ensure that it is rational in the long term for corporations to remove the hazard, rather than absorb the fine.[28] A further possibility is that, where an OHS offender accrues monetary benefits from an OHS offence, the OHS statute empowers the court to order the offender to pay, in addition to the fine, a further fine in an amount equal to the estimation by the court of the amount of the monetary benefits.[29]

However, any increases in fines are subject to the general criticisms of fines which we have outlined earlier in this Chapter, and in particular the risk of being passed on to innocent parties. Some proponents of fines[30] as a sanction have therefore responded to these assertions by suggesting that innocent parties can be protected by a variety of modifications to the fining process: equity fines,[31] instalment fines (i.e. collecting fines in instalments), pass-through fines (directly imposed on shareholders in proportion to their shareholdings), and superadded liability (which imposes pro rata liability on shareholders for fine amounts in excess of corporate assets). Metzger and Schwenk[32] argue that at best, 'these devices merely limit the classes of innocents subjected to punishment'.

One by-product of an increase in fines imposed upon OHS offenders is that the revenue from the fines might be diverted to fund the increased resource demands that will be placed upon courts and OHS agencies by the adoption of some of the new corporate sanctions that we outline below.

Tailoring fines to the offender's resources

OHS offenders come in all shapes and sizes. They range from sole proprietors running very small enterprises, to huge multinational corporations with thousands of employees, based in several jurisdictions. The OHS statutes tend to do little to deal with this complex issue. In outlining the maximum penalties, many of the statutes distinguish between individuals and corporations. But this distinction does not deal adequately with small corporations which are effectively a one or two person operation.[33] A related issue is the extent to which the courts actually take the defendant's resources into account in determining the level of fine to be imposed.

We suggest that an effective OHS sanctioning regime must have the capacity to tailor OHS penalties to the offender's resources,[34] to ensure that penalties not only serve as a deterrent, and promote OHS reform in the workplace, but are just in the actual circumstances of the case. One possibility is to modify a well-known European mechanism used to take into account the offender's resources when determining the level of a fine.[35] Some European jurisdictions have dealt with the issue of ensuring that the level of fine is tailored to the seriousness of the offence *and* to the resources of the defendant by introducing a *'day fine' system*.[36] For example, Sweden has an elaborate unit fine procedure for determining the means of an offender and relating the fine to the gravity of the offence, the culpability of the offender, and the means of the defendant.[37] The Swedish Criminal Code allocates to each category of offence under the code a maximum number of fine units (a number on a scale of 1 to 120 which reflects the relative gravity of the offence). The value of each unit, called the day fine, is assessed according to the means of the offender.[38] According to a report by the British Section of the International Commission of Jurists, Justice, the Swedish day fine scheme 'has removed the mystery attached to determining the amount of the fine, and virtually eradicated inconsistency and unfairness'. Pilot programmes for day fine systems have been run in the UK[39] and USA (Staten Island, Milwaukee, Arizona, Connecticut, Iowa, and Oregon), but generally were limited to minor offences. While it is probably still too early to form judgments, the results appear to be promising.

We note that the European schemes were introduced largely to deal with the problem of low-income offenders being too harshly penalized by fines which ignored their means; that they were developed as diversionary devices—instead of being prosecuted and fined, offenders agree to pay the fine in exchange for charges being dropped. We suggest that the principles of the day fine system might be adapted to an OHS sentencing regime so that the first stage in the sentencing process is that the courts conduct a careful examination of the actual and potential resources of the offender so that the court can determine: (i) the level of fine most likely to deter the particular offender; and (ii) whether there is likely to be a 'deterrence trap' in the case of the particular offender, in which case the court should be required to impose sanctions which avoid such a trap. The availability of a further range of sanctions, such as those which we discuss later in this Chapter, will mean that, where an offender is low on cash resources, the penalty imposed can utilize mechanisms such as community-based orders, publicity orders, or corporate probation which will ensure that the sanction acts as a sufficient deterrent, while at the same time leaving the offender in a position to improve its OHS policies, procedures, and practices. We will return to the issue of sanctions tailored to the defendant's resources when we examine sentencing guidelines later in this Chapter.

Introducing new corporate sanctions

In this section, we outline a range of new *corporate* sanctions which might be introduced into the OHS statutes to increase the range of sanction options currently available to the courts.[40] We have already noted earlier in this Chapter that the development of new corporate sanctions in the OHS arena will need to focus on objectives going beyond specific and general deterrence, to include retribution, denunciation, and organizational reform. But the ultimate purpose of corporate criminal sanctions for OHS offenders must be to reduce the contraventions of OHS obligations, and hence to reduce the incidence of work-related injury and disease.

We emphasize again that none of the sanctions we discuss in this Chapter is capable of functioning as a singly effective technique of control. As we argued earlier,[41] financial penalties aimed at general deterrence may not be effective in changing the behaviour of corporations which do not act as rational profit maximizers. Because of the potential of the 'deterrence trap', discussed in the previous section, they may even be ineffective for rational profit maximizers. In any event, it is difficult to ensure that the certainty and severity of any one sanction will provide sufficient incentives in all circumstances. Even sanctions which require organizations to restructure their OHS systems provide no incentive to organizations to comply proactively with the law before OHS contraventions have been detected.

If only the organization is punished, the behaviour of individuals within the organization may not be adequately controlled. But if only individuals are punished, the integrity of the organization itself will remain intact, and it will be free to continue to break the law in the future secure in the knowledge that sacrificial lambs will be offered when called for to appease the community. Corporate sanctions must, therefore, be used alongside sanctions against individuals.

To acknowledge the limitations of the different sanctions does not deny their value. Rather, it is a reminder that a combination of measures will yield the best results, particularly because, as is the case with individual malfeasance, corporate wrongdoing is often influenced by a range of variables, and is unlikely to be susceptible to any single mechanism of control. Fox[42] observes that 'what is clear is that no single technique provides a panacea. . . . [A] new eclecticism seems called for in relation to measures aimed at correcting corporate tendencies to breach the law.'

We now outline a few new sanctions which will bolster the top of the enforcement pyramid. Each has the potential to overcome at least some of the weaknesses of the fine identified earlier in this Chapter, and together provide courts with an eclectic array of sanctions which can be tailored to the circumstances of each case.

Publicity orders

We have stressed that general deterrence depends to a significant degree on business organizations and corporate officers being aware of the serious penal consequences of contravening OHS statutes. We have also pointed out that most business organizations jealously guard their reputations, and that these non-financial considerations often play a substantial role in the motivation of an organization, and can be important for financial success. There is some suggestion that 'it may be only large high profile organisations which are sensitive to adverse publicity. In contrast, small organisations thrive on their invisibility to the public at large . . . and so may be little affected by public impression of their activities.'[43] Nevertheless, this suggests that publicity given to OHS offenders is a significant component of a state strategy to control OHS offending.[44] Yet most OHS agencies do not appear to have a routine procedure for publicizing all OHS offences and offenders, and, even where they do, it is rare that the media reports on an OHS conviction, unless it arises out of an event that is deemed newsworthy in the first place. Haines[45] shows how media interest in OHS is usually confined to coverage of an event resulting in injury or a fatality, and that media interest was transient and focused on the human emotion surrounding the event.[46] As a result, media coverage of OHS is unlikely to have any worthwhile impact on a corporation's response to injuries and death at its workplace.

Court-ordered adverse publicity was originally introduced under the English Bread Acts, first enacted in 1815,[47] and endorsed in 1970 by the USA National Commission on Reform of Federal Criminal Laws, though this was never fully implemented.[48] It has been authorized as an optional condition of corporate probation under the USA *Sentencing Guidelines for Organizations*. Formal publicity has been used in administrative law as a penalty and in civil actions as a remedy.[49] Recently Nova Scotia amended its OHS statute to give the court the discretion to direct a convicted offender to publish the facts of an offence in a form prescribed by the court.[50] If the offender fails to comply, the Director of the OHS agency may publish the facts in compliance with the order and recover the costs of publication from the offender. French law provides for the publication of a sentence in the press.[51] Similarly, article 7 of the Dutch Economic Offences Act 1950 provides that certain verdicts (not OHS offences) of the court must be published.[52]

Court-ordered adverse publicity is an example of the use of shaming as a means of social control,[53] 'by bringing to public awareness the incongruity between criminal conduct and accepted models of behaviour and thereby coercing the offender's behaviour in the exemplary direction'.[54] Adverse publicity orders have the potential to be a powerful sanction against corporate wrongdoing, and could be ordered by a court once an OHS offence was proved. Depending on the circumstances of the offence, a formal court-ordered sanction of adverse publicity could accompany other forms of penalty, or could be a penalty sufficient in itself.[55] To avoid the problems raised by Haines earlier in this section, a publicity order should be carefully structured and targeted. The court can order the company to publicize the OHS offence in an appropriate manner to the whole community or the sectors of the community affected by the offence. For example, the court could order a company convicted of an OHS offence to take out an advertisement in a national newspaper in the case of a nationally-based company, or in a metropolitan paper in the case of a company operating in one of the large cities, or in the local paper for rural-based companies. The advertisement could be required to follow a format set out in the governing OHS statute (details of the hazard leading to the offence, including full details of the actual and potential harm caused; full details of the prosecution (including charges laid), conviction, and penalty imposed by the court; the top level executives involved; an indication of the remedial action taken, or ordered to be taken, by the company; and an apology). The company could also be required to pay for and have published remedial notices or warnings, a sanction which is important where the offence is a contravention of a general duty imposed upon a supplier or manufacturer of plant, equipment, or substances.

Adverse publicity orders provide a solution to virtually all of the criticisms of the fine which we rehearsed earlier in this Chapter. In particular, the order can be tailored to focus explicit attention on the internal reform of the corporation to avoid further OHS offences, and, as our example above shows, can be designed to have an impact only on managers and directors, and not employees, consumers, or shareholders. It will give other organizations notice of the consequences of OHS rule breaking and may trigger the shareholders to take action against the directors of the company. This sort of publicity may help to educate the public, and judiciary, about the seriousness of OHS crimes, and help to counter perceptions that these crimes are only 'quasi-criminal'. A significant advantage of publicity orders is that they do not require either the court or the OHS inspectorate to expend much by way of time or resources in monitoring the defendant organization's compliance with the order. Consequently, of all the new sanctions we suggest in this section, adverse publicity orders may be the easiest to implement in the short term.

Critics of this sanction have suggested that its use might lead to injustice, in that its impact (unlike, it is claimed, a monetary fine) is uncertain, and might be disproportionate to the severity of the offence.[56] Yet few other sanctions are assessed in terms of their impact upon the offender, and there is no reason why court-ordered adverse publicity should be treated differently.[57] While shaming through adverse publicity might be considered to be unacceptably cruel when applied to individuals,[58] corporations are unlikely to suffer adverse psychological consequences.[59]

We note that both the British Robens Report and the Australian

Industry Commission's final report recommended that adverse publicity be required in a particular form, in directors' reports and company annual reports.[60] These recommendations should be seen as additions to the sanction of adverse publicity, rather than as an alternative, because the recommendations focus on the overall OHS performance of the company, in addition to any specific offences which the company may have committed.

Supervisory orders and corporate probation

In the context of an OHS enforcement pyramid which aims to ensure that duty holders take a systems-based approached to OHS management, the corporate sanction which overcomes many of the deficiencies of the monetary fine is corporate probation.[61] Such supervisory orders remedy the lack of judicial power to order that corporate structures be remedied to prevent future OHS offences, and could go further to enable a court to require an OHS offender to develop and implement an appropriate SMS.

Under the Sentencing Reform Act 1984 corporate probation has been available to USA judges in relation to Federal crimes since 1987. The principal underlying goal of this sanction is deterrence, but, as the discussion of the sanction in this section shows, it contains important elements of rehabilitation (organizational restructuring). The sanction involves a court making a variety of orders, which might range from the court monitoring the activities of the organization for a limited period of time, requiring the senior management of the corporation to change the way it implements OHS measures (such as introducing a safety officer, altering work procedures, or substituting less hazardous substances), or making orders that require the implementation of internal disciplinary measures within the company. The operation of the company can be scrutinized and reformed. New policies and systems can be developed, and new personnel can be engaged. Senior management can be required to undergo OHS training, to undertake a programme to introduce safer work systems and to prevent future criminal conduct, or even to introduce a state-of-the-art SMS. 'The organizational defects of the company—its "psyche"—can be meddled with in ways which would be inappropriate in the case of an individual.'[62] The American Bar Association has recognized that these types of sanctions may be appropriate where there exists a 'clear and present danger' to the public health and safety.[63]

The sanction is an important one in an OHS regulatory model that, as we suggest, places great store on SMSs. It reflects at the prosecution stage the importance of SMSs, and the need to ensure that the whole enforcement pyramid is geared towards incentives for OHS duty holders to restructure their operations with OHS as a central concern. In the next few pages we briefly discuss three forms of supervisory orders: internal discipline orders, organizational reform orders, and punitive injunctions.[64]

Internal discipline orders[65]

In an internal discipline order, the court makes an order that an offending corporation must 'investigate an offence committed on its behalf, undertake appropriate disciplinary proceedings, and return a detailed and satisfactory compliance report' to the court which issued the order.[66] The investigation may be conducted by the organization itself, or by the organization engaging a suitably qualified third party, such as an OHS consultant. The aim is not to have managers responsible for the OHS contravention produced before the court, but rather to require the organization itself to carry out the disciplinary action, or face greater sanctions from the state. If the investigation and disciplinary action carried out by the corporate officers specified in the order is unsatisfactory, criminal sanctions are applied against those officers.

This form of sanction is a court order for an organization to activate its private justice system, rather than being aimed at reforming the OHS management system (see organizational reform orders, discussed below), and parallels our earlier discussion[67] of using prosecutorial discretion to give organizations the opportunity to carry out their own internal disciplining where OHS contraventions have occurred. If such a sanction is introduced, however, the authorizing statute must ensure that safeguards are introduced to ensure that the internal disciplining procedures accord substantive and procedural fairness, and do not shift blame onto scapegoats within the organization.[68]

It might be argued that this form of sanction requires courts to undertake supervisory functions for which they have no expertise or resources.[69] Further, courts generally tend to eschew advisory functions, and so corporate probation requires a shift in courts' activities. But the overall benefits of the various forms of corporate probation, including internal discipline orders, suggest that these issues of resources and court function should be faced and resolved, even if it means that corporate probation cannot be introduced in the immediate short term. It is cheaper for the organization to monitor its own activities, than for the state to do this.[70] The problem of court resources and expertise could be reduced by the court appointing the OHS inspectorate to oversee the supervisory activities. While this would increase the burden on the inspectorate, these supervisory activities could dovetail with a general move by OHS inspectorates to develop strategies to promote internal disciplining. Alternatively, the court could establish a panel of independent experts (separate from the consultants who might be engaged by the organization to assist in compliance with the order) who could supervise the order.[71] The cost of the expert could be borne by the defendant organization. Where the organization failed to comply with the order, the inspectorate (or the independent expert) could bring the matter to the attention of the court for a further order.

Organizational reform orders

Organisational reform orders[72] might be made when OHS contraventions by an organization are serious and repetitive and are a result of deficient OHS management systems. The sanction would be of a limited duration (say up to five years), and would require an organization to reform its internal OHS management system. It would involve monitoring of the OHS management system of the convicted organization, through reporting, record keeping, and auditing controls designed to increase the organization's internal accountability and to improve OHS programmes so that they comply with OHS statutory obligations, but stop short of the court reviewing legitimate 'business judgment' decisions.[73] Once again, the court could delegate its supervisory functions to the OHS inspectorate or a court appointed independent expert. We argue that any increased call on the OHS inspectorate's resources by the supervisory requirements of an organizational reform order are justified by the enhanced rehabilitative effects of the sanction, the signals it sends duty holders in general about the need for compliance programmes and OHS audits, and the expertise in monitoring the auditing of OHS management systems and compliance programmes which the inspectorate will develop. This expertise can be used by the inspectorate in its routine activities of advising duty holders about OHS management systems and compliance programmes.

As a example of an organizational reform order,[74] the court might order a convicted organization to develop and submit to the court, or an independent expert or the OHS inspectorate, a compliance programme (including details of how the programme will be implemented) to detect and prevent OHS contraventions.[75] OHS compliance programmes can be defined broadly as any programme or 'systematic organizational effort'[76] (for example, specific policies aimed at preventing employees from breaching OHS obligations, procedures to identify and remedy contraventions or potential contraventions of OHS statutes, and programmes aimed at ensuring compliance with OHS standards) that has the aim of ensuring a corporation's compliance with the law.[77] They are an example of self-policing through performance quality management systems.[78]

Compliance programmes are ultimately the responsibility of line managers who run the daily aspects of a business, and require managers and employees to be engaged in an ongoing process of identifying, facilitating, and encouraging lawful conduct by employees of the organization, and of monitoring compliance results.[79] Managers should indicate that it is company policy that all company activity comply with legal requirements for OHS, and should provide specific directions to employees about how they must carry out activities to follow those laws. The directions should be reinforced by appropriate training, corporate reward, and disciplinary arrangements. In effective compliance programmes all employees understand the level of performance demanded by the programme; the demands are effective to satisfy legal requirements; and the programme aims to prevent offences rather than detect them after the event. There should be regular inspections of organizational practices to ensure that they comply with OHS requirements, using a variety of monitoring and auditing practices. These should be supplemented by procedures encouraging employee suggestions in relation to matters where compliance falls short of requirements. Employee representatives should play a major part in developing and implementing the compliance programme.[80] The results of compliance inspections should be documented in reports to be used by management to evaluate compliance levels, develop responses to deficiencies, and monitor the operation of compliance systems.

Once the court approved the programme, the organization would have to advise its employees and shareholders of the conviction, and the programme proposed to avoid further contraventions. If the organization failed to produce a satisfactory compliance programme, the court could engage outside experts, funded by the organization, to prepare the programme. The organization would make regular progress reports to the court, and might be subjected to unscheduled inspections. The OHS inspectorate might be involved in the supervision of the programme. This sanction would enable the court to ensure that the OHS offender not only remedied the actual contraventions which gave rise to the prosecution, but introduced an ongoing and systematic programme to ensure at least minimum compliance with the obligations set out by the OHS statute.

Of course, as we have already mentioned in Chapter 3, in the context of whether SMSs should be legislatively mandated, and Chapter 5 in the context of auditing 'Track Two' SMSs, we cannot be sure that compliance programmes and SMSs will be anything more than tokenistic 'paper systems', particularly if they are imposed upon organizations. It may be that business organizations will feel a greater commitment to implementing compliance programmes and SMSs properly if they know that their efforts are being closely scrutinized by the court, and serious fines will be imposed if their implementation efforts are considered to be inadequate. Certainly, especially in relation to punitive injunctions and SMSs, the court must stress that SMSs are not simply 'add ons', but will require a total revamping of an organization's management systems.[81] In Chapter 5 we discussed a number of methods of auditing SMSs, and most of those suggestions would be worth considering in the context of compliance programmes in corporate probation and SMSs in the context of punitive injunctions. We suggest that whatever solution is adopted, it should focus on internal audits by managers and employee representatives familiar with their own organization, programmes, and employees and not rely too heavily on expensive external monitors and auditors.[82]

Punitive injunctions

Organizational reform orders are relatively soft sanctions, focusing more on rehabilitative or organizational reform effects than deterrent or retributive punishment.[83] A tougher form of sanction, to be used in the most serious cases where deterrence is required, is the punitive injunction. This would go beyond requiring the organization to remedy deficient OHS management systems, and might require the organization to introduce innovative measures, such as a SMS. 'The punitive injunction could thus serve as both a punishment and super-remedy.'[84] For example, the French Labour Code[85] empowers a court to order an employer, within a specified time period, to implement a plan of safety measures, under the supervision of the OHS inspectorate, and subject to heavy fines for non-implementation.

The punitive injunction's focus on innovative methods lends itself well to an OHS enforcement regime which has as one of its primary aims the development of SMSs. It may well be that the punitively demanding nature of the punitive injunction, particularly if it is used to require organizations to adopt innovative SMSs, requires the punitive injunction to be a sanction confined to the upper reaches of the enforcement pyramid, namely prosecution in a higher level court.

The three forms of corporate probation discussed above provide possible criminal sanctions which remedy a glaring deficiency in the criminal law to authorize disciplinary or structural intervention in corporate activities, or to demand innovative and proactive remedies, such as the introduction of SMSs. An important issue in the supervision process would be finding personnel or bodies who could monitor the corporation to determine whether its internal reporting, restructuring, or implementation of a SMS was being appropriately carried out. The body must not be capable of being lobbied by the business organization, or influenced by a changing political climate. Possible candidates for appointment by the courts to supervise a probation order are OHS management consulting firms, or university-based consultants,[86] or independent solicitors. The corporation would pay the fees of the appointee. As discussed in Chapter 5, we envisage a major role for trade union representatives in the monitoring process.

The order will be discharged upon proof by the offender of satisfactory compliance. Such proof might be satisfied by certification by the offending company's board of directors that the order has been complied with, as long as it was a statutory offence to provide false certification of compliance.[87] The court can ask for a report from the court appointed supervisor, and give trade union representatives the opportunity to present evidence of non-compliance with the probationary order.

Corporate probation would overcome many, if not most, of the disadvantages associated with the traditional monetary fine. In its different forms it acknowledges the complexity of corporate behaviour, and enables an examination of factors within the organization which have led to contraventions, and the development and implementation of programmes to prevent further contraventions. Probation can be used against corporate offenders with cash flow difficulties. It minimizes 'spillover' effects and addresses non-financial values within the organization.[88] Some authors argue that as corporate probation can include conditions which lower the company's reputation in the eyes of the public, the sanction can have severe deterrent or retributive effects.[89] Punitive injunctions, for example, clearly indicate society's disapproval of the crime and have significant deterrent effects.[90] Probation is also most effective in relation to government instrumentalities, where the imposition of a fine would simply involve recycling money among government agencies.

The major disadvantage of corporate probation might be its level of intrusiveness into the affairs of the corporation, steering 'the criminal justice system into a political conflict with the economic ideology of *laissez faire*'.[91] The Australian Law Reform Commission[92] noted that this level of intrusion was minor in comparison to the level of intrusion involved in imprisoning individuals for serious offences. It also indicated that where it is used the order should minimize the level of intrusiveness involved in its execution, and, if practicable, should only be imposed after the offending corporation has had an opportunity to specify what disciplinary and structural reform measures it intended to take in response to the commission of the offence.

Information about the suitability of corporate probation might be gathered in a routine corporate inquiry report, which we discuss below. Where the corporation's response was inadequate, the OHS agency (perhaps with the assistance of consultants) might be required to give a detailed report about the corporation's internal management processes leading to the contraventions, and the measures required to avoid future contraventions.[93] The USA Sentencing Reform Act 1984 enables a convicted corporation to provide a critique of terms proposed by outside experts, and to propose alternatives of its own.

In conclusion, there is much to suggest that corporate probation is a sanction which should have a central place at the top end of an enforcement pyramid which focuses on providing incentives to organizations to improve the OHS performance, and which encourages the adoption of SMSs. For this reason, corporate probation along the lines that we have discussed in this section should be available as a sanction for all OHS prosecutions. We are mindful of the increased resource demands that all forms of corporate probation place on courts and OHS inspectorates, but argue that corporate probation is so important to an OHS sanctioning regime that, even if corporate probation is not introduced in the immediate short term, in the long term resources should be allocated to ensure that courts and OHS inspectorates are able to implement the sanction. We reiterate that the severe demands of the punitive injunction might require it to be available only in a higher level court.

Community service orders

Community service orders are an important sentencing option for the criminal courts when dealing with more traditional individual criminal offenders. These orders have rarely been considered to be appropriate as a sanction for business organizations. Yet a community service order can be modified[94] for application against a corporate offender.[95] It can act as a deterrent without bankrupting the organization, and can stimulate internal disciplining within the offending organization, particularly if the organization decides to allocate managers involved in the OHS offence to the project.[96] Recently, the Nova Scotia OHS statute introduced community service orders as a possible sanction for an OHS offence.[97]

Community service orders can require a corporate offender during normal working hours to carry out research or socially useful projects which utilize the resources and special skills of the corporation. They can also be used against subunits or individuals within the organization where the defendant organization identifies such a subunit or individual as being responsible for the OHS contravention, as for example under Fisse and Braithwaite's accountability model discussed in Chapter 6, or where the court determines that a particular subunit or individual has special responsibility for the offence. The court must ensure that the community service project is seen by the community as part of a penalty the corporate offender is paying for an offence, rather than as an altruistic contribution.

The form of community service can be determined by the court alone, or proposed by the offender and approved by the court, and should be performed within a specific time. The project must have some relationship with the offence.[98] The order should be tailored to ensure that it does result in organizational change so that the offence, or other OHS offences, are prevented. For example, rather than requiring the OHS offender to build a park for the local community, or contributing to its favourite charity, a community service order might require the offender to divert resources into developing and testing new methods for reducing hazards emanating from the products it produces. The offender could research and develop innovative OHS management systems or control measures appropriate for the defendant's industry; sponsor, organize, and run industry conferences or committees to investigate causes of and solutions to hazards prevalent in the industry; run a community awareness campaign on a certain type of workplace hazard and ways of preventing it; pay for public education on OHS;[99] build a rehabilitation centre for victims of a particular type of workplace hazard; produce mechanisms to modify workplaces to accommodate injured workers; or manufacture artificial limbs or robots which could benefit workers suffering particular forms of injury.

Legislation establishing community service orders for OHS offenders should specify that the amount of community service ordered by the court should 'be quantified in terms of the actual net cost of materials, equipment and labour to be used for the project'.[100] The maximum cost of the community service order should be the same as the maximum amount of the fine for the relevant offence. Unless the court otherwise certifies, the project must be performed by employees of the offender, and should include managerial and executive officers. The court might appoint independent auditors, funded by the offending company, to supervise the execution of the order, and to ensure that offending companies do not falsify compliance reports; engage in projects which they have already completed in the normal course of their business; or enlist 'second- or third-rate personnel for the required project'.[101]

Dissolution

At the top of the enforcement pyramid might be the corporate equivalent of capital punishment, the dissolution of the corporation, the revocation of its charter, or the seizure of its assets.[102] Unlike capital punishment, this sanction need not be permanent.[103] It might be invoked where the company's OHS record is so bad that the dissolution of the company is warranted in order to protect the community.[104] The court should also consider imposing restrictions on the ability to be involved in other companies of the directors and managers most responsible for the offences, lest they simply take their bad OHS practices to other organizations. French OHS law provides for heavy fines for OHS breaches. The penalties gradually become more severe against employers and others who repeat an offence within twelve months, and can lead to 'the total or partial, permanent or temporary closure of the establishment or site on which the necessary health and safety measures have not been taken'.[105]

The disadvantage of dissolution as a sanction is that it is a blunt instrument, causing severe economic dislocation in some cases, and will affect the interests and livelihoods of innocent parties, such as shareholders, suppliers, consumers, and employees (whose health and safety is being protected by the sanction). The sanction could even result in a distortion of the competitive structure of the industry in which the offender is operating. Accordingly this sanction should only be used by a high-level court as a very last resort with an irrational[106] or intransigent, recidivist organization, after all other sanctions have failed to have an impact upon the offender. For example, if the court finds that the offender can only operate by engaging in activities which endanger its workers and the public then the company should be disbanded.[107]

Box[108] has proposed that offending companies should be nationalized for a limited period of time, with public appointed directors installed to make structural reforms to the company. We note that such forms of intervention are not totally alien to corporate life: companies are often run by administrators or receivers. Nevertheless, in the existing capitalist corporate culture, this would be an extreme sanction, but may be worth considering, in a mild form, for middle size to large corporations which have consistently bad OHS records. For example, the 'draft' USA environmental sentencing guidelines provided that where an offender repeatedly violated the conditions of probation, the court could appoint a trustee to ensure compliance with court orders.[109]

Other possible corporate sanctions

For the sake of completeness we note that there are other possible corporate sanctions which might be considered as part of an OHS enforcement pyramid.

Coffee has suggested that a sanction which overcomes some of the limitations of fines is the *equity fine*, also known as stock or share dilution.[110] Instead of a severe fine being imposed on the corporation in cash, it is imposed via the securities of the corporation. The court can order a convicted OHS offender to issue to, say, the OHS regulatory agency, shares calculated to be the equivalent value of the appropriate fine. The agency would then be able to liquidate the shares as it wished. This would ensure that more funds are available for prevention-based activities, particularly a larger inspectorate, better resourced prosecutors, and more resources for research into preventative strategies. Alternatively, shares might be issued to other organizations involved in OHS, workers compensation, or rehabilitation.

A major factor in the attraction of equity fines is that management's interests will be affected by reducing the value of their share options and holdings in the company. This rationale would be weakened if management did not hold shares or share options in the company, or restructured salary packages to exclude shares or share options if there is a possibility of an equity fine. Equity fines might have a significant impact on the convicted corporation because, as more shares are created, the corporation might face a greater prospect of a takeover. The costs of the equity fine fall mainly on shareholders, rather than on workers and consumers, and may result in market analysts counselling against the purchase of shares in companies with poor OHS management systems. The device of the equity fine avoids the deterrence and retribution traps, because the court can tailor the sanction to include not just the liquid assets of the company, but also fixed assets, future earnings, and owner-ship rights against the company's investments in plant and equipment. But equity fines do not promote internal disciplinary or reform measures, and do not affect the non-financial values of the organization.

Despite these misgivings we suggest that the equity fine is a useful mechanism to include in a mix of new OHS sanctions. Because of the impact that the sanction will have on shareholders, it should only be available as a sanction in higher level courts, in prosecutions brought at the top of the pyramid. It should be used with measures like corporate probation which require OHS systems to be restructured.

A sanction probably best used to supplement other penalties if and when they prove ineffective, is the *injunction*.[111] An injunction in this sense is quite different to the punitive injunction we discussed earlier in relation to corporate probation. It might be sought by the government to prevent continual breaches of public rights, and granted when the ordinary threats of the criminal law, at least in relation to lesser offences, have failed to deter the offender. If the offender disobeys the injunction, it can be fined an amount left to the discretion of the court[112] or the breach can be converted into civil or criminal contempt proceedings. Some Canadian OHS statutes authorize OHS agencies to apply to obtain a court-ordered injunction to take prompt action to address potential or existing harm.

Injunctions may not be as significant in OHS enforcement as in other areas of protecting public rights, because many jurisdictions, as we out-lined in Chapter 4, give their OHS inspectorates the power to issue prohibition notices, which have the effect of halting dangerous work. Injunctions can be seen as a more formal, court sanctioned, version of this administrative remedial measure, which can be used for more sym-bolic purposes to express public disapproval of dangerous work practices. It may be worth giving OHS agencies a similar power to seek an injunc-tion where an OHS offender is continuing a system of work which contravenes the OHS legislation, and where prohibition notices are being ignored by the company.[113]

Another possible sanction is the *disqualification of OHS offenders from government contracts*, or from tendering for government contracts.[114] In some jurisdictions OHS agencies are already utilizing this sanction in an administrative process, rather than as a court imposed sanction. The great advantage of this sanction is that it provides a competitive advantage to law-abiding companies. If used as a court imposed sanction, as is the case in France,[115] the disqualification order might impose OHS criteria for requalification, such as the adoption of an OHS systems-based approach, or an accepted OHS management approach. There might be objections to this sanction being used as a court imposed penalty on the basis that it is inappropriate for courts to be seen to be distributing government contracts. It seems to us, however, that this sanction merely involves removing companies from the tendering pool, rather than deciding who is to be awarded the final tender. Nevertheless, it is probably best used as an administrative measure to provide an incentive to organizations to adopt SMSs, as we discussed in Chapter 1.

New sanctions in the pyramid

All of the major corporate sanctions outlined above should be available to higher level courts where prosecutions are taken at the top of the enforcement pyramid. For lower court prosecutions further down the enforcement pyramid, corporate sanctions might be restricted to fines, community service orders, internal discipline orders, and organizational reform orders, because it would undermine the possibility of escalating sanctions within the enforcement pyramid to have all these sanctions available at a lower level in the pyramid. OHS offenders must be clear that not only will an escalating enforcement response involve prosecution in a higher level of court, but that a higher level prosecution will result in more serious sanctions. Sanctions such as adverse publicity orders, punitive injunctions, and dissolution are very big sticks which should be available only at the top of the pyramid.

Structuring sentencing discretion

In most jurisdictions, courts have a very broad discretion in sentencing OHS offenders, and receive little guidance. In his study of OHS prosecutions in Victoria, Australia, Johnstone argues that most OHS prosecutions tend to be taken where there has been an injury or fatality, and that instead of the case focusing on the defendant's failure to establish and maintain a safe system of work, the evidence examines the intricate details of the events leading up to the incident causing injury or death.

This 'event focus' of most OHS prosecutions enables the defendant's representatives to decontextualize OHS offences and to make them appear far less serious than they actually were when presenting arguments in mitigation of penalty. The fact that the prosecution is built around an 'event' immediately focuses the court's attention on the incident itself, rather than on the underlying system of work, the defendant's OHS management systems, or potential hazardous outcomes. The decontextualization is exacerbated by the fact that, in the typical case, the defendant pleads guilty, and the prosecution involves the prosecutor or OHS inspector giving a brief summary of the events giving rise to the prosecution and details of the defendant's previous convictions, followed by the defendant's plea in mitigation which usually raises a combination of factors designed to reduce the defendant's culpability.[116]

Johnstone's study examines in some detail the mitigating factors raised by the defendant, and the consequences of the court's lack of information in countering these factors. Once the event is drawn out of its context, the defendant can focus the court's attention on the minute details of the event (rather than on the underlying system of work) and can use a number of very common arguments to further isolate the event from its OHS context. For example, it is common for the defendant to argue that the injured worker contributed to the accident which led to the injury, or that the machine upon which the worker was injured was supplied without guards, so that in each of these cases the court should find the employer less culpable and impose a lower penalty. A further argument, which is put in the vast majority of cases, is that the defendant is a 'good corporate citizen', with an unblemished record and good attitudes to OHS. Generally it is very difficult for the prosecutor to challenge the 'good corporate citizen' argument because the law gives the prosecutor a very limited role in the sentencing process, and, even if the prosecutor does intervene to counter this argument, the prosecutor needs to have detailed information about the defendant's OHS record, information which is difficult to assemble for each prosecution. The English Law Commission[117] and Johnstone's research suggest, too, that very often the 'good corporate citizen' plea is a distortion of the true position, because the focus is usually on the absence of prior convictions, rather than on an absence of previous injuries, fatalities, or near misses, and ignores the fact that often low prior convictions are due to low OHS prosecution rates in general. The 'good corporate citizen' argument provides the basis for another common argument in mitigation of penalty, namely that the event giving rise to the prosecution was a 'freak accident' or 'one-off' event. In other words, the defendant has an unblemished approach to OHS, but, on this occasion, the exceptional, the unforeseeable, and hence unpreventable, had occurred.

A final set of common arguments in mitigation ensure that the event leading to the injury is isolated in the past, and so the court has no cause to impose a penalty to deter, rehabilitate, or punish the defendant, because the wrong has been corrected. The point here is that once again attention is drawn away from the defendant's OHS management systems and performance, and focused on the event leading to the accident, which is then pushed into the past with an assurance that it will not happen again, or that all the damage has been patched up to the best of the defendant's ability.

The implications of this analysis are very important, because the range of commonly used mitigating factors will generally ensure that the court assesses the defendant's offence to be at the lower end of the sentencing range.[118] The international data also suggests that there are serious inconsistencies between the courts, particularly the lower level courts, in the penalties imposed on OHS offenders. A further OHS sentencing issue arises if a wider range of sanctions, as outlined above, is introduced into the OHS statutes. How are courts to be guided in the use of these sanctions? The issue of disparity is likely to become more acute if sentencing courts not only have discretion to impose differing levels of fine, but also have a range of sanctions from which to choose.

In the remainder of this Chapter we examine a range of possible measures which can be used to ensure that, during the sentencing process, the courts remain focused on the essence of the systems-based OHS contravention before them, and impose penalties which are sufficient to deter OHS offences, provide incentives to OHS duty holders to introduce appropriate OHS management systems, and are consistent across the board.

What can be done to improve sentencing OHS offenders? The empirical studies referred to earlier in this section suggest that one of the problems with sentencing OHS offenders, at least in adversarial systems such as the UK and Australia with a tradition of prosecutorial passivity at the sentencing stage of the trial, is that the defendant corporation tends to control the information available to the court. Bergman, discussing the UK system,[119] points out that, while in mainstream criminal prosecutions the court has access to any relevant information about sentencing, which will be gained from the police evidence and the probation officer's social inquiry report, none of these processes is followed in relation to corporate offenders. The court remains ignorant about crucially important aspects of the company, such as its annual profits, turnover, health and safety record, compliance record, and history of co-operation with the OHS agency. Bergman suggests that this makes it impossible for the court 'to calculate an appropriate and just fine'.[120] As Johnstone's research suggests, the court also has very little information to help it to respond to the isolation techniques raised regularly by defendants.

OHS agencies can play a role in developing detailed profiles of each OHS offender, so that the prosecutor can inform the court of the offender's resources, OHS record (injuries, claims, incidents, notices issued, and previous convictions), record of involving employees in OHS policy development; and OHS systems generally. OHS inspectors should routinely gather such information which can then be used to determine the appropriate enforcement approach in a particular case in accordance with the enforcement pyramid; and inform the court at the sentencing stage, or challenge the veracity of factors argued in mitigation of penalty by the defendant.[121]

Further, in Anglo-Australian jurisdictions, prosecutors in all forms of criminal trial do not generally play an active role in the sentencing process. In some jurisdictions, prosecutors tend to see their role as answering questions asked by the courts about relevant sentencing principles and to correct errors of fact in defence submissions.[122] One way of overcoming the tendency to decontextualize OHS offences in the sentencing process is to ensure that OHS prosecutors, wherever possible, seek to play a more active role in the sentencing process. In particular, prosecutors should ensure that a complete picture of a defendant's OHS record is before the court, especially where defence counsel has argued in mitigation that this record is exemplary.[123] OHS prosecutors might also try to include in the material presented to the court a corporate defendant's prior convictions for OHS offences in other jurisdictions, and prior convictions for offences similar to OHS offences—for example, environmental offences.

Bergman[124] has argued that one solution to the problem of inadequate sentencing information would be the introduction of a procedure such as the 'corporate inquiry' report, which would be compiled by a court official and presented to the sentencing court. An example of this procedure is to be found in the USA's Sentencing Reform Act 1984, which requires a Federal probation officer to conduct a pre-sentence investigation into each convicted company, to determine the relevant sentencing factors, the appropriate fine to be imposed, and which of the other sanctions discussed earlier in this Chapter might be imposed. In particular, the corporate inquiry report might be supplemented by the more detailed report providing information about the suitability of the corporate probation sanction, again discussed earlier in this Chapter.

This corporate inquiry report might easily be introduced in some jurisdictions by adapting existing mechanisms. For example, in New South Wales, Australia, judges and magistrates can request pre-sentence reports to assist them in the sentencing process. The court sentencing an OHS offender could ask for an independent pre-sentence report to be prepared by an OHS consultant with expertise in the defendant's industry. The consultant would be given specific guidelines to put together a detailed analysis of the defendant's financial resources; OHS record (prior convictions, injuries, fatalities, OHS incidents, near misses, penalty, improvement, and prohibition notices); and an assessment of the defendant's general OHS management systems, and an indication of when these systems were implemented. The defendant would have an opportunity to dispute (with evidence) the statements made in the pre-sentence report. The court could then determine the appropriate penalty from its own assessment of the severity of the contravention.[125]

A controversial innovation, perhaps even more controversial in the context of 'preventive', inchoate, regulation such as OHS legislation, is the use of victim impact statements, which give victims of offences the right to participate in the sentencing process.[126] Some jurisdictions have statutory provisions enabling victims of offences to make, and be cross-examined on, oral or written statements to the court about the nature and significance of the offence.[127] Critics argue that the procedure does little to reduce sentencing disparities; it encourages the introduction of irrelevant or prejudicial matters; and it inflicts additional economic and physical burdens and further psychological damage upon the victim and the victim's family.[128] From an OHS perspective, victim impact statements have the advantage of helping the court to appreciate the seriousness of the injury which befell an injured worker. The process, however, increases the 'event' focus of the prosecution, and is only of use in prosecutions resulting from injuries to workers.

Although corporate inquiry reports and victim impact statements will go some way towards providing the sentencing court with appropriate sentencing information, these devices do not provide clear guidance to the court as to what it should do with the information, and do not provide transparent processes which indicate in advance to OHS duty holders the likely sentencing consequences of contravening the OHS statute. These effects are more likely to be achieved by sentencing guidelines.

Developing sentencing guidelines for OHS prosecutions utilizing a wider range of sanctions

Earlier in this Chapter we argued that perhaps the most significant problem with sentencing OHS offenders is that during the sentencing process OHS issues are decontextualized, transformed, and individualized, so that defendants appear to the court to be at a low level of culpability for the offence. One solution to this phenomenon is to develop sentencing guidelines for OHS offenders.[129]

OHS sentencing guidelines should aim to ensure that the sentencing court stays focused on the essence of the offence, namely the offender's departure from the OHS standards laid down in the relevant OHS statute, particularly general duty provisions which emphasize the need for a *safe system* of work. This requires, at the minimum, that the OHS sentencing discretion needs to be guided (even reshaped) in a *qualitative* way to ensure that the court is not diverted from the essence of the offences by mitigating factors raised by defence counsel which decontextualize and individualize the offences.

In our model of OHS regulation sentencing guidelines would perform a second function. Such guidelines would make the court's sentencing processes more transparent, and would signal clearly to OHS duty holders that a co-operative approach to compliance and attention to OHS programmes would not only reduce the likelihood of their being prosecuted in the first place, but would mitigate any penalty if they were to be prosecuted.

There are very few models of sentencing guidelines for corporate defendants upon which OHS sentencing guidelines can be based. Fortunately, we are provided with considerable assistance by the recently introduced USA Sentencing Commission's fourth draft of its *Sentencing Guidelines for Organizations*, which became law in 1991,[130] and the controversial *Draft Sentencing Guidelines for Environmental Crimes* presented to the

Sentencing Commission in 1991 by an advisory group.[131] While these are very recent and relatively untested innovations, they model an approach to sentencing that is consistent with our overall model of the enforcement pyramid, particularly the 'carrot and sticks' approach to encouraging voluntary compliance in the shadow of serious sanctions for non-compliance.[132] We will briefly outline the salient characteristics of these guidelines, and then sketch our own model of OHS sentencing guidelines.

Both the guidelines for organizations and environmental guidelines set out mandatory sentencing guidelines. The guidelines considerably increase the penalties that corporations will face for crimes, and endorse a range of corporate sanctions beyond fines, some of which (for example, corporate probation) institutionalize in the sentencing process prevention of crime through internal monitoring and rehabilitation through corporate restructuring.[133]

Both sets of guidelines specify that the calculation of a corporate *fine* is based on two factors: a 'base fine' (reflecting the seriousness of the type of offence), and a multiplier (the 'culpability score', which reflects the culpability of the organization), which together produce a 'guideline range fine'. Once calculated, the fine range can be increased or decreased using the organization's culpability score, which can go up or down depending on any aggravating or mitigating factors. Aggravating factors can include matters such as a prior history of similar misconduct by the organization, and participation in or condonation of the offence by senior members of management. The draft environmental guidelines made it an aggravating factor for the organization *not* to have a compliance plan. Mitigation factors aim to give organizations the incentive to 'maintain internal mechanisms for preventing, detecting, and reporting criminal conduct',[134] and might include the maintenance of an effective compliance programme to prevent and detect breaches of the law, self-reporting of offences, and full co-operation by the organization in the investigation of the offence. The draft environmental guidelines increase the discount for compliance plans, but try to eliminate 'paper' compliance programmes by requiring the organization to satisfy several exacting requirements before it can qualify for the discount.[135]

The guidelines also indicate how sanctions such as corporate probation, adverse publicity, and dissolution should be implemented. For example, under the 1991 Guidelines for Organizations a court must impose probation if, amongst other things, the organization is unable to pay a monetary fine; or it has more than fifty employees and it did not have an effective programme to prevent and detect violations. If the organization does not have an effective programme to detect and prevent contraventions, it may be required to submit one for the court's approval.

To date there is little empirical evidence as to the operation of the guidelines in the USA. Data up to the end of June 1995 suggested that the 208 corporations[136] by then sentenced under the guidelines had not been very successful in mitigating their penalties under the guidelines.[137] Thirteen per cent received no mitigation at all. Almost all of the remaining cases accepted responsibility for their actions, and 64 per cent co-operated with the investigation. Only three organizations (each already under investigation) voluntarily reported their offences. In just over half of the 208 cases, fines were increased because of the knowledge or involvement of high-level executives. Sixty-one per cent of convicted organizations were placed on corporate probation, and 14 per cent were ordered to implement compliance programmes. Compliance programmes have also been key features of settlement agreements between alleged corporate offenders and enforcement agencies in a number of industries. Some commentators suggest that compliance programmes are 'proving to be more burdensome and counterproductive than originally anticipated'.[138]

Of major interest to OHS regulators is the experience of the Guidelines for Organizations in inducing organizations to develop effective compliance programmes.[139] Only four organizations claimed to have effective compliance programmes, and still only one (an unusual case unlikely to serve as a model) was successful.[140] This low use of compliance programmes in mitigation occurred despite survey evidence that nearly two-thirds of 300 organizations surveyed in 1995 had either developed compliance programmes or improved existing programmes in response to the guidelines.[141] Measures taken included teaching employees about the organization's compliance programme, including compliance in employee's performance appraisal (two-thirds of organizations surveyed), reporting violations internally, internal audits (40 per cent of organizations surveyed), and enforcement of compliance programmes. This data is disappointing, and suggests that OHS policy makers intending to develop OHS sentencing guidelines should pay careful attention to ensuring that there are clear incentives in the guidelines to business organizations to adopt effective compliance programmes and OHS management systems.

Even though the guidelines have been strongly criticized, in particular for having overly severe penalties, for inadequate flexibility in sentencing, for ignoring organizational efforts which should mitigate penalties,[142] both of these sets of guidelines provide very useful models upon which to develop OHS sentencing guidelines. They provide examples of incentives to entice business organizations to comply with the law, through a mix of new sanctions (including corporate probation, adverse publicity, community service, and dissolution) and potentially high fines, ensuring that organizations are unlikely to simply consider fines to be a cost of doing business. Of particular importance for OHS crime is the potential place of the defendant's OHS compliance programme and OHS management system in the sentencing regime. The environmental sentencing guidelines show how sentencing can be used to provide strong incentives to organizations to develop SMSs, while at the same time ensuring that efforts to develop these systems are authentic, and not simply 'paper systems'.[143] In Chapter 3 we suggested that a major incentive for organizations to adopt SMSs is a reduction in the potential fine that they would face if for some reason things went wrong, and they were prosecuted for OHS contraventions.[144] Both sets of guidelines show how the offender's SMS could be a major factor in mitigating the penalty a Track One or a Track Two offender receives for a contravention of an OHS statute. The guidelines also show how organizations can be provided with incentives to monitor their own OHS management systems for deficiencies, remedy those deficiencies, and then report on the deficiency and the remedy to the OHS inspectorate. Organizations can use this proactive and internal self-regulation to mitigate penalties imposed in the event of a later prosecution. But, as the empirical data illustrates, there is a delicate balance to be struck between using guidelines to encourage compliance programmes, or SMSs, and making sure that such programmes and systems are sufficiently effective to warrant a sentencing discount.

As a starting point, OHS sentencing guidelines might emphasize exactly what the essence of the offence was: for Track One firms the failure to provide and maintain a system of work, in a very broad sense, so that it is safe and without risks to health. The essence of the offence for Track Two firms which had failed to implement SMSs would be both the failure to introduce SMSs and the breach of the trust placed in them by the OHS agency. The guidelines should then provide broad principles to guide the court in maintaining the essence of the offence in the sentencing process, indicating what the purposes of sentencing are, what factors are relevant to increase or decrease the penalty imposed, and the range of penalties which might be appropriate in each case. As our discussion of OHS sentencing in Australia and the UK earlier in this Chapter suggests, two crucial issues in sentencing OHS offenders are: (i) giving proper credit (for example, in reduced penalties) to organizations which have acted in good faith and have effective systems for ensuring compliance with OHS standards, and who go further and introduce SMSs; and (ii) ensuring that organizations' claims that they are 'good corporate citizens' in that sense are properly scrutinized by the courts.

We suggest that OHS sentencing guidelines might:

- clearly specify the purposes of sentencing OHS offenders—a mixture of specific and general deterrence, rehabilitation (in the guise of internal restructuring within organizations), with sub-goals including retribution (particularly in the case of manslaughter prosecutions) and denunciation;
- indicate that the sentencing process is to be seen in the context of the two-track approach, and of the 'enforcement pyramid', which we outlined in earlier Chapters;
- focus the attention of the courts on 'the essence of the offence', by including, in a schedule to the OHS statute, judicial training, or court manuals, illustrations of what Parliament considers to be a good approach to OHS;
- specify which factors are *not* to be raised by the defendant, and considered by the court, in passing sentence;
- set out the kinds of *mitigating* and *aggravating* factors which the court *can*, and should, consider in reducing or increasing the penalties imposed in the particular case; and
- set out *presumptive* guidelines which:

  — indicate when fines, corporate probation, adverse publicity orders, community service orders, and dissolution are to be used, and how;
  — indicate to the court the presumed range and severity of sanctions in each case;
  — specify that the court can depart from the presumed range if there are good reasons for doing so; and
  — ensure that the sanctions are tailored to the resources of the defendant, along the principles of the 'day fine' system which we discussed earlier in this Chapter,[145] so as have the desired impact

on the offender but not bankrupt the offender. Sanctions other than the fine should be used where there is danger of a deterrence trap.

We do not suggest that sentencing guidelines introduce the detailed and complex 'base fine' and 'culpability score' approach of the USA guidelines, as these impose restrictions on courts which would be far too inflexible in, for example, Anglo-Australian jurisdictions. But penalties should be tailored to the defendant's resources. The OHS legislation should require all convicted offenders to produce income statements and balance sheets for the preceding five years, so that the court is able to assess accurately the organization's resources. Once the level of fine reflects the offender's resources, the starting point for the fine for a mid-range offence[146] should be, say, 50 per cent of the maximum fine for an organization with the offender's resources. The starting point should be higher for a serious contravention, and lower for a relatively minor contravention.

We suggest that the *impermissible* sentencing factors should include the role of the injured worker or other workers, unless there is evidence that a superior manager was solely responsible for the offence;[147] reference to previous inspections by OHS inspectors in relation to which the defendant argues that the inspectors omitted to point out the hazard to the defendant; a failure of the supplier of plant or substances to comply with the provisions of the OHS statute; and unsubstantiated assertions that the offender was a 'good corporate citizen' with an excellent OHS record.

Acceptable *mitigating* factors (which will reduce the fine to below 50 per cent for a mid-range offence) should include clear evidence of an effective[148] OHS management system or a compliance programme to prevent and detect breaches of the OHS Act, which were in existence before the offence took place (a lesser reduction might be made when the programme is not effective, but nevertheless demonstrates efforts made in good faith to comply with OHS obligations); evidence of full co-operation by the organization in the investigation of the offence; evidence that the organization accepted responsibility for its criminal conduct and immediately rectified the contraventions and introduced measures to ensure that were not repeated; and evidence that on previous occasions the offender had identified OHS breaches and had notified the OHS inspectorate of these breaches and the steps the offender had taken to remedy the breaches.

Aggravating factors (which would increase the fine) should include situations where the defendant came to court without having remedied the circumstances which led to the offence; evidence that the OHS offence was committed after the offender's management was notified of the circumstances which led to the contravention (for example, from complaints by workers, improvement or prohibition notices issued by inspectors, previous incidents of a similar nature which indicated that there was a potential hazard, or where there were codes of practice covering the particular hazard); evidence that the offender has a prior history of similar misconduct over the previous ten years; evidence that the organization had not in the past addressed its mind to developing any form of OHS management system; and evidence that the OHS contravention took place in order to save money and to make a profit.

The guidelines should give the court discretion, in appropriate cases, to substitute a fine with an order that the offender perform community service, the financial value of which amounts to the level of the appropriate fine, in an approved community project.

For most OHS contraventions, the guidelines should include a presumption that the court will make an order of an appropriate form of corporate probation, which will include internal disciplining procedures and the development of an effective compliance programme to prevent and detect breaches of the OHS statute. The defendant can present evidence to the court that such an order is unnecessary because: (i) an internal investigation has already taken place; and (ii) an effective OHS management system and a compliance programme to prevent and detect breaches of the OHS statute is already in place. Where there is evidence that the offender has a deficient OHS management system, a higher level court should, after seeking advice (in the form of a corporate inquiry report) from an expert in OHS management, have the discretion to order the offender to introduce an appropriate OHS management system.

The guidelines should also specify that there is a presumption that the higher level court can order the company to publicize the offence in an appropriate manner to the whole community or the sectors of the community affected by the offence. Where offenders have a bad history of OHS offences, and where there is strong evidence of a continuing danger to employees and to the public as a result of the offender's neglect of OHS management principles, the court might, in its discretion, order that the offender cease trading until appropriate OHS management systems are introduced, or that a court appointed trustee supervise the development of appropriate OHS management systems at the expense of the convicted organization. In the most extreme case, a higher level court can make an order that the offending company be wound up.

A court which decides not to follow these guidelines must give cogent reasons for departing from the guidelines. If the court fails to follow the guidelines in imposing sentence, and fails to provide cogent reasons for doing so, there should be a right of appeal to a superior court.

We have placed great store, in these guidelines, on encouraging organizations voluntarily to develop their own OHS compliance programmes and SMSs. One potential problem is that an organization may be unclear as to what form of OHS compliance programme will mitigate the penalty.[149] Gruner and Brown note that little work has been done on developing tests for identifying reasonable law compliance efforts in the USA's *Sentencing Guidelines for Organizations*.[150] This problem is not easily remedied because the form and content of a compliance programme will depend on a host of factors, including the size of the organization, and the industry in which it operates.[151] There are a number of possible sources of guidance for organizations.[152]

First, OHS agencies may introduce their own guidance notes or codes of practice, which outline the fundamental elements of an OHS compliance programme or a SMS.[153] The OHS agency could work with industry groups, with representatives from employers and trade unions, to develop 'best practice' approaches to compliance programmes and SMSs. Different guides can be produced for large, medium-sized, and small organizations within each industry. This may require a large investment of resources, and will not be able to be done in the short term. In the longer term, however, as OHS agencies gain greater knowledge and experience of OHS management systems and compliance programmes, and develop their own internal training packages and manuals, they should be able to make their expertise available to organizations.

Second, OHS prosecutors may produce prosecutorial directives (for example, the USA Department of Justice Prosecutorial Directives for Environmental Offences) which encourage self-auditing, self-policing, and self-reporting, and which show how prosecutors take factors such as compliance programmes and OHS management systems into account when exercising prosecutorial discretion.[154] Again, this is not a short-term solution, but in the longer term a body of prosecutorial directives may become available to guide organizations.

Third, best practice approaches to OHS compliance programmes and SMSs can be developed within industry, to guide parties to design and implement the best possible compliance programmes.[155] Finally, organizations might ask the OHS agency to review their compliance programmes or SMSs before any alleged contravention occurred.[156] Many OHS agencies may not have the resources to do this. One solution might be to outsource this function to private consultants, on a fee for service basis. The private consultant would conduct the review accompanied by worker representatives. The OHS agency would reserve the right to prosecute if a subsequent OHS contravention occurred. The organization would not be able to argue that the approval of the OHS agency or consultant was a 'defence',[157] but could bring evidence of the compliance programme and its approval in mitigation of penalty.

Guidelines also need to ensure that organizations are not reluctant to introduce a compliance programme or an OHS management system because of a fear that the results of internal audits, self-reporting, and internal investigations may be used against them if a prosecution is actually conducted, or in fact to gather evidence to bring a prosecution in the first place.[158] Self-monitoring and self-reporting is a crucial aspect of any effective compliance programme, and the use of compliance programmes may need to be supported by selective amnesties or privileging information gathered from internal investigations, internal audits, or self-reporting (self-evaluative privilege).[159]

We conclude with a note of caution in relation to the political complexities of introducing sentencing guidelines. Overseas experience suggests that sentencing guidelines should be developed and introduced with the full support of all parties affected by the guidelines. Tonry's survey of the American literature suggests that standards for sentencing can change the sentences courts impose and can reduce disparities, but unless they are flexible enough to allow judges to 'treat different cases differently', they are likely 'to drive discretion underground as judges and lawyers try to achieve sentences that those involved agree are reasonable'.[160] Tonry further observes that the empirical literature on sentencing policy in Western legal systems suggests that planning is crucial to the effective implementation of sentencing guidelines which will change judges' behaviour and sentencing outcomes, and that, as different jurisdictions have different kinds of problems, policy makers should avoid crude transplants of sentencing regimes from one system to another.[161]

A major difficulty facing policy makers in countries without a tradition of sentencing guidelines (for example, the UK and Australia) is that there is likely to be concerted judicial resistance to the notion of guidelines in the first place. But, at the same time, we discern a desire amongst judicial officers for some help in grappling with the new issues raised in sentencing OHS offenders. Guidelines in those countries should be introduced only after consultation with the judiciary, and could be discretely introduced in the OHS statutes themselves, in, for example, a schedule to the OHS statute, where principles are set out to guide courts in the calculation of fines, and the circumstances under which new sanctions should be used. In countries with a tradition of judicial education, the introduction of the guidelines could be incorporated into the ongoing educational process. Another alternative is to work with the judiciary to incorporate the guidelines into the judiciary's own internal guidance notes given to judges to assist in sentencing.

Conclusion

In this Chapter we have outlined a series of proposals for tough sanctions at the top of the enforcement pyramid. In Chapter 4 we argued that an OHS enforcement strategy should be based on an assumption that organizations are virtuous, and that incentives and persuasion will be sufficient to ensure voluntary compliance with OHS standards. But where such compliance is not forthcoming, strong deterrent, remedial, and punitive measures are required.

We have also outlined a series of measures aimed at providing the courts with a flexible range of sanctions which can ensure that OHS offenders are sufficiently deterred from committing OHS offences, and remedy those offences when they do commit them. On occasion, the court can go further, to require an offender to adopt a SMS. We have suggested that on occasion equity fines and community service orders may provide better justice in a particular case. Extreme measures, such as dissolution or suspension of trading, may be taken by a high-level court at the very top of the pyramid. We recognize that, if the top of the pyramid is to act as a general deterrent, there needs to be greater publicity given to successful OHS prosecutions. Such publicity can be ensured by OHS agency publications and, at the top of the pyramid, publicity orders.

Finally, we have argued that, in order to ensure that penalties consistently reflect the essence of the offence, sentencing guidelines be developed by all parties involved in OHS prosecutions. The guidelines will also ensure the transparency of the tough penalties at the top of the pyramid, which will increase their deterrent value.

NOTES

1 See N. Gunningham, *Safeguarding the Worker* (Law Book Company, Sydney, 1984), Ch. 4; R. Johnstone, *The Court and the Factory: The Legal Construction of Occupational Health and Safety Offences in Victoria*, unpublished Ph.D. thesis, The University of Melbourne, 1994, Chs. 1 and 2; and Industry Commission Australia, *Work, Health and Safety* (AGPS, Canberra, 1995), Vol. I, Ch. 7.

2 D. Bergman, 'Corporate sanctions and corporate probation' (1992) 25 *New Law Journal* 1312 at 1312.

3 B. Fisse, 'Sentencing Options against Corporations' (1990) 1 *Criminal Law Forum* 211 at 212–13. For a brief discussion of the costs of workplace injuries and disease, see Health and Safety Executive, *The Costs of Accidents at Work* (HMSO, London, 1993); F. Van Waarden, J. den Hertog, H. Vinke, & T. Wilthagen, *Prospects for Safe and Sound Jobs: The impact of future trends on costs and benefits of occupational safety and health* (Dutch Ministry of Social Affairs and Employment, Gravenhage, 1997), Ch. 2; National Board of Occupational Health and Safety in Sweden, 'The Costs of Ill-health' (1992)(2) *Newsletter* 1; Industry Commission Australia, *Work, Health and Safety* (AGPS, Canberra, 1995), Vol. II, Appendix C; and R. Johnstone, *Occupational Health and Safety Law and Policy: Text and materials* (LBC Information Services, Sydney, 1997), pp. 19–21.

4 See also our brief discussion of the disqualification of directors in Ch. 6.

5 C. D. Stone, 'Corporations and Law: Ending the Impasse', in J. M. Johnson and J. D. Douglas (eds.), *Crime at the Top: Deviance in Business and the Professions* (J. B. Lippincott Company, Philadelphia, 1978), pp. 357–8. See also R. Fox, 'Corporate Sanctions: Scope for a New Eclecticism' (1982) 24 *Malaya Law Review* 26 at 27; and M. B. Mentzger & C. R. Schwenk, 'Decision Making Models, Devil's Advocacy, and the Control of Corporate Crime' (1990) 28 *American Business Law Journal* 323 at 326.

6 See Australian Law Reform Commission, *Sentencing Penalties*, Discussion Paper No. 30 (AGPS, Canberra, 1987), pp. 168–70.

7 See J. Braithwaite & P. Grabosky, *Of Manners Gentle: Enforcement Strategies of Australian Business Regulatory Agencies*, (Oxford University Press and the Australian Institute of Criminology, Melbourne, 1986); and Industry Commission Australia, *Work, Health and Safety* (AGPS, Canberra, 1995). See also Robens Committee (Committee on Safety and Health at Work), *Report of the Committee on Health and Safety at Work 1970–1972* (HMSO, London, 1972), p. 81.

8 In 1990, the British Health and Safety Commission referred to the 'undue

lenience' of the courts in imposing fines (see Health and Safety Commission, *Health and Safety Commission Annual Report 1988/89* (HMSO, London, 1990)). Since then, there have been some exceptionally high fines in individual cases, a few as high as £250,000, a couple of £200,000, and a few over £100,000. See S. Tombs, 'Law, Resistance and Reform: "Regulating" Safety Crimes in the UK' (1995) 4 *Social and Legal Studies* 343 at 354. See also European Foundation for the Improvement of Living and Working Conditions, Walters, D. (ed.), *The Identification and Assessment of Occupational Health and Safety Strategies in Europe*, Vol. 1 (Office for the Official Publications of the European Communities, Luxembourg, 1996), pp. 196 and 201. Late in 1997 the level of fines imposed by the lower courts in the UK was described as 'derisory' by Angela Eagle, the Labour government minister responsible for health and safety (A. Eagle, 'Raising the Profile of Health and Safety at Work' (1997) Dec. *Health and Safety* 8).

9   T. Wilthagen, 'Reflexive Rationality in the Regulation of Occupational Health and Safety', in R. Rogowski and T. Wilthagen, *Reflexive Labour Law* (Kluwer, Deventer, 1994), p. 366, n. 48. For Sweden, see Swedish National Board of Occupational Health and Safety (1995) 2 *Newsletter* 15.

10 European Foundation for the Improvement of Living and Working Conditions, 'Danish building and construction sites are hazardous workplaces' (1997) Apr. *Euroline*.

11 S. Shapiro & R. S. Rabinowitz, 'Punishment versus Cooperation in Regulatory Enforcement: A Case Study of OSHA' (1997) 49 *Administrative Law Review* 713 at 743.

12 T. H. McQuiston, R. C. Zakocs, & D. Loomis, 'The Case for Stronger OSHA Enforcement—Evidence from Evaluation Research' (1998) 88 *American Journal of Public Health* 1022 at 1022.

13 The following criticisms of the fine are drawn from B. Fisse, 'Sentencing Options against Corporations' (1990) 1 *Criminal Law Forum* 211 at 214–29; B. Fisse, 'Individual and Corporate Criminal Responsibility and Sanctions Against Corporations', in R. Johnstone (ed.), *Occupational Health and Safety Prosecutions in Australia: Overview and Issues* (Centre for Employment and Labour Relations Law, The University of Melbourne, 1994), p. 100, at pp. 103–4; and Australian Law Reform Commission, *Sentencing Penalties*, Discussion Paper No. 30 (AGPS, Canberra, 1987), pp. 168–70. See also M. B. Mentzger & C. R. Schwenk, 'Decision Making Models, Devil's Advocacy, and the Control of Corporate Crime' (1990) 28 *American Business Law Journal* 323 at 327–32; and Note, 'Criminal Sentences for Corporations: Alternative Fining Mechanisms' (1985) 73 *California Law Review* 443. See also A. Cowan, 'Note: Scarlett Letters for Corporations? Punishment by Publicity Under the New Sentencing Guidelines' (1992) 65 *University of Southern California Law Review* 2387 at 2391.

14 B. Fisse, 'Individual and Corporate Criminal Responsibility and Sanctions Against Corporations', in R. Johnstone (ed.), *Occupational Health and Safety Prosecutions in Australia: Overview and Issues* (Centre for Employment and Labour Relations Law, The University of Melbourne, 1994), p. 103.

15 A. Norrie, *Crime, Reason and History: A Critical Introduction to Criminal Law* (Weidenfeld & Nicolson, London, 1993); and also see L. H. Leigh, *The*

Criminality of Corporations in English Law (Weidenfeld & Nicolson, 1969), pp. 140–3, 160–1.

16 We note that the tendency of fines to penalize innocent shareholders has been a major reason cited by courts for not penalizing corporations heavily: see, for example, *R* v. *Wattle Gully Gold Mines N.L.* [1980] VR 622. See also R. Fox, 'Corporate Sanctions: Scope for a New Eclecticism' (1982) 24 *Malaya Law Review* 26 at 33; and B. Fisse, 'Reconstructing Corporate Criminal Law: Deterrence, Retribution, Fault and Sanctions' (1983) 56 *Southern California Law Review* 1141 at 1237 and the sources he cites. Bergman suggests that shareholders may not be as innocent as workers and consumers because (i) they 'own' the company, and should take responsibility for its decisions; (ii) they benefit from corporate profits gained by diverting resources away from OHS; and (iii) there might be a significant deterrent effect in that fear of reduced dividends may induce shareholders to demand more information about the company's OHS performance when making investment decisions, which might lead directors to devote resources to improving OHS performance. See D. Bergman, 'Corporate sanctions and corporate probation' (1992) 25 *New Law Journal* 1312 at 1313. See also M. B. Mentzger & C. R. Schwenk, 'Decision Making Models, Devil's Advocacy, and the Control of Corporate Crime' (1990) 28 *American Business Law Journal* 323; and A. Cowan, 'Note: Scarlett Letters for Corporations? Punishment by Publicity Under the New Sentencing Guidelines' (1992) 65 *University of Southern California Law Review* 2387.

17 Workers and consumers might conceivably suffer a 'double injury'—the ill effects from hazards, and bearing the cost of the penalty. Economists would argue that the process of passing on costs to consumers is appropriate, because it makes the product more expensive and reduces demand for it, thus affecting company profits.

18 M. B. Mentzger & C. R. Schwenk, 'Decision Making Models, Devil's Advocacy, and the Control of Corporate Crime' (1990) 28 *American Business Law Journal* 323 at 331–2.

19 See J. C. Coffee, '"No Soul to Damn: No Body to Kick": An Unscandalized Inquiry into the Problem of Corporate Punishment' (1981) 79 *Michigan Law Review* 386 at 389–92; and A. Norrie, *Crime, Reason and History: A Critical Introduction to Criminal Law* (Weidenfeld & Nicolson, London, 1993), p. 103.

20 See, generally, B. Fisse, 'Sentencing Options against Corporations' (1990) 1 *Criminal Law Forum* 211 at 215–19.

21 See, generally, N. Gunningham, *Safeguarding the Worker* (Law Book Company, Sydney, 1984), p. 327; Industry Commission Australia, *Work, Health and Safety* (AGPS, Canberra, 1995), Vol. I pp. 116–17; and R. Laing, *Report of the Inquiry into Operations of the Occupational Health and Safety and Welfare Act 1984* (Western Australian Government, Perth, 1992), p. 220.

22 Some Canadian statutes (for example Ontario since 1991 and North East Territories) have maximum penalties for corporations of C$500,000, but others have maxima of $50,000 and lower. In *R* v. *Cotton Felts Ltd.* (1982), 2 CCC (3d) 287 the Ontario Court of Appeal intimated that fines should be set at levels to achieve a deterrent effect.

23 Pursuant to a 1989 Review of OHS legislation.

24 Pursuant to a 1995 Review of its OHS legislation.

25 Report (1997) rec. 41, p. 86.

26 Roughly £230,000 and £45,000, or US $350,000 and US $70,000.

27 Industry Commission, Australia, *Work, Health and Safety* (AGPS, Canberra, 1995), Vol. I, p. 117.

28 N. Gunningham, *Safeguarding the Worker* (Law Book Company, Sydney, 1984), p. 328.

29 See Occupational Health and Safety Act 1996 (Nova Scotia), s. 74.

30 Note, 'Criminal Sentences for Corporations: Alternative Fining Mechanisms' (1985) 73 *California Law Review* 443.

31 See below pp. 275–76.

32 M. B. Mentzger & C. R. Schwenk, 'Decision Making Models, Devil's Advocacy, and the Control of Corporate Crime' (1990) 28 *American Business Law Journal* 323 at 331.

33 Some courts (for example, in New South Wales) have tried to deal with the issue by looking through the corporate veil to determine whether the corporation is truly a corporation within the meaning of the OHS statute, or whether in fact it was an individually run business incorporated for the convenience of bookkeeping or tax purposes. See, for example, *Mauger v. Kremar Engineering Pty. Ltd.* (1993) 47 IR 359. Recent inquiries into OHS regulation in New South Wales have strongly criticized the penalty differentiation between corporate and non-corporate penalties, and have suggested, instead, that a system of graduated penalties be developed. See R. C. McCallum (Chair), *Review of the Health and Safety Act 1983: Final Report of the Panel of Review* (New South Wales Government, Feb. 1997), p. 104; and Parliament of New South Wales, Legislative Council, Standing Committee on Law and Justice, *Report of the Inquiry into Workplace Safety*, Interim Report, Report No. 8 (New South Wales Government, Dec. 1997), pp. 58–9.

34 Most courts are required to take into account the means of the defendant when imposing a penalty. We suspect, though, from our observations of the sentencing practices of Australian lower courts, that this process is often, if not generally, cursorily undertaken when courts sentence OHS offenders, and that the information about the defendant's resources before the court is often patchy, and untested.

35 See R. Fox, 'Corporate Sanctions: Scope for a New Eclecticism' (1982) 24 *Malaya Law Review* 26 at 38.

36 For example, Sweden, Finland, Denmark, Germany, Portugal, Greece, Austria, and Hungary: see J. Junger-Tas, *Alternatives to Prison Sentences: Experiences and Developments* (RDC—Dutch Ministry of Justice/Krugler Publications, Amsterdam, 1994), pp. 22–5. Since 1974, s. 153a of the German Code of Civil Procedure has provided for a day fine system as in Sweden. The legislation sets out maximum and minimum daily amounts which can be imposed, and multiplies this figure by the gravity of the offence and the defendant's culpability. A major difference from the Swedish system is that in Germany accurate information about income is not available as it is in Sweden, and so assessments of means are approximate only, and are based on statements of offenders and details of employment given in police reports.

37 See, generally, Australian Law Reform Commission, *Sentencing of Federal Offenders*, Report No. 15 Interim (AGPS, Canberra, 1980), pp. 235–6; Justice, *Sentencing: A Way Ahead* (Justice, London, 1989), pp. 21–2; M. Tonry, *Sentencing Matters* (Oxford University Press, New York, 1996), pp. 125–6; and J. Junger-Tas, *Alternatives to Prison Sentences: Experiences and Developments* (RDC—Dutch Ministry of Justice/Krugler Publications, Amsterdam, 1994), pp. 22–5.

38 The day fine is generally calculated at one-thousandth of the offender's annual gross national income, less expenses incurred because of employment. Further adjustments are made for, amongst other things, people with high incomes (reductions because of progressive taxation); dependent spouses and children; for offenders with capital above a certain value (the size of the unit is increased); and offenders with large debts. All this information is collected by the police during their investigation of the offence, and then verified by the judge with the offender in court. The information can also be checked with income tax returns and the register of incomes (both of which are accessible public documents in Sweden).

39 In 1989 a successful pilot scheme was run by the Home Office Research and Planning Unit in selected towns in England and Wales to assess the benefits of a 'unit fine system' based on assessment of the offender's weekly disposable income, and with calculations based on a weekly, rather than daily, income. See Home Office, Research and Planning Unit, *Unit Fines: Experiments in Four Courts*, Paper 59 (Home Office, London, 1990); Justice, *Sentencing: A Way Ahead* (Justice, London, 1989), p. 22; and M. Tonry, *Sentencing Matters* (Oxford University Press, New York, 1996), p. 126. A national system of unit fines was introduced in 1992, but was soon abandoned after a series of media stories about offenders receiving ridiculously high sentences for minor offences, which discredited the whole system. M. Tonry, *Sentencing Matters* (Oxford University Press, New York, 1996), p. 127, notes that the problems experienced in Britain are all soluble, but that the British experience poses important warnings for other jurisdictions wanting to establish such a system.

40 A broader, more innovative range of sanctions was recommended by the Industry Commission, Australia, *Work, Health and Safety* (AGPS, Canberra, 1995), Vol. I, rec. 19.

41 See above p. 189–90.

42 R. Fox, 'Corporate Sanctions: Scope for a New Eclecticism' (1982) 24 *Malaya Law Review* 26 at 32–3.

43 F. Haines, *Corporate Regulation: Beyond 'Punish or Pursuade'* (Clarendon Press, Oxford, 1997), p. 222.

44 See A. Cowan, 'Note: Scarlett Letters for Corporations? Punishment by Publicity Under the New Sentencing Guidelines' (1992) 65 *University of Southern California Law Review* 2387.

45 F. Haines, *Corporate Regulation: Beyond 'Punish or Pursuade'* (Clarendon Press, Oxford, 1997), pp. 188–90 and 221.

46 See also Parliament of New South Wales, Legislative Council, Standing Committee on Law and Justice, *Report of the Inquiry into Workplace Safety*, Interim Report, Report No. 8 (New South Wales Government, Dec. 1997), pp. 64–5.

47 B. Fisse, 'Sentencing Options against Corporations' (1990) 1 *Criminal Law Forum* 211 at 240; and B. Fisse, 'The Use of Publicity as a Criminal Sanction Against Business Corporations' (1971) 8 *Melbourne University Law Review* 970.
48 Ibid.
49 See A. Cowan, 'Note: Scarlett Letters for Corporations? Punishment by Publicity Under the New Sentencing Guidelines' (1992) 65 *University of Southern California Law Review* 2387.
50 Occupational Health and Safety Act 1996 (Nova Scotia), s. 75.
51 M. Mialon, 'Safety at Work in French Firms and the Effect of the European Directives of 1989' (1990) 6 *The International Journal of Comparative Labour Law and Industrial Relations* 129 at 143.
52 Publicity orders have recently been recommended in New South Wales (see Parliament of New South Wales, Legislative Council, Standing Committee on Law and Justice, *Report of the Inquiry into Workplace Safety*, Interim Report, Report No. 8 (New South Wales Government, Dec. 1997), pp. 55–6).
53 See J. Braithwaite, *Crime, Shame and Reintegration* (Cambridge University Press, Cambridge, 1989).
54 B. Fisse, 'Sentencing Options against Corporations' (1990) 1 *Criminal Law Forum* 211 at 241, referring to P. French, 'Publicity and the Control of Corporate Conduct', in B. Fisse & P. French (eds.), *Corrigible Corporations and Unruly Law* (1985).
55 See, generally, National Commission on Reform of Criminal Laws, *Study Draft of a New Criminal Code* (Washington, 1970), § 405(1)(a); B. Fisse, 'The Use of Publicity as a Criminal Sanction against Business Corporations' (1971) 8 *Melbourne University Law Review* 107; B. Fisse, 'Reconstructing Corporate Criminal Law: Deterrence, Retribution, Fault and Sanctions' (1983) 56 *Southern California Law Review* 1141, at 1229–31; B. Fisse & J. Braithwaite, *The Impact of Publicity on Corporate Offenders* (State University of New York Press, Albany, New York, 1983); and Australian Law Reform Commission, *Sentencing Penalties*, Discussion Paper No. 30 (AGPS, Canberra, 1987), pp. 176–7.
56 J. C. Coffee, ' "No Soul to Damn: No Body to Kick": An Unscandalized Inquiry into the Problem of Corporate Punishment' (1981) 79 *Michigan Law Review* 386 at 427.
57 B. Fisse, 'Reconstructing Corporate Criminal Law: Deterrence, Retribution, Fault and Sanctions' (1983) 56 *Southern California Law Review* 1141 at 1230–1; and B. Fisse, 'Sentencing Options against Corporations' (1990) 1 *Criminal Law Forum* 211 at 243–4.
58 J. Braithwaite, *Crime, Shame and Reintegration* (Cambridge University Press, Cambridge, 1989), p. 60.
59 B. Fisse, 'Reconstructing Corporate Criminal Law: Deterrence, Retribution, Fault and Sanctions' (1983) 56 *Southern California Law Review* 1141 at 1230.
60 Industry Commission, Australia, *Work, Health and Safety* (AGPS, Canberra, 1995), rec. 35, Vol. I pp. 193–4; Robens Committee, *Report of the Committee on Safety and Health at Work 1970–72* (HMSO, London, 1972), p. 24.
61 See, generally, M. B. Mentzger & C. R. Schwenk, 'Decision Making Models, Devil's Advocacy, and the Control of Corporate Crime' (1990) 28 *American Business Law Journal* 323.

62 D. Bergman, 'Corporate sanctions and corporate probation' (1992) 25 *New Law Journal* 1312 at 1313.

63 American Bar Association, *3 Standards for Criminal Justice* (Little, Brown, Boston, 1980).

64 Australian Law Reform Commission, *Sentencing Penalties*, Discussion Paper No. 30 (AGPS, Canberra, 1987), p. 172.

65 Internal disciplinary orders were recommended in 1977 by the South Australian Criminal Law and Penal Methods Reform Committee. The proposal was based on an earlier proposal by the Canadian Law Reform Commission (1976). See also J. C. Coffee, ' "No Soul to Damn: No Body to Kick": An Unscandalized Inquiry into the Problem of Corporate Punishment' (1981) 79 *Michigan Law Review* 386 at 455–6; American Bar Association (1980), pp. 162 and 369–70; and further references in B. Fisse, 'Reconstructing Corporate Criminal Law: Deterrence, Retribution, Fault and Sanctions' (1983) 56 *Southern California Law Review* 1141 at 1223.

66 South Australia, Criminal Law and Penal Methods Reform Committee, *The Substantive Criminal Law*, Fourth Report (South Australian Government Printer, Adelaide, 1977), pp. 361–2.

67 See pp. 217 and 225–27 above.

68 B. Fisse & J. Braithwaite, *Corporations, Crime and Accountability* (Cambridge University Press, Sydney, 1993), pp. 169–77, 182–7. See our discussion of the dangers of scapegoating at pp. 220 and 226 above.

69 See Parliament of New South Wales, Legislative Council, Standing Committee on Law and Justice, *Report of the Inquiry into Workplace Safety*, Interim Report, Report No. 8 (New South Wales Government, Dec. 1997), p. 56.

70 Canada, Law Reform Commission, Working Paper 16, *Criminal Responsibility for Group Action* (Information Canada, Ottawa, 1976), p. 31.

71 We discuss this issue further below at pp. ••.

72 These orders have been recommended by a number of bodies, including the United States, Sentencing Commission, Preliminary Draft, *Sentencing Guidelines for Organizational Defendants* (United States Government Printing Office, Washington DC, 1991); and the American Bar Association, *3 Standards for Criminal Justice* (Little, Brown, Boston, 1980), and can be found in s. 19B of the Australian Federal government's Crimes Act 1914 (Cth.). The Australian Law Reform Commission, Report No. 68, *Compliance with the Trade Practices Act 1994* (AGPS, Canberra, 1994), has recently recommended the introduction of corporate probation as a sanction to enforce the Commonwealth Trade Practices Act (see R. Gruner, 'To Let the Punishment Fit the Organization: Sanctioning Corporate Offenders through Corporate Probation' (1988) 16 *American Journal of Criminal Law* 1; C. Stone, 'A Slap on the Wrist for the Kepone Mob' (1977) 22 *Business and Society Review* 4; J. C. Coffee, R. Gruner; & C. Stone, 'Standards for Organisational Probation: A Proposal to the United States Sentencing Commission' (1988) 10 *Whittier Law Review* 77; and other references cited in B. Fisse & J. Braithwaite, *Corporations, Crime and Accountability* (Cambridge University Press, Sydney, 1993), p. 42, n. 124). For cases in which this provision has been applied, see B. Fisse & J. Braithwaite, *Corporations, Crime and Accountability* (Cambridge University Press, Sydney, 1993), p. 42, n. 124.

73 American Bar Association, *3 Standards for Criminal Justice* (Little, Brown, Boston, 1980), para. 18.2.8(a)(v).

74 This model is similar to the proposed conditions for probation in the draft United States environmental sentencing guidelines, described below. See also B. Fisse, 'Individual and Corporate Criminal Responsibility and Sanctions Against Corporations' in R. Johnstone (ed.), *Occupational Health and Safety Prosecutions in Australia: Overview and Issues* (Centre for Employment and Labour Relations Law, The University of Melbourne, 1994), p. 100 at p. 105, describing the proposals for corporate probation in the Australian Trade Practices legislation.

75 The Occupational Health and Safety Act 1996 (Nova Scotia), s. 75, enacts a mild version of a compliance programme and may make an order requiring a convicted company 'to comply with such conditions as the court considers appropriate and just in the circumstances for the securing of the offender's good conduct and for preventing the offender from repeating the same offence or committing any other offence'.

76 R. S. Gruner & L. M. Brown, 'Organizational Justice: Recognizing and Rewarding the Good Corporate Citizen' (1996) *The Journal of Corporation Law* 731 at 734.

77 See United States, Sentencing Commission, *Guidelines Manual* (1995) §8A1.2; and K. B. Huff, 'The Role of Corporate Compliance Programs in Determining Corporate Criminal Liability: A Suggested Approach' (1996) *Columbia Law Review* 1252 at 1252–3.

78 R. S. Gruner & L. M. Brown, 'Organizational Justice: Recognizing and Rewarding the Good Corporate Citizen' (1996) *The Journal of Corporation Law* 731 at 737.

79 The discussion of compliance programmes in this paragraph is drawn from R. S. Gruner & L. M. Brown, 'Organizational Justice: Recognizing and Rewarding the Good Corporate Citizen' (1996) *The Journal of Corporation Law* 731. Gruner and Brown give a detailed analysis of indicators of due diligence in law compliance programmes at 749–65.

80 See again our discussion of employee representatives' role in audit teams in Ch. 5.

81 See our discussion of these issues at p. 72 above.

82 Note, 'Growing the Carrot: Encouraging Effective Corporate Compliance' (1996) 109 *Harvard Law Review* 1783 at 1791–2; and B. Fisse & J. Braithwaite, 'The Allocation of Responsibility for Corporate Crime: Individualism, Collectivism, and Accountability' (1988) 11 *Sydney Law Review* 469.

83 B. Fisse, 'Sentencing Options against Corporations' (1990) 1 *Criminal Law Forum* 211 at 236–7; and B. Fisse & J. Braithwaite, *Corporations, Crime and Accountability* (Cambridge University Press, Sydney, 1993), p. 42, n. 126.

84 American Bar Association *Standards for Criminal Justice* (Liffle, Brown, Boston, 1980) para. 18.2.8(a)(v). See B. Fisse, 'Sentencing Options against Corporations' (1990) 1 *Criminal Law Forum* 211 at 237.

85 Article L 263-8–263-11. See M. Mialon, 'Safety at Work in French Firms and the Effect of the European Directives of 1989' (1990) 6 *The International Journal of Comparative Labour Law and Industrial Relations* 129 at 143.

86 See J. C. Coffee, ' "No Soul to Damn: No Body to Kick": An Unscandalized Inquiry into the Problem of Corporate Punishment' (1981) 79 *Michigan Law Review* 386 at 448–55.

87 Australian Law Reform Commission, Report No. 68, *Compliance with the Trade Practices Act 1994* (AGPS, Canberra, 1994), para. 10.10.

88 See B. Fisse, 'Individual and Corporate Criminal Responsibility and Sanctions Against Corporations', in R. Johnstone (ed.), *Occupational Health and Safety Prosecutions in Australia: Overview and Issues* (Centre for Employment and Labour Relations Law, The University of Melbourne, 1994), p. 100 at pp. 105–6; Australian Law Reform Commission, *Sentencing Penalties*, Discussion Paper No. 30 (AGPS, Canberra, 1987), pp. 173–4; and B. Fisse & J. Braithwaite, *Corporations, Crime and Accountability* (Cambridge University Press, Sydney, 1993), p. 43.

89 D. Bergman, 'Corporate sanctions and corporate probation' (1992) 25 *New Law Journal* 1312 at 1313.

90 B. Fisse, 'Sentencing Options against Corporations' (1990) 1 *Criminal Law Forum* 211 at 237.

91 A. Norrie, *Crime, Reason and History: A Critical Introduction to Criminal Law* (Weidenfeld & Nicolson, London, 1993), p. 104.

92 Australian Law Reform Commission, *Sentencing Penalties*, Discussion Paper No. 30 (AGPS, Canberra, 1987), p. 174. See also B. Fisse & J. Braithwaite, *Corporations, Crime and Accountability* (Cambridge University Press, Sydney, 1993), pp. 43–4.

93 D. Bergman, 'Corporate sanctions and corporate probation' (1992) 25 *New Law Journal* 1312 at 1313.

94 South Australia, Criminal Law and Penal Methods Reform Committee, *The Substantive Criminal Law*, fourth Report (South Australian Government Printer, Adelaide, 1977), p. 364.

95 Community service orders can be ordered under the corporate probation provisions of the United States Sentencing Reform Act 1984. For examples of the use of community service orders against corporations in the United States, see B. Fisse, 'Sentencing Options against Corporations' (1990) 1 *Criminal Law Forum* 211 at 244–5.

96 See B. Fisse, 'Community Service as a Sanction Against Corporations' [1981] *Wisconsin Law Review* 970; B. Fisse, 'Reconstructing Corporate Criminal Law: Deterrence, Retribution, Fault and Sanctions' (1983) 56 *Southern California Law Review* 1141; and Australian Law Reform Commission, *Sentencing Penalties*, Discussion Paper No. 30 (AGPS, Canberra, 1987), p. 175.

97 Occupational Health and Safety Act 1996 (Nova Scotia), s. 75. The court can impose conditions in the order. See also Parliament of New South Wales, Legislative Council, Standing Committee on Law and Justice, *Report of the Inquiry into Workplace Safety, Interim Report*, Report No. 8 (New South Wales Government, Dec. 1997), p. 56, where community service orders are endorsed.

98 For a discussion of this and other potential pitfalls of this sanction, see generally B. Fisse, 'Community Service as a Sanction Against Corporations' [1981] *Wisconsin Law Review* 970 at 983–9 and 1008–16; B. Fisse, 'Recon-

structing Corporate Criminal Law: Deterrence, Retribution, Fault and Sanctions' (1983) 56 *Southern California Law Review* 1141 at 1228–9; and Australian Law Reform Commission, *Sentencing Penalties*, Discussion Paper No. 30 (AGPS, Canberra, 1987), p. 176.

99 See Occupational Health and Safety Act 1996 (Nova Scotia), s. 75.

100 B. Fisse, 'Community Service as a Sanction Against Corporations' [1981] *Wisconsin Law Review* 970 at 983–9.

101 B. Fisse, 'Reconstructing Corporate Criminal Law: Deterrence, Retribution, Fault and Sanctions' (1983) 56 *Southern California Law Review* 1141 at 1229.

102 See D. Yellen & C. J. Mayer, 'Coordinating Sanctions for Corporate Misconduct: Civil or Criminal Punishment?' (1992) 29 *American Criminal Law Review* 961, especially at 969–71. Yellen and Mayer note that this sanction has been proposed as a sanction for a number of crimes, but never actually imposed.

103 See L. H. Leigh, *The Criminality of Corporations in English Law* (Weidenfeld & Nicolson, 1969), pp. 157–8.

104 Most of our interviewees in our original New South Wales project who discussed this issue thought that there might be rare and extreme cases where dissolution was justified.

105 M. Mialon, 'Safety at Work in French Firms and the Effect of the European Directives of 1989' (1990) 6 *The International Journal of Comparative Labour Law and Industrial Relations* 129 at 143. Other penalties include a ban on practising one's profession for up to five years.

106 I. Ayres & J. Braithwaite, *Responsive Regulation: Transcending the Deregulation Debate* (Oxford University Press, Oxford, 1992), p. 30.

107 See P. J. Devine, 'The Draft Organization Sentencing Guidelines for Environmental Crimes' (1995) 20 *Columbia Journal of Environmental Law* 249 at 255.

108 S. Box, *Power, Crime and Mystification* (Tavistock, London, 1983), Ch. 2.

109 See the discussion of the guidelines later in this Chapter.

110 See J C Coffee, ' "No Soul to Damn: No Body to Kick": An Unscandalized Inquiry into the Problem of Corporate Punishment' (1981) 79 *Michigan Law Review* 386 at 413–24. See also Australian Law Reform Commission, *Sentencing Penalties*, Discussion Paper No. 30 (AGPS, Canberra, 1987), pp. 171–2; B. Fisse, 'Reconstructing Corporate Criminal Law: Deterrence, Retribution, Fault and Sanctions' (1983) 56 *Southern California Law Review* 1141 at 1233–7; and B. Fisse, 'Sentencing Options against Corporations' (1990) 1 *Criminal Law Forum* 211 at 230–3.

111 R. Fox, 'Corporate Sanctions: Scope for a New Eclecticism' (1982) 24 *Malaya Law Review* 26 at 41–2.

112 Under British common law, see *Attorney-General* v. *Harris* [1961] 1 QB 74; *Attorney-General* v. *Sharp* [1931] 1 Ch. 121.

113 But see the problems raised by R. Fox, 'Corporate Sanctions: Scope for a New Eclecticism' (1982) 24 *Malaya Law Review* 26 at 42.

114 See, generally, Australian Law Reform Commission, *Sentencing Penalties*, Discussion Paper No. 30 (AGPS, Canberra, 1987), p. 171. This sanction was recently endorsed by the Parliament of New South Wales, Legislative

Council, Standing Committee on Law and Justice, *Report of the Inquiry into Workplace Safety*, Interim Report, Report No. 8 (New South Wales Government, Dec. 1997), p. 56.

115 See M. Mialon, 'Safety at Work in French Firms and the Effect of the European Directives of 1989' (1990) 6 *The International Journal of Comparative Labour Law and Industrial Relations* 129 at 144.

116 See R. Johnstone, *The Court and the Factory: The Legal Construction of Occupational Health and Safety Offences in Victoria*, unpublished Ph.D. thesis, The University of Melbourne, 1994, chs 7 and 9; Law Commission, Published Working Paper No. 30, *Codification of the Criminal Law: Strict Liability and the Enforcement of the Factories Act 1961* (Law Commission, London, 1970); and D. Bergman, 'Corporate sanctions and corporate probation' (1992) 25 *New Law Journal* 1312 at 1312. See also E. Tucker, 'The Westray Mine Disaster and its Aftermath: The Politics of Causation' (1995) 10 *Canadian Journal of Law and Society* 91 at 115–16, who confirms that this event focus occurs in Canadian prosecutions.

117 Law Commission, Published Working Paper No. 30, *Codification of the Criminal Law: Strict Liability and the Enforcement of the Factories Act 1961* (Law Commission, London, 1970).

118 For further details of this argument, see R. Johnstone, *The Court and the Factory: The Legal Construction of Occupational Health and Safety Offences in Victoria*, unpublished Ph.D thesis, The University of Melbourne, 1994, Ch. 14; and R. Johnstone, 'The Legal Construction of Occupational Health and Safety Offences in Victoria : 1983–1991', in R. Johnstone (ed.), *Occupational Health and Safety Prosecutions in Australia* (Centre for Employment and Labour Relations Law, The University of Melbourne, Melbourne, 1994), pp. 95–7.

119 D. Bergman, 'Corporate sanctions and corporate probation' (1992) 25 *New Law Journal* 1312 at 1312.

120 Ibid.

121 The role of such information in sentencing is discussed below at pp. 283–89.

122 See R. Fox & A. Freiberg, 'Silence is not Golden: The functions of prosecutors in sentencing in Victoria' (1987) 61 *Law Institute Journal* 554; and I. Temby, 'The Role of the Prosecutor in the Sentencing Process' (1986) 10 *Criminal Law Journal* 199. In relation to OHS prosecutors, see R. Johnstone, 'The Legal Construction of Occupational Health and Safety Offences in Victoria : 1983–1991', in R. Johnstone (ed.), *Occupational Health and Safety Prosecutions in Australia* (Centre for Employment and Labour Relations Law, The University of Melbourne, Melbourne, 1994), pp. 377–82.

123 See the discussion of the 'good corporate citizen' submission in the at 278 above.

124 D. Bergman, 'Corporate sanctions and corporate probation' (1992) 25 *New Law Journal* 1312 at 1312.

125 See N. Gunningham, R. Johnstone, & P. Rozen, *Enforcement Measures for Occupational Health and Safety in New South Wales: Issues and Options* (WorkCover Authority of New South Wales, Sydney, 1996), p. 163. See also J. Hickey & C. Spangaro, *Judicial Views About Pre-Sentence Reports* (Judicial Commission of New South Wales and the New South Wales Probation Service, Sydney, 1995).

126 In Anglo-Australian common law, the impact of the offence on the victim is a relevant consideration in the sentencing process. Victim impact statements are increasingly being used in OHS prosecutions in Australian jurisdictions, particularly in New South Wales and Victoria. See Parliament of New South Wales, Legislative Council, Standing Committee on Law and Justice, *Report of the Inquiry into Workplace Safety*, Interim Report, Report No. 8 (New South Wales Government, Dec. 1997), pp. 61–4; and New South Wales Law Reform Commission, *Sentencing*, Discussion Paper 33 (Sydney, 1996), p. 436.

127 See D. Mitchell, 'Victim Impact Statements: A brief examination of their implementation in Victoria' (1996) *Current Issues in Criminal Justice* 163; and G. Hall, 'Victim Impact Statements: Sentencing on Thin Ice' (1992) 15 *New Zealand Universities Law Review* 143.

128 See D. Mitchell, 'Victim Impact Statements: A brief examination of their implementation in Victoria' (1996) *Current Issues in Criminal Justice* 163; and R. Douglas & K. Laster, *Victim Information and the Criminal Justice System: Adversarial or Technocratic Reform?* (La Trobe University, Melbourne, 1994).

129 Sentencing guidelines were recommended by the Industry Commission, Australia, *Work, Health and Safety* (AGPS, Canberra, 1995), Vol I, rec. 15. See also Parliament of New South Wales, Legislative Council, Standing Committee on Law and Justice, *Report of the Inquiry into Workplace Safety*, Interim Report, Report No. 8 (New South Wales Government, Dec. 1997), pp. 52–4.

130 See generally E. Zagrocki, 'Federal Sentencing Guidelines: The Key to Corporate Integrity or Death Blow to Any Corporation Guilty of Misconduct' (1992) 30 *Duquesne Law Review* 331; R. S. Gruner, 'Just Punishment and Adequate Deterrence for Organizational Misconduct: Scaling Economic Penalties Under the New Corporate Sentencing Guidelines' (1992) 66 *Southern California Law Review* 225; B. Baysinger, 'Organizational Theory and the Criminal Liability of Organizations' (1992) 71 *Boston University Law Review* 341; M. K. Block, 'Optimal Penalties, Criminal Law and the Control of Corporate Behaviour' (1992) 71 *Boston University Law Review* 395; J. Moore, 'Corporate Culpability Under the Federal Sentencing Guidelines' (1992) 34 *Arizona Law Review* 743; I. H. Nagel & W. M. Swenson, 'The Federal Sentencing Guidelines for Corporations: Their Development, Theoretical Underpinnings, and Some Thoughts on their Future' (1993) 71 *Washington University Law Quarterly* 205; P. H. Bucy, 'Organizational Sentencing Guidelines: The Cart Before the Horse' (1993) 71 *Washington University Law Quarterly* 329; P. E. Fiorelli, 'Fine Reductions Through Effective Ethics Programs' (1992) 56 *Albany Law Review* 403; E. F. Novak, 'Sentencing the Corporation' (1995) 21 (Summer) *Litigation* 31; R. S. Gruner, 'Towards an Organizational Jurisprudence: Transforming Corporate Criminal Law Through Federal Sentencing Reform' (1994) 36 *Arizona Law Review* 407; D. H. Freyer, 'Corporate Compliance Programs for FDA-Regulated Companies: Incentives for Their Development and the Impact of the Federal Sentencing Guidelines for Organizations' (1996) 51 *Food and Drug Law Journal* 225; and Note, 'Growing the Carrot: Encouraging Effective Corporate Compliance' (1996) 109 *Harvard Law Review* 1783.

131 The organizational guidelines expressly did not apply to environmental offences, on the grounds that environmental crime was different from the other types of offences included in the organizational guidelines, and environmental crime was too controversial to include in the organizational guidelines. The draft guidelines for environmental crimes was a draft of a suggested 'Chapter Nine' to the *Sentencing Guidelines Manual*. Although published in the Federal Register for public comment, and carefully considered by the Sentencing Commissioners, this advisory working group report has not been formalized as an amendment to the sentencing guidelines, nor has it been modified in any way since its original submission. For a general discussion and critique of these proposed guidelines, see J. M. Lemkin, 'Deterring Environmental Crime Through Flexible Sentencing: A Proposal for the New Organisational Environmental Sentencing Guidelines' (1996) 84 *California Law Review* 307; J. C. Coffee, 'Environmental Crime and Punishment', Thursday 3 Feb. 1994, *New York Law Journal* 5, 10, and 29; P. E. Fiorelli & C. J. Rooney, 'The Environmental Sentencing Guidelines for Business Organizations: Are There Murky Waters in Their Future?' (1995) 22 *Environmental Affairs* 481; K. Woodrow, 'The Proposed Federal Environmental Sentencing Guidelines: A Model for Corporate Environmental Compliance Programs' (1994) 25 *Environmental Reporter* 325; P. J. Devine, 'The Draft Organization Sentencing Guidelines for Environmental Crimes' (1995) 20 *Columbia Journal of Environmental Law* 249; R. L. Kracht, 'Comment: A Critical Analysis of the Proposed Sentencing Guidelines for Organizations Convicted of Environmental Crimes' (1995) 40 *Villanova Law Review* 513; and M. Harrell, 'Organizational Environmental Crime and the Sentencing Reform Act of 1984: Combining Fines with Restitution, Remedial Orders, Community Service and Probation to Benefit the Environment While Punishing the Guilty' (1995) 6 *Villanova Environmental Law Journal* 243.

132 Further guidance for developing guidelines for determining the level of penalties can be drawn from the discussion of civil penalties under the United States OSH Act, in the United States Department of Labor, Occupational Safety and Health Administration, *Field Inspection Reference Manual*, Sept. 1994.

133 United States, Sentencing Commission, Preliminary Draft, *Sentencing Guidelines for Organizational Defendants* (United States Government Printing Office, Washington DC, 1991), pp. 22786 and 22787.

134 United States, Sentencing Commission, Preliminary Draft, *Sentencing Guidelines for Organizational Defendants* (United States Government Printing Office, Washington DC, 1991), Introductory Commentary to Ch. 8.

135 For example, seven minimum factors *all* need to be satisfied before a sentencing discount is possible for having a compliance programme. These include: attention to environmental compliance by officers and managers; the existence of training and educational programmes for employees; the presence of auditing and reporting systems to uncover offences; continuing compliance evaluation and improvement; the existence of incentives programmes; and consistent and visible disciplinary procedures to deter environmental offences. This last factor includes a specification that the organization's sales and production programmes cannot be inconsistent

with its environmental compliance programmes. Further mitigation is possible if the seven minimum factors are satisfied and the organization can show that it has implemented 'additional innovative approaches to environmental compliance'. See P. J. Devine, 'The Draft Organization Sentencing Guidelines for Environmental Crimes' (1995) 20 *Columbia Journal of Environmental Law* 249 at 257–9; and P. E. Fiorelli & C. J. Rooney, 'The Environmental Sentencing Guidelines for Business Organizations: Are There Murky Waters in Their Future?' (1995) 22 *Environmental Affairs* 481 at 494–7.

136 The majority of offences committed were for fraud (average fine $465,967), with the other offences including antitrust (average fine $569,868), environmental and tax violations (average fine just over $50,000). Note, 'Growing the Carrot: Encouraging Effective Corporate Compliance' (1996) 109 *Harvard Law Review* 1783 at 1786–7.

137 The material in this and the following paragraph is drawn from Note, 'Growing the Carrot: Encouraging Effective Corporate Compliance' (1996) 109 *Harvard Law Review* 1783; and D. H. Freyer, 'Corporate Compliance Programs for FDA-Regulated Companies: Incentives for Their Development and the Impact of the Federal Sentencing Guidelines for Organizations' (1996) 51 *Food and Drug Law Journal* 225. See also E. F. Novak, 'Sentencing the Corporation' (1995) 21 (Summer) *Litigation* 31.

138 D. H. Freyer, 'Corporate Compliance Programs for FDA-Regulated Companies: Incentives for Their Development and the Impact of the Federal Sentencing Guidelines for Organizations' (1996) 51 *Food and Drug Law Journal* 225 at 226.

139 For a discussion of the benefits to organizations of corporate compliance programmes, see D. H. Freyer, 'Corporate Compliance Programs for FDA-Regulated Companies: Incentives for Their Development and the Impact of the Federal Sentencing Guidelines for Organizations' (1996) 51 *Food and Drug Law Journal* 225 at 228–30.

140 Note, 'Growing the Carrot: Encouraging Effective Corporate Compliance' (1996) 109 *Harvard Law Review* 1783 at 1787. In two of the cases the organization had an effective compliance programme, but they were deemed ineffective because corporate presidents were involved in the offences. In the other case, the conduct for which the organization was convicted took place at newly acquired facilities not covered by the comprehensive compliance programme (see D. H. Freyer, 'Corporate Compliance Programs for FDA-Regulated Companies: Incentives for Their Development and the Impact of the Federal Sentencing Guidelines for Organizations' (1996) 51 *Food and Drug Law Journal* 225 at 240).

141 Note, 'Growing the Carrot: Encouraging Effective Corporate Compliance' (1996) 109 *Harvard Law Review* 1783 at 1787, referring to a survey conducted by Andrew Apel, presented to a United States Sentencing Commission symposium on 7 Sept. 1995.

142 See M. Lemkin, 'Deterring Environmental Crime Through Flexible Sentencing: A Proposal for the New Organisational Environmental Sentencing Guidelines' (1996) 84 *California Law Review* 307 at 327.

143 See our earlier discussions of this issue in Ch. 5 at p. 282.

144 See, for example, P. E. Fiorelli & C. J. Rooney, 'The Environmental Sentencing Guidelines for Business Organizations: Are There Murky Waters in Their Future?' (1995) 22 *Environmental Affairs* 481.

145 We do not suggest that OHS statutes adopt the day fine system currently operating in Sweden or Germany. Certainly its diversionary aspects are inappropriate for OHS crime. Rather, we note that it provides a valuable example of the way in which a corporate defendant's resources (turnover, revenue, profits, and total assets) can be assessed and incorporated in an assessment of a fine after conviction, in a way that ensures that the fine can act as a deterrent on the offender, but that at the same time does not cripple the offender. It can, in other words, ensure effective specific deterrence.

146 By a mid-range offence we mean a contravention that was neither trivial (i.e. trivial in that it did not endanger workers or members of the public), nor very serious (serious in the sense of resulting in a number of fatalities).

147 If the company intended to run this argument in mitigation, it would have to give the OHS agency notice of this 'defence' as soon as prosecution proceedings were issued, so that the agency could investigate the manager's culpability, ensure that there was no scapegoating, investigate whether top management had given the manager adequate authority, time, and resources to develop appropriate OHS systems, and then bring proceedings if warranted.

148 The guidelines should provide the court with a checklist of factors to examine to ensure that the management system or compliance programme were effective and not merely a 'paper system'. Useful guidance is to be found in the factors required under the environmental guidelines in n. 135 above. See also our discussion below of the ways in which guidance can be given to organizations about establishing compliance programmes.

149 Note, 'Growing the Carrot: Encouraging Effective Corporate Compliance' (1996) 109 *Harvard Law Review* 1783 at 1786 and 1788–94.

150 R. S. Gruner & L. M. Brown, 'Organizational Justice: Recognizing and Rewarding the Good Corporate Citizen' (1996) *The Journal of Corporation Law* 731 at 734. See their analysis of United States' cases looking at the issue, at 742–9.

151 I. H. Nagel and W. M. Swenson, 'The Federal Sentencing Guidelines for Corporations: Their Development, Theoretical Underpinnings, and Some Thoughts on their Future' (1993) 71 *Washington University Law Quarterly* 205 at 230.

152 Note, 'Growing the Carrot: Encouraging Effective Corporate Compliance' (1996) 109 *Harvard Law Review* 1783 at 1788–94.

153 See our discussion of codes and safety management systems at p. 77 above, in Ch. 3.

154 Note, 'Growing the Carrot: Encouraging Effective Corporate Compliance' (1996) 109 *Harvard Law Review* 1783 at 1792–3. Freyer gives examples of cases in which US prosecuting authorities have exercised prosecutorial discretion in favour of an organization with a co-operative attitude and a compliance programme (see D. H. Freyer, 'Corporate Compliance Programs for FDA-Regulated Companies: Incentives for Their Development and the Impact of the Federal Sentencing Guidelines for Organizations'

(1996) 51 *Food and Drug Law Journal* 225 at 241). See also R. S. Gruner & L. M. Brown, 'Organizational Justice: Recognizing and Rewarding the Good Corporate Citizen' (1996) *The Journal of Corporation Law* 731 at 732.

155 Note, 'Growing the Carrot: Encouraging Effective Corporate Compliance' (1996) 109 *Harvard Law Review* 1783 at 1793. See, for example, the Australian Industry Commission's recommendation in relation to industry codes of practice: Industry Commission, Australia, *Work, Health and Safety* (AGPS, Canberra, 1995), pp. 90–4, rec. 11.

156 Note, 'Growing the Carrot: Encouraging Effective Corporate Compliance' (1996) 109 *Harvard Law Review* 1783 at 1798–9, where the suggestion is made that the agency would provide the organization with a letter indicating that the programme was acceptable, or indicating why it was not acceptable. These letters would be made publicly available so that organizations would become appraised of accepted ways of developing compliance programmes.

157 For a discussion of the admissibility of evidence of a compliance programme in proceedings against the corporation in United States courts, where the issue is whether a corporation should be found criminally liable for the acts of its employees, see K. B. Huff, 'The Role of Corporate Compliance Programs in Determining Corporate Criminal Liability: A Suggested Approach' (1996) *Columbia Law Review* 1252, especially at 1253–67. The dominant position seems to be that compliance programmes are legally irrelevant when determining whether to impute liability from the employee to the corporation.

158 R. S. Gruner, 'Towards an Organizational Jurisprudence: Transforming Corporate Criminal Law Through Federal Sentencing Reform' (1994) 36 *Arizona Law Review* 407 at 458; J. Arlen, 'The Potentially Perverse Effects of Corporate Criminal Liability' (1994) 23 *Journal of Legal Studies* 833 at 836–7; Note, 'Growing the Carrot: Encouraging Effective Corporate Compliance' (1996) 109 *Harvard Law Review* 1783 at 1794; and B. L. Williams & K. Kavanaugh, 'Compliance programs and Federal Organizational Sentencing guidelines' (1993) June *Res Gestae* 558 at 560.

159 See our discussion of this at the end of Ch. 6.

160 M. Tonry, *Sentencing Matters* (Oxford University Press, New York, 1996), p. 180.

161 Ibid. p. 176.

8   Conclusion: Recommendations for Reform

The main purpose of this book has been to argue the case for redesigning OHS regulation and its enforcement, so as to make it both more effective and efficient in addressing the OHS problems confronting advanced industrial nations. In doing so, it has identified a range of 'next generation' regulatory and enforcement instruments which, we believe, should be central to any future policy reform agenda.

In this final Chapter, we summarize the conclusions of the previous Chapters and identify their major implications, in terms of recommendations for reform. To a more limited extent, we point to issues where our conclusions are tentative or speculative, and suggest issues that might be the subject of future examination.

Compliance with what? The roles of specification and performance standards

In Chapter 2, we addressed an issue which has become central to contemporary OHS policy debates: the question of precisely what it is that has to be complied with, and, in particular, what sorts of standards are most appropriate to deliver successful OHS outcomes? To answer this question, we examined both the main types of standards which have been used to achieve OHS goals in the past, and new approaches which might be better suited to achieve 'best practice' OHS in the future.

We began by reviewing the debate between specification standards (which tell duty holders how to meet a goal by requiring compliance with detailed technical requirements) and performance standards (which define the duty in terms of problems to be solved or goals to be achieved). In particular, we noted the trend towards substantially replacing specification with both performance standards and principle-based standards (for example, very general 'duty of care' provisions) while complementing these latter with greater reliance on codes of practice rather than regulations. This approach has two considerable benefits. First, it establishes general goals but leaves duty holders with the flexibility and discretion of finding the most effective and efficient way to achieve them. Second, it provides duty holders with the practical guidance that some organizations, particularly SMEs, require through codes which, rather than being prescriptive, simply offer one acceptable means of achieving the principles set out in the performance-based standards. More recently, there has been a trend to complement performance standards with process-based standards: standards which address procedures for achieving a desired result.

From our analysis of the relative benefits of performance and specification standards, and based on the available evidence of their functioning in practice, we endorsed the current trend and recommended that in most circumstances performance and process standards are the most appropriate ways of providing guidance as to how to comply with the general duty requirements imposed by most jurisdictions. Education of duty holders as to the evidentiary implications of the codes of practice is the most appropriate way of overcoming the criticism that they have become *de facto* mandatory regulations (with clear indications being given to duty holders that compliance with the codes will be regarded by enforcement agencies as compliance with minimum performance and other standards). As a corollary, the weight of specification standards should be substantially reduced, such standards being reserved for exceptional situations: for example, where there is a high degree of risk and there are specific controls which are applicable to all circumstances where the risk occurs and which are essential to control the risk.

However, small employers and subcontractors, in particular, remain a problem under this approach because they often require much more concrete direction as to what is required of them. Such direction can continue to be given, but this is most appropriately done through technical guidance documents rather than through legally binding and prescriptive specification standards (though the former may include much of the information that was formerly contained in the latter). More broadly, hazard-based codes of practice have value in providing a set of principles and performance requirements and should continue to be adopted. Industry-specific standards are also useful in promoting best practice but should be developed on a tripartite basis rather than left to the voluntary efforts of industry associations, in order to avoid the risk of a lowest common denominator approach.

We also argued that specification standards and principle and performance standards (what we call 'phase one' and 'phase two' standards) both have their limits. They do not encourage enterprises to improve OHS over and beyond the legal limits prescribed. They do not encourage continuous improvement or industry best practice. Nor do they directly encourage enterprises to develop a safety culture or to 'build in' safety considerations at every stage of the production process. Certainly, for some firms, it may be unrealistic to expect performance much above the legal minima, and, for these firms, specification and performance standards (particularly the latter) will continue to play an important role. However, there are many other enterprises which could achieve far more than the legal minima. An important role of law is to encourage them to do so. For these enterprises, there is a considerable potential for developing a 'third phase' of standards design—one which provides not only considerable flexibility and enables enterprises to devise their own least-cost solutions, but one which gives them direct incentives to go 'beyond compliance' with the minimum legal standards prescribed under phases one and two.

Towards a systems-based approach: Voluntarism, legislation, or incentives?

This third phase (whose precursor was more limited process-based standards of the 'identify, assess, control' variety) involves standards which require the duty holder to implement an OHS management system (i.e. a system which involves the assessment and control of risks and the creation of an inbuilt system of maintenance and review). Such standards, appropriately designed, encourage enterprises to go beyond compliance by the combination of systematic management practice and a high level of commitment to a safety culture. They involve both management and workers in the development and maintenance of the system and integrate within it continuous measurement, benchmarking, the capacity for system self-correction and the commitment to continuous improvement.

While there are certainly pitfalls to be avoided in adopting such systems, and a real risk of 'implementation failure', nevertheless there is considerable evidence that, carefully designed, and with top management commitment and active worker participation, these can and do deliver very substantial and sustained improvements in safety performance.[1]

In broad terms, policy makers wishing to encourage a systems-based approach are faced with three options: leave it to the market (i.e. rely on the enlightened self-interest of enterprises in voluntarily implementing management systems); require by law that all or some enterprises implement OHS management systems (or at least make it optional for them to do so); or provide incentives (including subsidies) to enterprises to implement management systems but with no element of compulsion.

Of these three options, voluntary approaches are the least convincing. Voluntarism has a very disappointing record, and most studies have concluded that, even in the most co-operative of atmospheres, it is unlikely that industry will change its safety culture without considerable prodding or encouragement. There are a variety of reasons for this, not least of which is the sometimes conflict between safety and profit, which substantially inhibits rational investment in OHS improvements. But even in circumstances where no such conflict exists, and where firms might gain a competitive advantage from improved OHS performance, not all of them will realize this, or necessarily behave in an economically rational manner.

On this analysis (which assumes that the market alone will be insufficient to change industry behaviour substantially) there remains a legitimate role for governmental intervention. A crucial question then becomes whether this intervention should be by means of coercive regulation or whether an incentives-based approach alone could achieve the desired outcomes at less cost or in a less interventionist manner that is more acceptable to business.

Turning to the first of these options, could OHS legislation be modified so as to compel enterprises to develop an OHS management system? If so, would such an approach be effective? Based on past experience, the answer to the first question is almost certainly yes (and indeed in Norway this approach has already been adopted) but the answer to the second is much less clear. The practical obstacles to successfully implementing a legislative approach are considerable. In terms of effectiveness, the most substantial problem is that firms who are unwilling to develop a management system voluntarily may respond to compulsion by complying with the letter of the law rather than its spirit. That is, they may simply adopt 'paper systems' which appear to meet the legal requirement, but which in practice are little more than empty shells. Moreover, where enterprises are required to adopt a system, one is dealing with conscripts rather than volunteers, and many of the former will lack commitment to the legislative requirement. Nor would such a mandatory approach solve the problem of enterprises merely 'tacking on' a SMS rather than reworking the entire management system.

A less interventionist but potentially more effective means of inducing enterprises to adopt a systems-based approach is by means of incentives. Incentive-based approaches have considerable attractions. They can influence behaviour without direct intervention in the affairs of enterprises, they encourage them to seek out the most cost-effective (and often innovative) solutions to problems, they decentralize decision making to enterprises who often have better information on how to solve a problem than government, and they reduce government's enforcement costs. However, the effectiveness of incentives will depend largely upon the design and appropriateness of the particular mechanisms adopted. Following a review of the available options and evidence of their implementation, we recommend the introduction of the following incentives:

- administrative benefits, such as offering a partnership/co-operative approach to regulation to enterprises who agree to adopt a SMS, blitzing recalcitrants who choose not to adopt such a system, or abbreviating an inspection where inspectors are satisfied that a SMS is in place;
- logo or other publicity or public relations benefits to enterprises who adopt management systems;
- making implementation of system a condition for granting of self-insurer status;
- making implementation of a system a condition for tendering for major government contracts;
- providing 'up front' bonuses under workers compensation insurance (where applicable) to systems-based firms;
- subsidies to 'kick-start' a systems-based approach in firms which, by reason of their size, economic circumstances, or other factors, would otherwise be unlikely to adopt such an approach; and
- various forms of regulatory flexibility for those adopting a systems-based approach, including: reducing the likelihood of inspections and prosecutions, less prescriptive regulatory requirements, and reductions in penalties if prosecutions take place.

However, not all workplaces are alike and not all employers are similarly motivated. Many indeed, may be incapable or unwilling to introduce SMSs even if given substantial incentives to do so. For these reasons a 'one fits all' approach to regulation would be both inefficient and ineffective. From this we conclude that two very different types of regulation will be necessary: one for those enterprises which are ready, willing, and able to adopt a systems-based approach (and who have demonstrated a previous high standard of safety performance), and another for those which are not.

Our recommendation is that enterprises should be offered a choice between a continuation of traditional forms of regulation on the one hand (Track One), and the adoption of a SMS on the other (Track Two). The latter will put primary responsibility on employers and workers themselves to find optimal means of reducing occupational injury and disease subject to government and third party oversight, and will involve a partnership between the agency and the enterprise, from which both sides will benefit. Even the former, however, need not involve an adversarial relationship between regulator and regulated and we make a case for developing a more co-operative and cost-effective approach even under traditional regulation, for those who are motivated to comply voluntarily.

Regulating Track One enterprises

For enterprises regulated under Track One (being those who choose to remain on this Track or those who are not eligible for Track Two), we argue that a first and crucial issue is how to make the most efficient and effective use of scarce regulatory resources. Here, three questions are centrally important: (1) What principles should govern the allocation of scarce enforcement resources among different sizes of firms, and how could an agency improve its allocation under current laws with current resources? (2) What sorts of enforcement action should be invoked and what should be the nature of inspection? For example, should inspections be made more intensive, or should more firms be inspected with shorter inspections? (3) How should inspections be targeted or focused? For example, should inspections be random or targeted to worst performers?

In terms of the first of these questions, there is no 'off the peg' solution that is likely to be conveniently applicable to all agencies and all circumstances. Rather, much depends upon the existing distribution of inspections between small, medium, and large firms. However, we do argue that regulators should engage in a reassessment of their current deployment of resources with the likelihood (based on empirical evidence from the USA) being that a strategy that focuses substantially on the larger firms will be more efficient than the converse. However, there is some doubt about the extent to which these conclusions can be applied elsewhere. In particular, this evidence may not translate well to Europe, for example, where there is far more co-regulation and where the largest firms are usually the most willing to adopt a co-regulatory approach and with it to take greater responsibility for inspecting themselves.

In terms of the second of these questions, the evidence shows that the very fact of an inspector's visit, coupled with some form of enforcement action (for example, an improvement or prohibition notice or an 'on-the-spot' fine), may have a significant impact on behaviour and consequently on injury levels, even in circumstances where compliance costs will likely exceed the economic benefits to the employer of compliance. Essentially, this is because such action may serve to refocus employer attention on safety and health problems they may previously have ignored or over-looked. It follows that inspectors should be encouraged to more regularly invoke some degree of formal enforcement action sufficient to bring the problem forcibly to the employer's attention even if the latter is not substantial.

Related to this, there is also support for the proposition that shorter inspections provide an efficient use of scarce regulatory resources. In particular, the USA has experienced success with 'focused inspections' for employers with strong and effective safety programmes. The OSHA's intent is to work with targeted industries to identify the most serious hazards in those industries. These hazards will then be given focused attention during inspections, as a means of encouraging the adoption of effective safety and health programmes. If an OSHA inspector conducts an inspection and finds an effective safety programme operating on-site, then the remainder of the inspection will be limited to the top four hazards in that industry. Elsewhere, such a scheme might be applied as an incentive to larger companies to establish a SMS, or it might be applied to smaller companies who, even without a SMS, might be able to demonstrate a quality of safety performance sufficient to justify an in-spection limited to priority hazards—thereby both providing them with an incentive to reduce workplace injury and disease and enabling re-gulatory resources to be redeployed elsewhere.

In terms of the third question, the empirical evidence suggests that targeting plants with the highest level of risk, as indicated by consistently high injury rates, may provide the greatest incentives for those plants to comply voluntarily and therefore lead to more efficient enforcement. The success of the Maine 200 experiment described in Chapter 3 (and now being expanded to other USA states) also suggests the importance of concentrating on the most serious threats to OHS and that redeploying enforcement efforts in this manner will do most to protect workers. Our recommendation is that regulators target those employers who have serious OHS problems which they are unwilling to address voluntarily, and focus on worst cases and major hazards. In particular, they should apply a version of the Maine 200 scheme, while retaining some random component in order to keep other firms 'on their toes'.

Although effective targeting can play an important role in an efficient and effective enforcement policy, it is heavily dependent on the depth and accuracy of an agency's statistical database and other information sources. Only with adequate data collection and interpretation can a targeted inspection programme realize optimal results. An over-reliance on workers compensation data for these purposes may be dangerous, given the significant limitations of this data. It follows that there is a need to explore alternative sources of information and to establish databases which provide more accurate profiles of individual firms, hazards, and industries. The Queensland approach, where considerable work has gone into developing alternative database sources, suggests a number of possibilities. These include: using workers compensation data subject to recognized limitations by developing a database coded to Workers Compensation National Data Set requirements; using a random audit programme to collect compliance data which can predict levels of compliance in particular industries or workplaces and decide on targeting priorities; monitoring industries, workplaces, or work practices which have been identified in external research as likely to put persons at risk of development of injury or disease; establishing data collection systems in conjunction with general practitioners, hospitals, and other health providers; and conducting specific purpose samples, for example a rural fatalities study.

Enforcement strategies under Track One

Legislation that is not enforced seldom fulfils its social objectives, and effective enforcement is vital to the successful implementation of OHS legislation. A crucial issue is how the inspectorate should go about the enforcement task, in order to achieve maximum compliance. Experience suggests that neither a pure deterrence nor a pure compliance strategy will achieve the best results in terms of improved OHS performance. Most contemporary specialists agree that a judicious mix of compliance and deterrence is likely to be the optimal regulatory strategy.

Of particular value is the concept of an enforcement pyramid as set out by Braithwaite and Ayres (1992). Under this approach, regulators start at the bottom of the pyramid assuming virtue—that business is willing to comply voluntarily. However, they also make provision for circumstances where this assumption will be disappointed, by being prepared to escalate up the enforcement pyramid to increasingly deterrence-oriented strategies.

Increasingly, regulatory agencies are not only familiar with the concept of an enforcement pyramid, but also claim to apply it. However, even though the aggregate enforcement data might appear to confirm this claim, closer empirical scrutiny reveals that such a pyramid is not actually used by inspectors as a framework for an escalated enforcement response against individual employers. And where inspectors do in fact use what resembles an enforcement pyramid against individual employers, this enforcement response is informal and unstructured. Moreover, there may be little consistency across the inspectorate about the circumstances in which the various notices are issued, or prosecutions taken.

To rectify these anomalies, our recommendations are that regulatory agencies train and instruct inspectors to adopt an escalating enforcement response against individual duty holders. This should involve the establishment of a consistent and well-documented set of procedures incorporating the principles of pyramidal enforcement. Regulators should also be prepared to escalate to the top of the enforcement pyramid where action at the lower end fails to achieve results. Finally, they should introduce mechanisms which ensure greater consistency in the enforcement activities of individual inspectors.

Turning to enforcement more broadly, an optimal strategy involves both carrots and sticks. Most valuable at the base of the enforcement pyramid may be those approaches which actively encourage duty holders to regulate themselves, and which give them a positive incentive to do so. There may also be opportunities to assist small employers in particular, for example by developing guidance material collectively where they face a limited range of control measures, or by providing both subcontractors, and those who engage them, with material on the OHS hazards they face and how to deal with them, as well as on their broader legal responsibilities. Advisory standards, codes of practice, and guidance notes on the use of subcontractors will be particularly important.

For Track One enterprises, there are some innovative alternatives which are well worth exploring. For example, the VPP, described in Chapter 4, might be modified so as to encourage Track One firms to go beyond the minima prescribed by law, by recognizing and rewarding excellence in OHS programmes provided by employers, but in a less demanding form which did not require the implementation of a full SMS.

Similarly, there is evidence that, even with the severe resource constraints which inevitably limit the opportunities for regulating small enterprises, incentives-based innovations, such as the MPI initiative described in Chapter 4, can have considerable success. This initiative is based on the premise that a clear distinction should be made between companies that have adopted detailed company compliance and prevention policies and those that take little or no positive initiatives. By encouraging voluntary compliance, the scheme enables regulatory resources to be redeployed and refocused on those who are not responsive to voluntary initiative. Similarly, the Maine 200 and focused inspection programmes, referred to above, have particular application to SMEs.

Moving further up the enforcement pyramid, there is also some evidence that the system of improvement and prohibition notices and on-the-spot fines has been successful in those jurisdictions where it has been implemented, and that such a 'tap on the shoulder' approach can be very effective in focusing employers' attention on safety issues. However, where carrots at the base of the pyramid, and 'tap on the shoulder' approaches at its middle levels, both fail, then we argue for increasingly severe deterrent-oriented strategies, as more fully described below.

Finally, we argued the case for a lateral approach to regulation, taking advantage of the considerable leverage that large firms have over smaller trading partners. This would involve using the former as surrogate regulators, who would act if not to 'police', at least to oversee, the OHS performance of their suppliers and contractors, exercising control through the extended application of their management systems. A judicious combination of these various strategies would produce a substantial improvement in their OHS performance of many enterprises which never aspire to, and may never achieve, OHS 'best practice'.

Regulating Track Two enterprises

We have argued that the best strategy is to reward the efforts of good safety performers, and confine negative strategies such as deterrence to those situations where rewards are unlikely to be effective. The evidence suggests that the best results are likely to be achieved by those who adopt a SMS and we have argued that incentives should be provided to enterprises to adopt such systems.

In Chapter 5 we examined the regulatory implications of a SMS approach and of Track Two regulation, mindful that, if a systems-based approach is to be administratively viable, ways must be found to ease the burden on regulatory resources which might otherwise render this approach impracticable. There are a number of credible ways of addressing this problem. They fall into two broad categories: first, various strategies which encourage effective self-regulation on the part of enterprises that commit themselves to a SMS, and second, various forms of third party oversight. The former approach transfers a substantial part of the regulatory burden onto employers themselves, the latter onto third parties who can act as surrogate regulators.

Specifically, we recommend that regulatory agencies establish minimum criteria for SMSs, which enterprises wishing to adopt Track Two regulation must satisfy, and that these criteria should include performance indicators and other requirements which facilitate self-regulation through the SMS. We also recommend that Track Two enterprises, in return for the benefits which Track Two provides, agree: (i) to subject their SMS to initial accreditation and periodic review by an independent third party auditor nominated by the regulatory agency; (ii) that the cost of the audit will be borne by the audited enterprise; (iii) that a summary of the results of the audit will be provided to the regulator as well as to the enterprise itself; and (iv) that the regulatory agency shall have access to the full audit report of a small number of third party audits in order to audit the auditors.

The audit team should be required to include a worker representative (for example, a member of the OHS safety committee). Since such worker representatives will have already undertaken an approved OHS training course and have, in most cases, considerable practical experience of OHS, their presence on an audit team would make a useful practical contribution. It would also serve to maintain the credibility of the audit team, reduce the chances of it being 'captured' by the regulated enterprise, build in a valuable 'whistle-blower', and build the possibility that the partnership implied under a systems-based approach be extended to include workers and their representatives. A further attraction of this proposition is that it builds on an existing mechanism for worker participation in OHS, which could readily be adapted for this additional purpose.

Notwithstanding the important roles of self-regulation and third party oversight, there will remain a basic function which the inspectorate itself must perform. As we have indicated, there will be temptations on those who self-regulate, to cut corners and minimize costs in the short term.

Some enterprises, rather than genuinely implementing a systems approach in order to improve their OHS, may be tempted to simply devise cosmetic 'paper systems' to keep the regulators off their backs or to gain other perceived advantages.

To deal with these problems, the regulatory design we propose involves a tiered regulatory response. First, it is designed to encourage Track Two enterprises to regulate themselves (as indicated above, one of the prerequisites to enter into Track Two will be that the SMS is self-referential and self-correcting). Second, would be the role of third party oversight, both at the stage of accrediting the system when it is introduced, and through subsequent periodic audits. Thus the third party audit fulfils a substantial role as surrogate regulator. However, there will also be a third tier, involving an underpinning of government regulation which 'kicks in' as a back-up mechanism in circumstances where there is reason to believe that tiers one and two have not delivered the required outcomes in terms of system-effectiveness and improvements in OHS. Specifically, we recommend that an inspection by the government regulator be triggered: if the worker representative, after going through an internal complaints process, complains to the inspectorate that the audit was not, in his/her view, fair and accurate; if the third party audit report itself expresses serious reservations about the effectiveness of the SMS; if an inspectorate's verification of the third party audit (conducted randomly on a small minority of audits) suggests that the audit itself was not fair and accurate; and if there is a serious accident or incident, or a series of complaints from workers.

What action should a regulator take with a Track Two enterprise whose SMS is failing? We recommend different responses depending upon whether the enterprise is identified as incompetent or recalcitrant. Incompetents can be given advice as to how to improve, but where, after a period, it seems that the system is not leading to improvements, than that firm should be given a period of grace in order to rectify breaches of the law (punishment is not an appropriate short-term response for firms which have genuinely attempted a Track Two approach but failed) and then returned to Track One.

However, in the case of those who deliberately abuse Track Two, a different response is required. If Track Two is to maintain its credibility, a betrayal of the trust implicit in the accreditation and self-regulatory arrangements must not only lead to loss of accredited status but also to rapid escalation up the enforcement pyramid. This is a necessary response to discourage foot-dragging by firms that claim to be regulating themselves, but who use this as a guise to avoid their legal obligations. The justification for such rapid escalation is that the privileges bestowed by Track Two also imply obligations. Where industry is largely trusted to regulate itself, devising its own solutions in ways which best meet its own particular circumstances, so a betrayal of that trust must be seen to have serious consequences sufficient to deter the minority who might otherwise be tempted to do so. The potential for trade union and other third party prosecutions (as described in Chapter 5) would also serve as a deterrent to any betrayal of the trust implicit in the Track Two approach.

Finally, it is important to emphasize that the initial costs involved in participating in Track Two may be substantial, in terms of time, money, and vulnerability (for example, flowing from a commitment to greater worker participation). As such, it is crucial that the incentives to participate in Track Two are considerable and sufficient to demonstrably outweigh the costs. In the related field of environmental protection, it has been the inability of regulators to provide sufficient and tangible incentives that has led to the faltering of a number of innovative regulatory flexibility initiatives.[2] It is also important to remember that the costs of investing in a SMS and in participating in Track Two are mainly short-term whereas the benefits and substantially long-term. For all these reasons, it cannot be expected that the potential savings (in terms of reduced injury and increased profits), that participation in a SMS may bring, will themselves necessarily be sufficient to encourage more than a very modest number of firms to participate, and that a judicious combination of the incentives identified earlier will be fundamental if Track Two is to attract a substantial number of participants.

Special measures to assist small business

As we outlined in Chapter 1, one of the issues that continues to preoccupy governments, their regulatory agencies, and employer associations is the difficulty that small business operators have in keeping up with a plethora of regulatory requirements emanating from a wide range of sources, including OHS regulation. Governments are increasingly examining ways of streamlining the requirements on small businesses to ensure that small business operators are not paralysed by the regulatory requirements placed on them. In the area of OHS, small business people have pointed to three related problem areas in meeting their OHS obligations. The first is that the volume of legislation, including regulations, ordinances, decrees, and codes of practice, is huge and detailed, and duty holders find it difficult to keep up with the regulatory requirements imposed upon them. The second is that even where performance-, principle-, or process-based standards are known, the standards are so general that duty holders may not see their relevance to the situations the duty holders are actually confronting. The third is that, even if the regulatory requirements are known and applied, it is difficult to ascertain exactly what constitutes compliance with the statutory standards, and to be sure of whether the requirements have been met. In short, how do duty holders find out what standards apply to their businesses, how do they interpret them in the context of their own workplaces, and how can they be confident that they have complied with the requirements?

Much work still needs to be done to ensure that these problems are solved, without diluting the protection afforded employees and contractors who are exposed to workplace hazards in a small business setting. Many current initiatives focus on providing small business duty holders with advice and guidance material which explains in simple terms what managerial action needs to be taken to bridge the gap between the statutory enunciation of OHS obligations and the actual situations facing duty holders in their workplaces. Many of the solutions, however, lie in recognizing that compliance is socially constructed, and ultimately determined by OHS inspectorates and the courts. Small business operators' confusion and uncertainty when seeking to comply with OHS obligations can be reduced if the OHS enforcement agencies and the courts indicate clearly what it is that they are looking for when they make judgements about compliance. If inspectorates target their inspections to the most significant hazards in the industry, indicate very clearly to duty holders that that is where their priorities should lie, and follow a procedure similar to the OSHA focused inspections outlined in Chapter 4, then duty holders will be able to prioritize their compliance efforts. If these initiatives are reinforced by enforcement procedures, prosecution policies, and sentencing guidelines which give duty holders credit for documenting, implementing, and regularly evaluating and improving their OHS management systems, then the uncertainties of complying with OHS obligations will be reduced. In other words, if inspectorates and courts make it clear to duty holders how compliance will be constructed (that is, their enforcement procedures and sentencing principles are transparent), then duty holders will have a better idea of what they need to do to satisfy those responsible for the enforcement of OHS statutes.

Flexible enforcement measures for non-factory workplaces

In Chapter 1 we intimated that OHS regulation has developed around the assumptions that work is based at a permanent, ongoing workplace, typically a factory, and involves mainly physical hazards, and full-time employees. In recent times, particularly in Europe, the OHS regulatory model has developed to cover occupational health issues, the organization of systems of work, and job satisfaction. At the same time, the model has been modified to cover mobile work sites, such as those to be found in the construction industry, and less traditional work arrangements, such as home-working and outsourcing.

We suspect that, once again, developments in standard setting have outstripped progress in enforcement of these standards. The challenge in the future is for policy makers to come up with enforcement strategies which will ensure that the working conditions of employees, contractors, and others are monitored even though work is being conducted at fragmented venues distant from the employer or principal contractor's headquarters. This implies the development by OHS agencies of enforcement strategies which examine the existence, quality, and effectiveness of the OHS management systems of organizations engaging contractors, subcontractors, outworkers, and homeworkers, and which require the organization to show how its system of work ensures that the health and safety of these workers is safeguarded. Such strategies also imply the need to focus on the inspection and audit of the OHS management systems of the central organization, rather than relying on sporadic inspections of individual workplaces. Where organizations have OHS management systems which purport to ensure that the health and safety risks of workers are identified and controlled, inspectors will be able to conduct spot checks to verify that these systems are effective, and can encourage workers to notify inspectors where such management systems are not being fully implemented.

Where there are no such systems, inspectors will have to use the full extent of their inspection powers to find out where workers are located, and use administrative sanctions and on-the-spot fines or citations to require the development of appropriate measures to ensure the health and safety of workers. By doing so, the fact that many of the people conducting the work of the organization are widely dispersed will not render them incapable of being effectively monitored. This may require more sophisticated record keeping by the inspectorate, so that the OHS agency's records track an organization's health and safety performance from project to project, or scrutinize a sample of homeworkers' working environments.

Similarly, if, as some writers have argued, future business organizations are likely to make more frequent use of ad hoc structures and project teams, with the structure changing from project to project, policy makers will need to sharpen the focus of their regulatory efforts onto the overall OHS policies, procedures, and programmes of the head organization, rather than on the individual project units which will come and go before traditional enforcement measures will have any effect.

The role of criminal prosecution at the top of the pyramid

In Chapters Six and Seven we turned our attention to the role of criminal prosecution in enforcing obligations imposed upon duty holders by the OHS statutes. We note that, disappointingly, the use of criminal sanctions in OHS enforcement has barely been developed since the first attempts at OHS regulation in the early nineteenth century. It is no more successful in preventing work-related illness and injury than it was then. Indeed, it may be worse. As work-related hazards have become more complex with advances in technology, the decline in manufacturing, and the expansion of service and related industries, and as OHS regulators have increasingly replaced specification standards with performance-, process-, and principle-based standards, the criminal law appears to many to be less appropriate to punish those who fail to comply with their OHS obligations. Consequently OHS regulators in many countries have increasingly turned away from prosecution towards greater use of administrative sanctions and civil penalties.

Our overall argument is that the historical tendency towards decriminalizing OHS regulation, and underutilizing or bypassing criminal sanctions, should be arrested and reversed by a vigorous but carefully targeted, flexible, and highly publicized OHS prosecution strategy. We are not arguing for an enforcement policy which centres on prosecution, or which makes greater use of prosecution than of persuasive or administrative enforcement measures. Rather, we argue that OHS agencies should be able to encourage, facilitate, and negotiate compliance with OHS standards within the shadow of potentially large penalties and the stigma of criminal sanctions, and that to do so requires that OHS policy makers explore avenues to reinvigorate court-based criminal prosecutions by developing an enforcement strategy with escalating sanctions culminating with criminal prosecution at the top of the enforcement pyramid.

In consequence, we argue that, far from abandoning OHS prosecutions and adopting enforcement methods culminating in inspectorate-based civil penalties or citations (as in many European countries and in the USA), policy makers should introduce a more comprehensive and graduated, and hence more effective, enforcement pyramid with a significant number of court-based criminal prosecutions for contraventions of the OHS statutes, and some prosecutions for manslaughter and related offences under the mainstream criminal law. Most prosecutions would arise in the course of enforcing the OHS statutes against Track One firms, although we do envisage a few prosecutions for Track Two firms who either abuse the trust placed in them by the regulator, or where there were circumstances at the firm's workplace which resulted in serious injury, death, or the potential for serious disease. In the latter case the firm's accreditation as a Track Two enterprise would constitute a significant factor in mitigation of final penalty.

The virtue of a tall enforcement pyramid with many levels, and with criminal sanctions at its peak, is that it enables a greater scope for a nuanced escalation of enforcement responses by the regulator, who thereby is given much greater flexibility. A further advantage is that it assists regulators in dealing with respondents at the lower levels of the pyramid. Regulators can 'speak softly', knowing that their discussions are backed up by 'a big stick'. Accordingly, we argue for two levels of court prosecutions to increase the enforcement options open to the regulator— a lower level court (such as a magistrates' or local court) and a higher level court (a District, High or Supreme Court). To emphasize the criminal nature of OHS prosecutions, we suggest that these courts be located in the existing criminal court hierarchy. In order to ensure a more informed treatment of OHS prosecutions within the existing court system, magistrates and judges with specialist knowledge of OHS issues might be appointed to hear OHS prosecutions. Each level of court proceeding would have available a wider range of sanctions than the traditional fine, with the superior court having all of the sanctions we outlined in Chapter 7, and the lower level court being able to impose internal discipline orders, organizational reform orders, and community service orders. The majority of prosecutions would be conducted in the lower level court, but prosecutions for manslaughter, prosecutions in response to fatalities, serious injuries, or high risk contraventions, major 'pure risk' prosecutions, and important test cases will be conducted in the higher level court.

We argue, however, that to be successful such a strategy requires major reforms to the legal architecture shaping OHS prosecutions and a rethinking of prosecution strategies. Such reforms need to distinguish between the use of the criminal law for traditional crimes (theft, homicide, assault, and so on) and its use for inchoate crimes committed by organizations, and individuals directing and managing those firms, where the offence is a failure to establish and maintain a system of work that is safe and without risks to the health and safety of workers and others, even if the failure does not actually result in an injury. In short, not only should OHS offences be considered to be 'real crime', to use the terminology we adopted in Chapters 1 and 6, but it should be real crime which recognizes and accommodates the peculiarities of crimes committed by business organizations, and their managers and directors, in failing to prevent hazards to workers and others. In particular, we identify a range of major criticisms of the traditional mechanisms for OHS prosecution, and the prosecution strategies adopted by OHS enforcement agencies.

Criticisms of traditional approaches to OHS prosecution

The first criticism has been that, in those countries where prosecutions have been initiated for contraventions of OHS statutes, the emphasis has been on offences involving physical hazards (such as dangerous machinery, falls from heights, and similar matters), usually in response to an incident resulting in serious injury or death to a worker. In other words, prosecution has generally been reactive, and reactive in relation to physical hazards only, and injuries rather than disease. This has meant that whatever enforcement benefits have derived from prosecution have been confined to a narrow range of hazards, reinforcing traditional preoccupations with manufacturing sector accidents, having little value in preventative enforcement strategies for chemical hazards, or hazards emanating from psychosocial or organizational factors. Traditional OHS prosecutions are increasingly unsuitable for the types of workplace hazards which are now more prevalent as a result of structural changes in modern developed economies.

The reactive and event-focused emphasis of OHS prosecutions has resulted in OHS prosecutions that fail to see breaches of the OHS statutes in the context of systems of work or OHS management systems, but rather construct OHS contraventions as a chain of specific actions leading to a specific injury or death. Consequently, arguments in mitigation of penalty shift the sentencing court's attention away from an analysis of the failure of the OHS system, to scrutinizing the minute details of the events leading to the injury. This enables defendants to shift blame onto workers and others; and facilitates the argument that the accident was a 'freak' or 'one-off'. Further, the defendant is able to argue that the event has been superseded by measures to remedy the hazard, the introduction of new OHS systems, or the subsequent benevolent treatment of the worker. The event is isolated in the past (i.e., remedied the actual hazard; have a new management team; have introduced new OHS system; etc.).

Where courts have a largely unfettered discretion to determine the level of penalty imposed on those prosecuted successfully for OHS contraventions these mitigating arguments generally result in inadequate levels of penalty, a point we discuss below.

This traditional approach to prosecution is ill-suited to play a role in enforcing OHS standards which follow the principle-, process-, or performance-based standards model, and has little to contribute to incentive schemes inducing duty holders to go beyond the requirements of minimum standards, and to strive continuously to improve their OHS performance by adopting some form of SMS, as described in Chapter 2.

The second and third criticisms of the traditional model of OHS prosecution (and of civil penalties for OHS contraventions as favoured by the USA and much of Europe) pertain to the sanctions imposed on OHS offenders—traditionally a monetary fine. Since the first half of the nineteenth century, the most widely articulated and pervasive criticism has been that the maximum fines available in OHS statutes are too small to be an effective deterrent, and that the courts rarely approach the maximum penalty when imposing sanctions for contraventions. The consequence of this is that the full impact of criminal prosecution for OHS offences has never been explored and evaluated.

A third criticism is the weakness of the fine itself as a sanction. As we outlined in Chapter 7, a monetary fine signals to offenders that offences are 'purchasable commodities' rather than activities or omissions judged by the state to be intolerable. Fines do not require the offender to investigate the reasons for the contravention, to discipline those responsible, or directly to review and remedy the defects in the OHS management system. Monetary fines only affect the financial values of the organization, and on their own can have little impact on non-financial values (such as reputation and prestige). In some circumstances fines can be passed on to consumers, employees, or shareholders. Finally, the level of fine required to reflect the seriousness of the offence, to serve as a general deterrent, or as retribution, may exceed the firm's capacity to pay (the so-called deterrence or retribution trap).

A fourth criticism of the traditional approach to prosecution is the overemphasis on corporate employers. Despite the wide range of duty holders under modern OHS statutes, and the availability in many statutes of the possibility of prosecutions of responsible corporate officers, generally OHS enforcement agencies tend to focus their enforcement measures, and particularly their prosecution strategies, against corporate employers. In countries favouring OHS prosecutions, there is too much of an emphasis on corporate liability, and not enough scrutiny of the responsibilities of top-level managers and directors. The focus is also too heavily centred on those in control of the workplace itself, and fails adequately to examine the roles of those who produce plant, equipment, substances, work systems, plans, and designs that will be used at work. This narrow enforcement approach severely underutilizes the armoury of the OHS enforcement agency, and in particular relies too heavily on the assumption that the corporate employer is a coherent and integrated unit, acting in a uniform and rational manner. It also fails to put pressure on duty holders further up the chain from employers to design and manufacture plant, equipment, and substances in ways that reduce or remove work hazards.

Rethinking OHS prosecutions

In Chapters 6 and 7 we attempted to remodel the legal architecture and strategies for OHS prosecution so that prosecution becomes a more significant and effective punishment for OHS contraventions, is more visible as the ultimate sanction at the top of the enforcement pyramid, and can play a role more supportive of the systems-based approaches to OHS regulation outlined in Chapter 2.

In our model, prosecutions would primarily be initiated when deterrent methods are required in the face of the failure of enforcement methods based on persuasion and voluntary compliance. The primary purpose of prosecutions at the top of the enforcement pyramid should be instrumentalist—mainly general deterrence, so that duty holders are induced to make rational judgments that the cost of non-compliance, based on the perceived likelihood of detection and the severity of the penalty, outweighs the benefits of non-compliance. Nevertheless, the purposes of prosecution should be flexible enough to achieve other instrumentalist aims, such as specific deterrence (essential to ensure that enforcement measures are escalated where compliance is tardy), re-habilitation or organizational reform (to ensure that more direct meas-ures are used to ensure that offending organizations adopt good OHS management systems rather than simply see OHS offending as a cost of doing business), and incapacitation (in the rare cases where business organizations accord OHS such a low priority that society is prepared to prevent them from operating at all).

We are sensitive, however, to moral and political pressures (from members of the public, workers, their families, and their trade unions) for a number of prosecutions to be initiated for important moral, symbolic, and retributive purposes, to signal clearly society's disapproval of OHS contraventions, and to ensure that the most egregious offenders are punished in proportion to the seriousness of their offending. These symbolic purposes are more effectively served by criminal sanctions rather than civil penalties, and criminal sanctions have the further advantage of affecting the non-financial values of business organizations and their officers, particularly their prestige and reputation, which are very sensitive to the stigma of criminal prosecution.

A major caveat to the adoption of the deterrence model is the assump-tion that all OHS duty holders are rational actors. We seem to be on fairly safe ground in attributing to business organizations some degree of rational decision making, at least to the extent that they will be more likely to comply with statutory OHS obligations if they believe there is a high likelihood of inspection and prosecution resulting in significant penalties. However, beyond this, much remains unclear, not least the interface between deterrence theory and empirically-based and theoreti-cal understanding of the ways in which different business organizations operate. We recognize that the assumption of rationality will not hold in all circumstances, and that there is still much we do not know about the way in which business organizations work. One of the central issues facing OHS regulators is the diversity of organizational forms, the uncer-tainty about the degree of 'rationality' that can be assumed about corpo-rate decision making, and the difficulty of ascertaining 'pressure points' which can be targeted by OHS regulators aiming to promote compli-ance with OHS minimum standards, and eager to provide incentives for organizations to strive for continuous improvement in OHS manage-ment going way beyond prescribed minimum standards.

Much more research is required before we can develop sensitive indicators of which organizations will be susceptible to which types of incentives, and, in particular, which organizations are unlikely to embrace voluntarily the principles of exemplary OHS management; which require deterrence-based measures; and how such measures should be tailored to the idiosyncrasies of the organization in question. In the meantime, the concept of an enforcement pyramid provides a powerful strategic approach to enforcement, because it can operate effectually even where the regulator does not know whether she or he is dealing with the rational, irrational, or incompetent. In particular, the assumption of virtue, and the capacity to escalate to tougher penalties when the assumption is disappointed, serve to mitigate, if not overcome, our current lack of understanding of organizational behaviour in an environment where OHS agencies are unlikely to have greater resources, and indeed are likely to have to operate with a decreasing ratio of inspectors to workplaces.

Nine principles for more effective OHS prosecutions

Building on the pyramid approach, in Chapter 6 we outlined the basic principles which we consider to be essential for OHS regulators to ensure that OHS prosecutions serve an effective role as the 'big stick' at the apex of an OHS enforcement strategy. In line with our argument in the earlier Chapters of this book that OHS regulators should provide incentives for the adoption of OHS management systems which go beyond minimum compliance with OHS obligations, the principles aim to encourage business organizations to meet their OHS obligations by adopting effective OHS management systems and compliance programmes. The nine principles are aimed at ensuring that enforcement measures are flexible, responsive, fair, consistent, transparent, and, if necessary, tough. The enforcement strategy must also play a part in reintegrating OHS crime with mainstream criminal law.

First, the OHS agency's enforcement strategy[3] and prosecution policy must be aimed at ensuring *consistent* escalation up the pyramid. Inspectors and prosecutors must use common and transparent criteria and decision-making processes, so that the public perception is that all duty holders are treated equally, and that inspectors use the same criteria to work their way up the enforcement pyramid, so that only the most serious contraventions result in prosecution. This imperative of equal treatment leads to the second and third principles.

The second principle is that prosecutions must focus not just on

corporate employers, but on *all duty holders,* including manufacturers, suppliers, designers, self-employed persons, franchisors, principal contractors, subcontractors, and the individual managers and
directors of corporate employers and other corporate duty holders.
The OHS enforcement agency cannot assume that in each case the
corporate employer is the only entity responsible for the contravention, or that each business organization will have the same organizational structure. It must be able to target the prosecution at individuals or organizational entities which are most responsible for the
contravention. The most significant cause of the contravention may
be parties providing plant, equipment, substances, or services to the
employer. In order to ensure that workplace hazards are removed at
source, prosecution strategies should cover the design of plant and
equipment for use in workplaces, the manufacture and supply of
dangerous substances, and the design and planning of construction
projects. Where contraventions take place within forms of commercial
organization utilizing outsourcing and franchising, the agency should
ensure that prosecutions are targeted not only at the immediate employers of labour (the subcontractor or franchisee), but also at the parties
responsible for the overall design or coordination of the work system
(franchisors, head contractors, principal contractors, and planning
supervisors).

We anticipate that, in some circumstances, these attempts to broaden
the scope of OHS prosecutions may come up against jurisdictional
hurdles, particularly in relation to the duties of designers and suppliers.[4]
To some extent this issue is addressed in some jurisdictions by duties
being imposed on importers, but the problem remains that the source of
the unsafe plant or toxic substance may be beyond the jurisdictional
reach of the OHS agency. In Federal states such as Canada and Australia the jurisdictional restrictions will be quite daunting. Similar difficulties are likely to be accosted in Europe, where even though the EC
Directives have resulted in a more uniform body of national OHS
statutes, enforcement issues are still carried out at a national level.
Research is needed to explore the extent of the problem, and to devise
legal solutions to ensure that OHS sanctions can have an impact further
back up the contractual chain, so that the axiomatic OHS principle
of removing hazards at source, and placing responsibility for OHS
contraventions on those in control of the work system, can be implemented. Within Europe, Canada, or Australia jurisdictions may wish to
work co-operatively to identify cross-jurisdictional designers and suppli-

ers who are breaching OHS obligations, so that the OHS agency in the relevant jurisdiction can take the appropriate enforcement action. Such co-operative arrangements will be more difficult where unsafe plant or substances are imported into the EU or into Australia, the USA, or Canada.

Given the evidence that deterrence does not have a uniform impact on corporate actors, but, as noted in Chapter 6, does have a significant effect on corporate officers, an OHS prosecution strategy should ensure that prosecutions are brought against senior corporate officers or directors as individuals where there are egregious examples of these individuals failing to introduce or enforce OHS management systems. These prosecutions can be taken for manslaughter and related offences or for contraventions of OHS statutes. They should focus on directors and members of top management, because these are the people who are responsible for the OHS culture of an enterprise and for developing standard operating procedures, and who are in a position to 'do something' about OHS. But in designing a prosecution strategy which includes prosecuting corporate officers, the OHS enforcement agency must ensure that organizations do not 'scapegoat', and should also ensure that talented and hard-working people are not deterred from taking positions with responsibility for OHS management.

We note in Chapter 6 the problem of identifying the responsible officer in large organizations. We suggest that these difficulties can be overcome by: (a) presuming that the CEO is the responsible officer; or (b) requiring the organization to nominate a responsible officer and provide that officer with adequate training and resources. A further possibility[5] is for the OHS regulatory regime to be structured so that it provides incentives for organizations to activate their own private justice systems, and then monitor that process to ensure that it is fair and does not involve scapegoating. For example, the OHS enforcement agency foreshadows that an organization will not be prosecuted for an OHS contravention where it conducts internal self-disciplinary proceedings, publicly identifies those responsible for the contravention, and introduces a compliance programme to avoid future contraventions.[6]

The third principle is that, to remedy the 'split pyramid' which we discussed in Chapter 4, the pyramid must be *properly integrated*, in the sense that it must be clear to all duty holders that the OHS agency is willing and able to work its way up the pyramid when compliance with OHS standards is not satisfactory. In particular, we are concerned that efforts be made to ensure that prosecutions are not initiated for incidents giving rise to injury and death, and rarely for contraventions which do not result in any actual harm to employees. Consequently, there will need to be more 'pure risk' prosecutions (prosecutions where an injury or disease has not yet resulted from the contravention). Once again, this is a difficult issue for OHS agencies to pursue, and the difficulties in mounting a case alleging a defective health and safety environment over a period of time, and not specifically focusing on a particular incident, will differ from jurisdiction to jurisdiction. For example, in Anglo-Australian jurisdictions the rules of criminal procedure require the charge to be specific and to include particulars of the particular incident giving rise to the charge. Procedural rules also require each charge to include one offence only, and in some jurisdictions it has been held that the employers' general duty provision contains a number of specific offences, which means that the charge must include only one of these specific offences. Lawyers in each jurisdiction will need to consider carefully the legal impediments to mounting 'pure risk' prosecutions, and will need then to decide which impediments require amendments to the rules of criminal procedure and evidence, and which can be sidestepped by thorough investigation, and the use of expert witnesses.

The fourth principle requires the *mainstream* criminal law (for example, manslaughter and related crimes) to be *integrated* into the pyramid, so that enforcement officials initiate prosecutions for manslaughter and causing serious bodily injury where fatalities and injuries occur, but coupled with prosecutions under the OHS statutes which portray contraventions in terms of inadequate OHS management systems. This strategy will circumvent the dilemma facing OHS regulators contemplating the use of prosecutions for manslaughter and related offences. Non-use of mainstream criminal prosecutions, on the one hand, offends a key principle of the Rule of Law that the mainstream criminal law apply equally to all legal subjects. On the other hand, the use of mainstream criminal prosecutions for contraventions resulting in fatalities, in preference to prosecutions under the OHS statutes, suggests that OHS contraventions are minor offences, and not 'really criminal'. The use of both types of prosecutions in tandem will at least avoid the suggestions that corporate offenders are not subject to the mainstream criminal law, or that OHS offences are 'quasi-criminal', and will increase the deterrent and retributive effect of prosecution. Once again, in some jurisdictions the legal rules governing the requirements for corporate liability for these offences will need to be reformed, so that corporate liability can be attributed from the acts of upper and middle management, or from an overall 'corporate culture'.

In Chapters 2 and 3 we showed how OHS standard setting has developed from detailed, prescriptive specification standards to a more systems-based approach using principle-, process-, and performance-based standards. In Chapter 7 we suggested that the criminal law has traditionally focused more on incidents than systems, and that the monetary fine has been criticized as a sanction which fails to require direct corporate action to remedy OHS contraventions or defective OHS management systems. To reinforce our emphasis on systems-based approaches to OHS standard setting and enforcement, and to remedy the weakness of the fine as a criminal sanction, the fifth principle requires criminal sanctions to be consistent with, and to promote, *systems-based approaches* to OHS management and compliance. Sanctions should be redesigned to give the courts the option of requiring measures to be taken to implement the principles of sound OHS management.

Sixth, there must be *tough sanctions* ('big sticks') at the top of the enforcement pyramid—higher maximum fines, and a broader range of sanctions. Higher monetary fines, and the possibility of imprisonment for culpable corporate officers, are required to provide a greater general and specific deterrent to OHS duty holders, to ensure that the perceived costs of OHS contraventions exceed the likely benefits of ignoring or sidestepping statutory OHS obligations. Sanctions should also signal to OHS duty holders that, where they fail to introduce appropriate OHS management systems, courts will order them to do so.

Principles five and six require OHS statutes to be reformed to increase the maximum fines available (to increase the size of the stick), but ensuring that penalties should be tailored to the duty holder's resources. Revenue from increased fines should be made available to the OHS agency and to the courts, to offset the call on their respective resources likely to result from the introduction of the new sanctions outlined below.

In order to ensure that there are other tough sanctions over and above the possibility of a prosecution for manslaughter (described above in principle four), that greater publicity is given to OHS prosecution outcomes (see below), and that those sanctions are consistent with a systems-based approach to OHS regulation, a number of new sanctions might be introduced into OHS statutes. These sanctions might include:

- *court-ordered adverse publicity*, which enables a court to require the details of the contravention and the outcome of the prosecution to be publicized through the media to enhance general deterrence, and to affect non-financial values of the organization.
- *supervisory orders and corporate probation*, which include:

  — internal discipline orders requiring the organization to investigate the contravention, discipline those responsible, and return a compliance report to the court;
  — organizational reform orders, which require organizations to report regularly to the court on its efforts to develop a compliance programme and to reform its OHS management system; and
  — punitive injunctions, where the court requires the organization to introduce a specific OHS management system.

- *community service orders*, which require the duty holder to carry out an OHS-related project using the organization's resources, involving top management, during normal business hours.
- *dissolution*, where the most egregious offenders are required to cease their activities until their OHS management systems are reformed, or wound up permanently if the court decides that they are incorrigible.
- *equity fines*, where instead of being fined, offenders are required to issue new shares to the OHS agency which can be liquidated by the agency when it chooses. This sanction might result in a dilution of the offenders shares, might increase its susceptibility to a takeover, and might decrease the value of top management's shares in the organization.
- *disqualification from tendering for government contracts.*

Courts may be ill-equipped to make judgments about which sanctions will be appropriate in each situation. As we pointed out in Chapter 2, OHS management systems cannot simply be tacked onto pre-existing management structures, and so sanctions like corporate probation will need to be carefully tailored to each offending company. Similarly, the court will need assistance in structuring sanctions like adverse publicity orders, community-based orders, and equity fines to the circumstances of the offender. We suggest that the court be assisted by routine inquiries conducted by court appointed officials, resulting in something like the USA's corporate inquiry report, which might give the court information on the structure of the company, its financial position, and the suitability of the different sanctions outlined above. In each jurisdiction research is required to developed institutions which fulfil these functions in a way that is consistent with the existing institutions and traditions of the legal system.

Most of these new sanctions have been introduced somewhere in Europe or North America. In some jurisdictions there has been resistance to their introduction, either because they appear to go against deep-seated ideologies about corporate criminal sanctions (see Chapters 1 and 6) and the non-supervisory role of the courts, or on the grounds that they will be expensive to introduce and monitor. The ideological objections should be questioned and challenged, because they are not in themselves substantive objections, but rather rely on taken-for-granted, and largely outdated, assumptions about courts, the criminal law, and corporations. These assumptions need to be rethought before the criminal law can be developed to play a role in preventing corporate crimes. Such ideologically-based arguments are likely to appear from country to country in different forms, and, again, there needs to be a detailed examination of their substance in each jurisdiction, so that their false assumptions can be excavated. The objections based on the costs of proposed reforms also need to be challenged in each jurisdiction, on the basis that they are more than justified when measured against the cost of workplace injuries and illness and the benefits of stronger and more systems-based sanctions.

We argued in Chapter 5 for a role for workers in monitoring SMSs. There is a danger that this monitoring role can be merely symbolic or tokenistic. To enable workers to ensure that their monitoring role is taken seriously by employers and other duty holders, in general we support the notion that OHS prosecutions might be *initiated by parties affected by poor OHS* management (for example, trade unions and public interest groups) if the OHS agency does not, and if the alleged contravention falls within the OHS agency's prosecution guidelines. The OHS enforcement agency and the public interest groups with prosecution powers will need to negotiate an enforcement strategy which ensures that the OHS agency shares with the public interest groups the fruits of its earlier inspection and investigations of the alleged offender, and ensures that our other eight principles are maintained.

To the extent that business organizations are rational actors, to maximize the general deterrent effect of prosecution two further conditions need to be satisfied. Our eighth principle requires that the *outcomes* of successful prosecution receive the *maximum publicity*, so that OHS duty holders are fully aware that there is an active prosecution strategy resulting in formidable sanctions where contraventions are pursued successfully through the court system. Publicity will also enhance the stigma of criminal prosecution for OHS contraventions, and will amplify the effect of prosecution in shaming OHS offenders.

Our ninth principle requires not only greater publicity of prosecution outcomes (principle eight), but also that OHS duty holders are aware *in advance* of the likely approach that will be taken to their contraventions by the OHS agency and the courts. In other words, the operation of the pyramid, and the role of prosecution within it, must be *transparent.* If duty holders are aware of publicly available prosecution guidelines and sentencing guidelines, they are more likely to respond to these signals and incentives, promoting voluntary compliance, for example, through the development of satisfactory OHS management systems.

Transparent and well-structured sentencing guidelines can be particularly useful in ensuring that irrelevant or inappropriate sentencing factors are not considered by the court, and that convicted offenders have the exemplary or unsatisfactory aspects of their OHS performance considered by the courts. For example, in Chapter 7 we suggest a framework for sentencing guidelines which outlines the appropriate range of sanctions for OHS offences, guides the courts in their choice of sanction, and indicates factors to be ignored, to be considered in mitigation (for example, a proven compliance programme and/or OHS management system, a regular process of self-reporting of contraventions and so on), and in aggravation (a poor OHS record, proven top management involvement in the offence, lack of co-operation in investigation, and so on). Transparent and well-publicized sentencing guidelines will signal to duty holders that their investment in compliance programmes and OHS management systems will be rewarded if they are prosecuted for a contravention which occurs despite these measures.

We are conscious of the limited resources that agencies have available for prosecution. One of the benefits of a more credible and publicized use of criminal sanctions at the peak of the enforcement pyramid is likely to be that the enforcement measures at the lower levels of the pyramid are given more authority and operate more effectively, paradoxically reducing the need for an increased number of prosecutions. In short, we are arguing that OHS prosecutions be made more visible and effective, in the expectation that that strategy will reduce the need for prosecutions to be more numerous.

We observed in Chapter 6 that ideally our model of OHS prosecutions would be based on an empirical analysis of the effectiveness of OHS enforcement strategies escalating in seriousness and formality, and culminating in prosecutions targeted along the lines we advocate in earlier Chapters. We reiterate this point here. While we are confident that our model of OHS enforcement has a sound theoretical basis, its practical effectiveness needs to be tested by rigorous research in a jurisdiction which implements our model. We doubt that the optimal approach to enforcement and prosecution will be achieved from the outset, but we believe that with ongoing evaluation and reform we can move much closer to an optimal enforcement strategy in which prosecution plays a significant role.

Conclusion

Stepping back from our more concrete and detailed arguments, we conclude with some broader reflections on several of the critical challenges that have confronted us in writing this book and how we have sought to address them.

First, in relation to many of the issues that were raised in Chapters 2 to 5, not least the appropriate role and effectiveness of OHS SMSs, there is very limited empirical evidence from which to draw definitive conclusions. For example, the experience with SMSs is highly variable. At their worst, SMSs can be behaviourist, coercive, exacerbate power differentials between workers and management, and suffer from implementation failure. Yet, at their best, they hold out the promise of continuous improvement in OHS performance, cultural change, and of ensuring a greater coincidence between public and private objectives.

Against this backdrop, our approach has been essentially incrementalist. We have advocated *limited* experimentation with a two-track system of regulation, arguing for the provision of incentives for participation in a systems-based approach, but not for mandating it. In this way, we seek to explore the opportunities for policy reform implicit in a wider application of SMSs, while minimizing the likelihood they will be abused by applying them, at least in the first instance, only to the best players. To the extent that a systems-based approach is successful, it can then be extended (if necessary with modification) to the next-best players, and so on. At each stage, the opportunity to test it by results and to learn by experimentation seems to us the most responsible means of confronting a still somewhat confused and incomplete empirical picture.

Second, in seeking regulatory reform under a two-track structure our aim has been both to make better use of regulatory resources and to facilitate, reward, and encourage better OHS performance. In doing so, we emphasize that we do *not* advocate any removal or weakening of existing regulatory models (as our critics sometimes assume). On the contrary, we are conscious that some may seek to abuse the privilege of greater self-regulation implicit in a systems-based approach, and it is for this reason that we have stressed that those who abuse regulatory flexibility should be subject not only to existing controls, but also to rapid escalation up an enforcement pyramid.

Third, any strategy for regulatory reform of OHS must overcome the fact that one size most certainly does not fit all. Regulated enterprises include leaders, laggards, and many shades of grey in between. Strategies are needed to reward best performers, to deter and punish recalcitrants, and to educate and support incompetents. We have provided a range of strategies that confront this reality and which recognize that considerable flexibility in regulatory response is necessary to accommodate to as wide variety of circumstances as possible. The different approaches we suggest for Track One and Track Two firms, the specific strategies intended for SMEs, and the pyramidal enforcement strategy outlined in Chapters 6 and 7, are central planks of such a flexible regulatory response.

Fourth, in judging our recommendations for regulatory reform, it is essential to recognize that we are comparing grossly imperfect alternatives. As we are painfully aware, existing approaches to regulation are grossly inadequate on many levels, and in evaluating proposals for reform it is important to ask not: will they work perfectly? but rather: are they likely to achieve a substantial improvement on the status quo?

Fifth, the proposals we have made are likely to play out very differently in different cultures. Put briefly, the greatest opportunities for regulatory reform are likely to be found in Scandinavia, where a history of relative trust between the 'social partners', in conjunction with a substantial amount of direct worker participation in OHS issues, provides grounds for optimism about the chances for the success of the sort of initiatives (accompanied by the sorts of safeguards) we have advocated. Indeed, some elements of our proposals are already evident in certain Scandinavian countries. At the other extreme, the history of adversarial legalism in the United States, the extreme tensions and lack of trust between the main stakeholders, weak trade unions, antipathy to state intervention, and regulatory gridlock, suggest that any form of serious reform will have to confront and overcome serious and in some cases intractable problems. Australia, Canada, New Zealand, and the other developed countries in Western Europe fall somewhere between these two extremes.

Sixth, we are mindful of the difficulties that OHS regulators will encounter in seeking to ensure that the protection afforded by OHS legislation extends to precarious or contingent workers, that is outworkers, subcontractors, home-based workers, and other workers in temporary employment. There are no easy answers to this increasingly serious problem. In some instances regulators may need to legislate to create new categories of duty holders (for example, franchisors). In others, they may need to discourage or even limit the engagement of certain types of worker if it is not possible to develop regulatory mechanisms to reduce the risks they encounter at work. In still others, there are opportunities to take advantage of the disparities of power between large contractors and small subcontractors, and to encourage, facilitate, or even compel the former to extend their systems and preventative strategies to the latter.

Seventh, in an ideal world we would call for greater resources for the enforcement of OHS standards against recalcitrant and recidivist duty holders. Knowing that such requests are unlikely to be well received by national government, we have argued, nevertheless, for a more robust approach to enforcement. Such an approach will require targeted inspection strategies, more responsive enforcement, and meaningful penalties. To encourage voluntary compliance with OHS standards we have suggested new strategies for prosecutors, and greater and more varied sanctions.

Finally, we are conscious that, at best, this book can make only a partial contribution to an ongoing debate, the parameters of which are constantly shifting as it accommodates to changes in the economy, in the workforce, in power differentials, and in the empirical evidence concerning both the nature of the problem and its solution. At the very least, we hope that this contribution will have some direct input on that debate in a manner that mitigates one of the most intractable social problems to beset the developed and developing world since the onset of the industrial revolution.

NOTES

1 See Ch. 3 above.
2 R. Steinzor, 'Reinventing Environmental Regulation: The Dangerous Journey from Command to Self-Control' (1998) 22 *Harvard Environmental Law Review* 103–202.

3 See the discussion of enforcement in Chs. 4 and 5.
4 The degree of control that a franchisor exercizes over the operations of a franchisee, and the fact that most franchise arrangements require the franchisee to lease plant from the franchisor or a subsidiary of the franchisor, tends to reduce problems of criminal jurisdiction in respect of prosecutions of a franchisor based outside the country or state of the franchisee. Nevertheless, similar jurisdictional issues may arise in relation to enforcement of OHS obligations against franchisors.
5 Following I. Ayres & J. Braithwaite, *Responsive Regulation: Transcending the Deregulation Debate* (Oxford University Press, New York, 1992); and B. Fisse & J. Braithwaite, *Corporations, Crime and Accountability* (Cambridge University Press, Sydney, 1993).
6 In order to encourage the organization to scrutinize its own OHS performance in this way, the OHS statute might include some form of 'self-evaluative privilege' which ensures that communications generated by an organization while it monitors its own OHS performance are not able to be used by a prosecutor to prove an OHS contravention.

Appendix   The Legal, Institutional, and Industrial Relations Environment Governing OHS in the USA, the UK, Sweden, Denmark, and Australia

This Appendix reviews the legal, institutional, and industrial relations environment in selected jurisdictions in order to identify methods of setting, legislating, and enforcing OHS standards. The countries considered are the USA, the UK, Sweden, Denmark, and Australia. The principal reasons for the selection of these jurisdictions is as follows: the two Scandinavian countries have developed particularly innovative approaches to OHS; the UK is the system which first adopted the recommendations of the highly influential British Robens Report;[1] and Australia, while following a broadly similar approach to the UK, has introduced some important innovations of its own. It is also the jurisdiction in which the initial empirical work for this book was conducted. The USA is included not only because it is a major jurisdiction in its own right but also because it has adopted a radically different approach which provides a valuable contrast to the other jurisdictions under consideration.

The principal legislation in these countries is: the Occupational Safety and Health Act 1970 (USA)—hereafter referred to as the OSH Act; the Health and Safety at Work Act 1974 (UK)—hereafter HSW Act; the Work Environment Act 1977 (Sweden); and the Working Environment Act 1975 (Denmark). Australia, for constitutional reasons (see below), has not adopted comprehensive national legislation. The regulation of workplace health and safety remains principally a responsibility of the individual states and territories, with Federal statutes covering Commonwealth government employees and the maritime industry.

Division of responsibilities between national and state governments

The UK, Sweden, and Denmark are examples of unitary systems where OHS is addressed at the national level and all significant legislation is national legislation, although it should be noted that many of the recent developments in these jurisdictions have been in response to EC Directives (see below). In contrast, both the USA and Australia have had to grapple with defining state-federal roles in OHS, and their experience will be considered below.

In the USA, until 1970, OHS in industry was regulated principally at state level. The Federal government's involvement was very modest, being limited to Federal employees and to a small number of other issues. However, the manifest inadequacy of state legislation and in particular of state enforcement, finally led Congress to introduce Federal OHS legislation.

In debating the 1970 reforms, Congress found that 'personal injuries and illnesses arising out of work situations impose a substantial burden upon, and a hindrance to, interstate commerce in terms of lost production, wage loss, medical expenses and disability compensation payments'. Under its powers to regulate interstate commerce, Congress enacted the OSH Act to encourage 'employers and employees in their efforts to reduce the number of occupational safety and health hazards at their places of employment, and to stimulate employers and employees to institute new and to perfect existing programs for providing safe and healthful working conditions'. A primary purpose of the Act was to ensure *uniformly* applied standards across the country.

Despite the clear constitutional capacity to regulate OHS federally under the interstate commerce power, the approach adopted in implementing the 1970 OSH Act is a form of co-operative federalism. This involves a clear demarcation between standards, which are to be set by the Federal government, and the administrative responsibility for enforcement and monitoring of those standards, which by and large is delegated to the states. The inducement for state involvement in monitoring and enforcement is financial: the states receive substantial Federal grants if they do participate, whereas, if they do not, they lose significant control over their local industry.

Pragmatic reasons, including the fiscal and administrative difficulties of implementing the legislation at Federal level, largely explain the Federal government's preference for delegation to the states. Nevertheless, the continuing involvement of the states has been severely criticized in some quarters. For example, Ashford,[2] in his major study of the OSH Act, concludes:

The states' previously poor record in OHS enforcement does not suggest that sufficiently increased worker protection will result from the return of the programs to the states, even though the total occupational health and safety compliance force will be increased somewhat. State takeovers could severely limit the ability of labor unions to act in a watchdog role and hamper their efforts to improve working conditions. Manufacturing concerns with operations in several states may find it difficult to conform uniformly to different state requirements.

Partly in recognition of this problem, reforms proposed by the Clinton Administration early in its tenure included provision whereby the OSHA would be required to investigate complaints against state plans. It is also proposed to modify the procedures for withdrawal of approval of a state plan. However, it is significant that these proposals were limited to improving state administration, rather than replacing it with direct Federal intervention.

Except where the Federal government clearly intends to 'pre-empt' state regulation, the states are at liberty to develop standards which are more stringent than those prescribed under Federal legislation. The proposed reforms would leave states free to impose additional safety and health requirements to protect the general welfare.[3]

In Australia, legislative power is divided between the Federal government, and the state and territory governments. Unless the Commonwealth Constitution expressly confers legislative power on the Federal government, legislative power can only be exercised by the states. The Commonwealth constitution contains no express power in relation to OHS, and the Commonwealth to date has declined to use other potential heads of power to introduce national OHS legislation. These would include the power to deal with 'trade and commerce with other countries and amongst the States', 'foreign corporations, and trading or financial corporations formed within the limits of the Commonwealth', commonwealth employees, the territories, and 'external affairs' which gives the Federal government the power to legislate to implement internal conventions, including International Labour Organisation (ILO) conventions. The Federal government has introduced legislation establishing the National Occupational Health and Safety Commission (NOHSC), and OHS legislation for Commonwealth employees and the maritime industry. Since 1990 the NOHSC has been charged with overseeing a concerted attempt to harmonize OHS standards in Australia, although this process came to a standstill in 1997. We discuss this process later in this Appendix. The important point is that, apart from Federal regulation of OHS in Federal government employment and the maritime industry, OHS regulation in Australia, both in terms of the enactment of standards and their enforcement, is the responsibility of state and territory governments.

The coverage of the OHS legislation

In the past, the scope of the OHS legislation was not comprehensive. For example, in the UK the law was inconsistent and incoherent, and the legislation (itself contained in a variety of statutes) only applied to designated premises, processes, or activities. As a result, the scope of the legislation was partial and inadequate and many workers had little or no protection from workplace hazards. This was clearly an undesirable state of affairs, and more recent legislation, not only in the UK, but also in Scandinavia, the rest of Europe, North America, and Australia, has adopted the principle that all workers and all workplaces should be covered by OHS legislation.

The Robens Report, on which both British and Australian contemporary legislation is based, called for the development of a unified and integrated body of law, and the UK HSW Act of 1974 largely implemented this objective. In particular, it established a comprehensive set of duties relating to the basic and overriding responsibilities of employers, the self-employed, employees, manufacturers and suppliers of plant and substances, and others. These standards protect any person present at or near a workplace, including visitors and members of the public. Some pre-existing legislation with differential coverage did, however, remain, and the 1974 Act exempted some sensitive establishments under government control.

In the USA, the OSH Act of 1970 applies to any employer who is engaged in a business affecting commerce, either in the USA or any territory administered by the USA. In doing so, it provides coverage to many employees who were not previously protected by any legislation. State and local government employees are now the only major group not given legislative protection, although they must be covered under plans submitted by each state as required by its law.[4] Reforms proposed by the Clinton Administration would rectify this anomaly. Specifically, the Comprehensive Occupational Safety and Health Reform Bill of 1993 would extend the OSH Act provisions to state and local government employees.

Similarly, the approach taken in Denmark and Sweden is to achieve comprehensive OHS coverage for the entire workforce and in all workplaces. In Sweden, the Work Environment Act covers 'every activity in which an employee performs work for an employer's benefit' subject only to exceptions involving work performed on ships and in the employer's home. The Act also applies to students in the compulsory school system. Draftees, prisoners, and patients are not normally regarded as employees, although they nevertheless receive some protection under the Act. In 1994, the National Board of Occupational Safety and Health was empowered to extend the applicability of the Act to one-person and family undertakings. There is a general philosophy that all areas of working life, both in the private and in the public sector, should be covered comprehensively by OHS legislation. This includes not only physical hazards, but also job content, working hours, work organization, employee participation in decision making, equal opportunity, and the psychosocial aspects of the workplace.

In Denmark, the Working Environment Act 1975 covers all work undertaken for an employer except some work in the aviation, fishing, and shipping industries (which are regulated by other Acts). Certain other types of work such as military service and self-employment are only partially covered. The Act's requirements are perceived as broad and much more comprehensive than previous legislation.

Australian jurisdictions, following the recommendations of the Robens Report and the model of the 1974 UK legislation, have also enacted legislation which (with a few exclusions, significantly mining) achieves comprehensive coverage of all persons in the workforce, and which applies to all workplaces.[5] This approach has considerable merits and has never been seriously challenged. While there is still scope to improve its implementation, the principle of comprehensive coverage is itself unlikely to become an issue in any future debate on reform of OHS legislation in Australia.

In summary, reasons of equity and efficiency dictate that uniform legislation should be implemented in such a way as to ensure comprehensive OHS protection for all workers and in respect of all workplaces. This has been substantively achieved in all the jurisdictions in this study although the drafting techniques utilized for this purpose differ somewhat between countries.

Legal responsibilities of employers and employees

In the UK, the HSW Act imposes broad-ranging general duties on both employers and employees.[6] The principal duty imposed on the employer is to ensure 'so far as reasonably practicable the health safety and welfare at work of all his employees'. This duty expressly includes the provision and maintenance of safe plant and systems of work; safe use, handling, storage, and transport of articles and substances; provision of information, training, instruction, and supervision; safe access and egress; and the provision of a working environment that is safe, without risks to health, and has adequate welfare facilities.

All these duties are qualified in the HSW Act by the phrase 'reasonably practicable'. This qualification has been interpreted to mean that the cost of preventive action in the workplace should be weighed in the balance against the probability of personal injury occurring, and the severity of the injury likely to occur. That is, it implies some form of cost-benefit analysis (see further below).

The employer under the UK HSW Act is also required to prepare a written statement of its general OHS policy in relation to its employees, and the organization and arrangements in force for carrying out this policy. This provision is intended to dispel apathy (believed by Robens to be the major cause of accidents at work), and to focus employers' minds directly on OHS issues and how to address them. The Robens Report explained that the policy statement will be 'a frame of reference for positive safety and health activity within the firm, and a stimulus to interest and participation by all personnel'. The Management of Health and Safety at Work Regulations 1992 implement the provisions of the EC Framework Directive (see below) which impose more detailed obligations on employers and self-employed persons, requiring them to plan, organize, control, monitor, and review all preventive and protective measures to remove or minimize workplace hazards, and to encourage the development of an active health and safety culture in their organizations.

The HSW Act similarly imposes a general duty upon each employee to take reasonable care for the health and safety of him or herself, and of other persons who may be affected by their acts or omissions at work, and to co-operate with other persons to enable them to carry out their statutory duties in respect of health and safety at work.

Over and beyond the general duty provisions, there remain in the UK a substantial number of specific and detailed regulations dealing with particular types of hazard (for example, fencing of dangerous machinery), with particular types of operations (for example, construction sites—see the Construction (Design and Management) Regulations 1994) or substances (Control of Substances Hazardous to Health and Health and Safety (General Provisions) Regulations 1992). These regulations may be supplemented by codes of practice (which are dealt with further below). The employer's duties in respect of consultation with employees, safety representatives, and safety committees are also addressed below.

In all the Australian jurisdictions, the British approach to general duty provisions has been followed closely. While there are some significant differences in drafting technique, the outcome, with one exception, is essentially the same as is achieved under the UK HSW Act.[7] The exception relates to the employer's obligation to provide a written OHS policy statement. Not all Australian jurisdictions have adopted this provision.[8] Given the potential for policy statements to focus management attention on OHS issues, and to encourage the development of OHS management systems, there may be merit in extending their use to all jurisdictions.[9]

Since the early 1990s, many Australian jurisdictions have adopted provisions which require duty holders to identify, assess, and control workplace risks. Many of these requirements are to be found in hazard specific regulations and codes of practice (see below). General risk assessment provisions are to be found in the Queensland Workplace Health and Safety Act 1995 (section 22), and in general OHS regulations in New South Wales, South Australia, and the Northern Territory.

In the USA, under the OSH Act 1970 each employer has a specific duty to comply with occupational safety and health standards promulgated under the Act. These standards comprise a myriad of extremely detailed prescriptions concerning most aspects of the workplace. In all cases not covered by specific standards, the employer has a general duty to 'furnish to each of his employees, employment and a place of employment which are free from recognized hazards that are causing or are likely to cause death or serious physical harm to his employees'. This requirement, known as the 'general duty clause', is an important 'backstop'. It has been the basis of many citations issued against employers (see below).

Since 1993 there have been a number of unsuccessful proposals before Congress which would require all employers to provide a written safety and health programme.[10] The purpose of the programme would be to identify and control hazards before injury or illness results. Health and safety programmes are required in some OSHA standards, most notably for the construction industry. In addition, some OSHA inspection programmes (such as Maine 200, the recent proposal for CCPs, and the focused inspection programme, all of which are discussed in Chapter 4 of this book) abbreviate inspections if employers can show that they have effective safety and health programmes. The OSHA is currently proceeding with a new standard which would require *all* employers to develop comprehensive safety and health programmes.

The OSH Act also imposes a duty on the employee to comply with OHS standards and all rules, regulations, and orders issued pursuant to the Act. However, no mechanism exists under the Act to force compliance by employees, with the result that, for practical purposes, ultimate responsibility for compliance rests with the employer.[11]

In Sweden, most of the legislative standards are also very broadly stated, the 1977 Work Environment Act being essentially 'framework' legislation that sets goals, and outlines systems, techniques, and allocation of responsibilities, for the working environment in general terms. For example, Chapter 2 of the 1977 Act, states that the 'work environment shall be kept in a satisfactory state having regard to the technological progress occurring in the community at large', that 'work must be planned and arranged in such a way that it can be carried out in healthy and safe surroundings', and that 'working premises must be arranged and equipped in such a way as to provide a suitable working environment'. It then goes on to specify requirements in respect of such matters as industrial hygiene, machinery, hazardous substances, personal protective equipment, and related matters. As indicated below, regulations (ordinances) may be promulgated to specify in more detail what is required of employers or others under the Act.

In the early 1990s changes were made to the Work Environment Act in response to the EC Framework Directive (see below), in order to emphasize the employer's responsibility for OHS and to define the work environment more broadly to include work systems and psychosocial problems. For example, employees must be given an opportunity to participate in designing their own working situation, and in processes of change. Technology, work organization, and job content must be designed so as to avoid physical or mental strain. 'Forms of remuneration and the distribution of working hours shall also be taken into account in this connection'; and 'efforts must be made to ensure that work provides opportunities of variety, social contact and co-operation, as well as coherence between different working operations'. Working conditions are to be adapted to people's individual aptitudes. The employer is made responsible for systematically planning, directing, and inspecting working environment efforts ('internal inspection'), as well as continuously investigating hazards and work injuries. These provisions have been supplemented by an Ordinance of the National Board of Occupational Safety and Health[12] requiring all employers to have a work environment policy, action plans, a clear allocation of duties and routines for charting risks and for investigating accidents and ill health ('internal control'). The employer must also carry out job modification and rehabilitation at the workplace.[13] Since 1994, under Chapter 4 of the Work Environment Act, the National Board of Occupational Safety and Health can require employers, and others responsible for safety, to compile health and safety-related documents, such as environment, activity, or health and safety plans. The provisions in relation to 'internal control' are discussed further in Chapters 3 and 5 of this book.

The 1977 Act also requires employers to provide an occupational health service. Most large companies have their own occupational health services, while joint occupational health service centres provide services to small and medium-sized organizations. The service should place special emphasis on preventive work, and should contribute to job adaptation and rehabilitation. The service is integrated, and includes occupational physicians, safety engineers, nurses, physiotherapists, and behavioural scientists. Until 1993 the costs of the service could be reimbursed through the Work Environment Fund. Since 1993 the costs have been borne by employers.[14] This has given rise to competition between joint occupational health service centres for companies' work.[15]

The responsibility of the employee under the Swedish legislation is to co-operate with the employer in establishing a good working environment, to observe current safety regulations, to use safety devices, and exercise the caution required for the prevention of ill health and accidents.

In Denmark, the 1975 Working Environment Act imposes broad duties on employers to ensure safe and healthy working conditions, and to ensure that work is performed safely and without risk to health. The more specific duties of the employer include effective supervision, informing employees of any risks of accidents and diseases and providing them with the necessary instruction and training to avoid danger or risk, informing safety representatives and shop stewards of any written communications (for example, improvement notices) issued by the inspectorate, and carrying out tests, examinations, and surveys at the request of the inspectorate.[16]

Supervisors are required, on behalf of the employer, to contribute towards ensuring health and safety within their field of activity, to check the effectiveness of measures to protect health and safety, to take steps to avert danger when they know of faults which may involve risks of accident or disease, and to inform the employer if danger arises which cannot be prevented by their intervention on the spot.

At the end of May 1997 the Danish Parliament amended the Act to include the workplace assessment requirements of the EU Framework Directive,[17] discussed above in relation to the UK and Sweden. By the end of 1998 all Danish companies with five or more employees will have to produce written workplace assessments of the health and safety conditions (Arbejdspladsvurdering) in their companies.[18] The assessments must be in written form, and must follow the general steps set out in the legislation: identification of working environment problems; description and evaluation of the identified problems; development of priorities and action plans for resolving problems; and implementation of procedures for following up the action plans. For companies with fewer than five employees, such assessments must be produced by the end of 2000.

The 1975 Act obliges employers engaged in hazardous work and in other specified industries to participate in occupational health services. Occupational health services play an important role in Danish OHS arrangements, and have been provided with greater resources in recent years. A local and private system, with its structure determined by ministerial decree, they operate at three levels, providing individual company services, industry-wide services, and providing services from local centres. The services operate as consultants to all levels of the company, although there has been criticism that they have inadequate contact with employees.[19]

The main duty of employees is to co-operate in ensuring that working conditions are safe and without risk to health, and to check the effectiveness of measures taken to promote health and safety. In particular, they must wear protective clothing supplied, obey work procedures and hygiene conditions, and attend training courses.

It should be noted that general duties are also imposed in most jurisdictions (apart from the USA) on other persons at the workplace such as controllers or occupiers of premises, and manufacturers and suppliers of substances or equipment.[20]

In Chapter 2 of this book, we provide a detailed critique of the general duty provisions used in the OHS statutes.

Role of industrial relations institutions in setting, legislating, and enforcing OHS standards

Neither the voluntary efforts of employers, nor the creation and enforcement of legislation, have succeeded in reducing occupational injury and disease to acceptable levels. Workers, as the group most directly affected by work injuries, have an important contribution to make: in identifying the hazards; in co-operating with employers and governments to bring about improved conditions; and, when necessary, in taking direct action to protect themselves from imminent danger. The law can play an important role in ensuring that workers and their organizations do make those contributions, by giving them enforceable rights, both in setting and legislating standards, and in enforcing them.

In terms of setting and legislating OHS standards, most jurisdictions now adopt a tripartite approach. For example, in the UK, the HSC is the body charged with carrying out the general purposes of the HSW Act, including the replacement of existing enactments and regulations by a new system of regulations and codes of practice. The Commission includes employer and employee representatives (and local authority and professional body representatives). The HSE, which advises and assists the Commission, is entrusted with the enforcement function under the Act, is also tripartite in nature.

The Robens Report placed worker involvement high on its list of ways to dispel apathy and recommended that workers 'must be able to participate fully in the making and monitoring of arrangements for safety and health at their workplace'. The UK legislation makes specific provision for worker participation in OHS through the mechanisms of safety representatives and safety committees. In particular, the Safety Representatives and Safety Committees Regulations 1977 gave trade union appointed safety representatives the right to investigate dangerous hazards and occurrences, to investigate worker complaints, to make representations to employers, to carry out workplace inspections, to represent employees in consultations with inspectors, to receive information from inspectors, to attend meetings of safety committees, and to paid time off for training and to perform other functions. Employers are required to establish a safety committee within three months of being requested to do so, but no specific powers were given to safety committees by the Regulations.

The relationship between OHS and industrial relations contemplated by the UK legislation is complex. The Robens Report asserted that there was far greater community of interest between employers and employees on OHS than on most other workplace issues, and therefore that there was no legitimate role for collective bargaining on OHS. However, trade unions never accepted this aspect of the Robens philosophy, and in industrial reality there has certainly been collective bargaining on OHS issues in the UK.[21] Although this reality was barely recognized by the 1974 Act itself, the 1977 Regulations, promulgated following the election of a Labour government, took a different approach.

In particular, those Regulations gave trade unions the sole right to appoint safety representatives, and also gave them a series of rights (described above) far more extensive than Robens would have believed necessary to achieve consensus solutions to OHS problems. The backdrop to this change of approach was the overall government commitment to the encouragement of collective bargaining, and union recognition, and the rights associated with it, as the preferred means of regulating relations between employers and employees. However, under a subsequent government, the exclusive rights given by the Regulations to trade unions have resulted in a large (non-unionized) component of the workforce being effectively excluded from the benefits of this aspect of the legislation. The problem has been exacerbated by the decreasing levels of union membership, particularly amongst male workers since 1979.

This weakness in the UK system was at least partially remedied in 1996, when the Health and Safety (Consultation with Employees) Regulations were introduced, to 'top up' the 1977 Regulations. Employers are required to consult employees not covered by safety representatives appointed under the 1977 Regulations. The consultations must encompass the same subject matter as those required under the 1977 Regulations. The employer can choose to consult employees directly, or though elected representatives. If the employer chooses to consult employees directly, the employer can choose whichever method is preferred by all parties involved. If the employer decides to consult employees through an elected representative, then employees have to elect one or more representatives to represent them.[22]

The UK experience of health and safety representatives is admirably summarized by James[23] as follows:

In workplaces where union membership is high, trade unions are well organised and have member support, and management adopts a supportive approach, safety representative and safety committee structures have been found to work well. In other situations a far less satisfactory situation has been found, with only very partial application of the regulations being discovered.[24]

The Scandinavian countries have gone further in terms of involving workers directly in the decision-making process. In Sweden, the Work Environment Act 1977 enshrines the concept that worker control is a critical aspect of a healthy working environment. Employee safety delegates are elected for a period of three-years by local unions (and by employees in non-unionized workplaces) at workplaces with five or more employees. Large workplaces have several safety delegates, one of whom is elected as a chief safety delegate.[25] Safety delegates are given the right to halt dangerous processes pending an investigation by an inspector. They are also given the right to participate in the planning of new premises, devices, work processes, work methods, and the use of substances liable to cause ill health or accidents, while employers are required to inform delegates of any changes having significant bearing on conditions in the areas they represent.[26]

Employers are also required to respond to representations made by safety delegates without delay and, failing a satisfactory resolution, an issue can either be referred to the inspector or to the joint employer-employee safety committee. Additional rights of safety delegates to time off with pay, for training, to freedom from harassment or discrimination, and rights to information are also provided by the 1977 Act.

The local trade union organization may appoint a regional safety delegate, sponsored by the Institute of Occupational Health. The regional delegate's principal duty is to stimulate safe work practices in small companies without safety committees.[27]

Joint committees are required in workplaces with fifty or more employees or where employees demand one. Their role is to plan and supervise company safety and health activities. Employee members are elected by all relevant unions at the workplace. In general, safety committees are expected to strive for consensus in decision making, but, in the event of disagreement, any member may choose to refer an issue to the government inspectorate.[28]

Following the enactment of the 1977 legislation, a new Working Environment Agreement was negotiated by the main government, trade union, and employer representative bodies. This provided more detailed rules and guidelines for implementing the law. In particular, workers were given a majority of one on safety committees, and at least one employer member of the committee was to hold a management position in the firm. The committee was to be treated as a decision-making and as an advisory body, and was given authority over company health services. Unanimous decisions of the committee on budgetary matters were made binding on the company and, if unanimity is not achieved, any member can refer a matter to an inspector.[29] The Swedish Employers' Federation abandoned this agreement in 1991, and to date no new agreement has been signed on a national level. The strategy of the employers has been to decentralize OHS arrangements as much as possible, so that individual companies develop co-operative arrangements with the local union. Over 80 per cent of the workforce are trade union members. Local agreements remain in operation, and 'most Swedish enterprises still have good industrial relations in the workplace'.[30]

As part of this strategy, nearly all collective agreements at industry level have been renegotiated in the past few years, and have become shorter and more general. While collective agreements still determine much of the local administration of safety, the basic framework remains that set up by the 1977 Act.[31] The history of co-operation between unions and employers has been influential in shaping health and safety policy and in limiting tensions between the legislation and the outcomes of collective bargaining.

In Sweden, safety delegates and safety committees play an important role in influencing work conditions and, in general, that authority has been well utilized. At present there are about 100,000 safety delegates. Evidence shows that the right of safety delegates to stop dangerous work is exercised with restraint and very rarely abused. This right was used most frequently in the years following its introduction, and the decline in the number of cases coming to the attention of the inspectorate is probably due to joint decisions by both employers and employees to suspend work.[32] The effectiveness of the safety delegate system can to a large extent be attributed to the education, training, and information received by delegates from the Joint Industrial Safety Council, the Joint Work Environment Council, and the Swedish Working Environment Association. However, there are concerns that committees, though active, may exercise little influence on production decisions and that they lack the power and authority necessary to assume the responsibility for the safety of the work environment.[33]

An unforeseen and undesirable side effect of this system has been to encourage and facilitate many employers avoiding direct responsibility for safety issues, and passing the burden instead to the safety delegate.[34]

Given that the employer has both the major responsibility for work accidents and the greatest ability to prevent them, this trend is a matter of serious concern. Some efforts have been made to reverse it (for example, legislative amendments to make clear that the committee's role is limited to 'participating' in the planning of the safety programme) but it is unclear to what extent they have been successful. There remains the fundamental issue of ensuring that health and safety concerns are integrated into the highest levels of the firm's decision-making processes.[35]

Sweden had also adopted a tripartite approach to the enactment and implementation of health and safety standards, operating through a number of Boards with tripartite operating structures. However, this too has been weakened in recent years, as employers have sought more decentralized arrangements. Until 1992, the Directorate of the National Board of Occupational Health and Safety comprised the Director-General, ten other members and two employee representatives. Seven of the members represented the labour market parties. However, following a strategy of reducing formal representation on government bodies, in an effort to reduce their support for 'corporatist' arrangements, private employers withdrew from the Directorate in 1992,[36] and by mid-1993 the government had terminated formal union representation on the Board, and other tripartite Boards dealing with OHS issues. Board membership is now dominated by government appointments, including appointments from the ranks of management personnel and retired unionists. Consultation and negotiation in relation to the development of standards continues, but at a more informal level.[37]

Denmark is in many respects very similar to Sweden. One of the main aims of the 1975 legislation is to create a system whereby workplaces themselves can solve health and safety problems, and where the 'social partners' work co-operatively to achieve this.[38] To this end, from July 1998 all workplaces with more than five employees are required to have safety representatives, safety groups, and safety committees.[39] Previously the threshold was companies with ten or more employees, and companies with twenty employees in the case of office-related workplaces. Failure to appoint safety representatives as required by the Act renders the employer liable to an order or fine by the Labour Inspectorate (see below). The Danish Employers' Confederation strongly resisted these changes, and, after withdrawing from tripartite arrangements (see below), agreed to a compromise in which the Minister of Labour offered employers a three year transition period.[40]

Safety representatives have inspection, information, and

representation rights, are protected from discrimination, and have rights to time off for training and to submit questions to the inspectorate.

Safety groups (which form the core of the internal safety organization) must be set up for each department or field of activity. Their role is to evaluate the acceptability of working conditions, to ensure that necessary instructions are given to employees, to ensure that employees observe safety regulations, to take action against risks arising out of working conditions, to participate in planning, and to establish action plans for problem solving.[41] The safety groups must also check compliance with safety regulations and report and investigate occupational injuries. Each safety group consists of the foreman/supervisor and the employees' safety representatives. The Group has power to stop work, or work processes, so far as is necessary to avert imminent and serious risk to health and safety.

Safety committees are composed of employees and members of management, and they co-ordinate the activities of the safety groups. In effect, safety groups are local branches of the safety committee. The role of safety committees is to plan, manage, advise on, and supervise activities concerning OHS at the workplace. The duty of the safety committee is purely advisory, and it has no power to make decisions as regards the execution of safety and health measures.

In 1996, before the threshold for implementation of the safety committee and representatives provisions was reduced (see above), it was estimated that 90 per cent of businesses in Denmark under an obligation to have health and safety representatives and committees had established these structures.[42] Overall it appears that 'this system provides an effective force for maintaining and improving the standard of health and safety within the undertaking' and that 'overall the standard of co-operation between employers and the trades unions is good'.[43] This system is successful because about 80 per cent of employees are union members, and most unions are affiliated to the Danish Confederation of Trade Unions. It enables the Working Environment Service (WES) inspectorate (responsible to the Ministry of Labour) to concentrate on higher risk areas of industry, and the less safety conscious employers.[44]

Again there is a substantial degree of tripartism at the national level, with both the Danish Federation of Trade Unions and Danish Employers' Confederation (the peak employer organization, which represents about 50 per cent of employers in the private sector) nominating members to sit on the Work Environment Council (which recommends policy changes under the Act), and on the twelve Trade Safety Councils (each of which covers one specific industry sector or group of industries, and provides detailed guidance on the application of OHS laws to the specific conditions of the particular industry or group of industries). This is consistent with the strong Danish tradition for 'involving the social partners in the formulation of new work environment rules'.[45]

From June 1997 until June 1998, the Danish Employers' Confederation withdrew from the Work Environment Council and the new Branch Work Environment Council, in response to major changes[46] in tripartite arrangements at the national level announced by the Minister of Labour. The Confederation was reacting to the Minister's reforms to reduce the threshold for the appointment of safety representatives and committees, and proposals to increase the Ministry's role in OHS regulation and in goal setting, particularly in relation to work carried out by the new 'Branch Work Environment Councils'. The Confederation alleged these measures, and the manner of their introduction, were contrary to the tradition of tripartism in Danish OHS regulation. The proposed changes would result in increased centralization, and bureaucracy, and a hierarchical policy and operating system.[47] The changes were made because of the Minister's belief that the labour market parties had not been fulfilling their roles under the legislation effectively. A compromise was reached in which the Minister, without changing the statutory provisions, agreed that the branch work environment councils would be given greater autonomy and independence, and that the Minister would only intervene if the work of the councils was in conflict with the government's action plan for OHS.[48]

Overall, the Danish system strongly emphasizes the resolution of health and safety problems at both national and workplace level by negotiation between the social partners. As Martin, Linehan, and Whitehouse[49] observe:

The legislature, the Government and the WES may give a lead, and be asked to approve and support the outcome, but will often look towards employers and trades unions to develop and implement solutions. This is illustrated by the way the problem of 'repetitive and monotonous work' was recently handled. The Danish Parliament asked the Government to prepare a plan of action to deal with the issue. In the event the response was prepared by negotiation between the social partners who, although they sought approval for the solution, undertook to implement the necessary arrangements without further Government action by way of regulation or enforcement.

Recently, however, there has been criticism of the operation of the Danish safety representatives and safety groups.[50] The main criticisms appear to be that these institutions are not part of the other co-operative systems in Danish workplaces, that they contribute towards maintaining a hierarchical structure which is now considered outdated in companies striving for greater flexibility, and that workers tend to use these institutions rather than seeking their own solutions to issues.

In Australia, worker involvement in OHS issues and the powers and rights of safety representatives extend beyond that provided by the existing UK approach and are much closer to the Scandinavian model. In most Australian jurisdictions, there is a substantial degree of tripartism in the setting and legislating of OHS standards. For example, most jurisdictions now have an Occupational Health and Safety Commission (or similar body), which usually consists primarily of representatives of employer and employee organizations,[51] and, at the Federal level, the NOHSC is also tripartite in nature. Victoria is the exception in recently departing from this model.

Queensland has the most developed institutions at an industry level. Until 1998 the Minister had power to establish tripartite industry workplace health and safety committees which gave advice on promoting and protecting workplace health and safety in their industry. In 1998 these committees were replaced with industry sector standing committees of the state-wide tripartite Workplace Health and Safety Board. The standing committees give advice and make recommendations to the Board in relation to health and safety in the industries in which they were established. Queensland has also made provision for the compulsory appointment of health and safety officers, responsible to management for health and safety issues, in workplaces with more than thirty employees.

Of far greater day-to-day significance are the roles of worker safety representatives and worker members of safety committees. Here, there is very little uniformity between the different Australian jurisdictions.[52] For example, while New South Wales and the Northern Territory only make provision for worker participation on safety committees, the large majority of jurisdictions provide for the election or appointment of worker safety representatives. When it comes to the rights and powers of safety representatives, there is again considerable disparity between different jurisdictions. Here, the main contrast is between the far-reaching Victorian approach, the weaker versions of this model to be found in South Australia, the Commonwealth, and the Australian Capital Territory, and the purely consultative role given to health and safety representatives in Queensland, Western Australia and Tasmania.

The Victorian approach is illustrative of what might be done and is particularly important since most other Australian legislation approximates a diluted version of it. It can be characterized as having six components:[53]

1. employers and worker representatives are required to consult about all health and safety issues that arise at a workplace, and shall attempt to resolve such issues;
2. health and safety representatives are given specific legal entitlements: to inspect the workplace; to accompany an inspector on her/his inspection of the workplace; to require the establishment of a health and safety committee; and to be present at an interview between an inspector and an employee;
3. employers have obligations to permit a safety representative to have access to specified OHS information in the possession of the employer; to permit the safety representative to be present at any interview between the employer and an employee on an OHS issue; to consult on proposed changes to the workplace, plant, or substances that may affect OHS; to permit the safety representative to have paid time off for training or for the performance of OHS duties; to provide facilities and assistance; and to supply prescribed medical information;
4. safety representatives are entitled to seek outside assistance from people with expertise in OHS to enable them to perform their functions;
5. safety representatives are empowered to issue improvement notices similar to those which inspectors issue, described below. However, these notices are 'provisional' in that the person to whom the notice is issued can dispute the notice by requiring an inspector to attend the workplace, in which case the effect of the notice is suspended and the inspector will make his or her own independent determination and take appropriate action; and
6. where there is an immediate threat to health and safety such that it is not appropriate to follow agreed procedures, the health and safety representative may, after consultation, direct that work shall cease. This last power is clearly both the most potent and the most contentious, although in practice the evidence suggests that it is used sparingly and responsibly.

Separate provisions address the role and powers of safety committees. These provisions are facilitative rather than prescriptive. That is, although safety representatives have the right to require the establishment of a safety committee in prescribed circumstances, once the committee is in operation there is considerable flexibility about its functioning. Its main roles are envisaged to be facilitating co-operation between employer and employees in instigating, developing, and carrying out measures designed to ensure OHS, and formulating, reviewing, and disseminating the standards, rules, and procedures relating to OHS which are carried out or complied with at the workplace.

In Australia, as in the UK, there have been various attempts to divorce OHS and industrial relations (not least the Robens Report itself). However, the provisions in Victoria and elsewhere, which provide extensive powers for safety representatives, significantly erode managerial prerogative, and thereby substantially reconnect OHS and industrial relations.[54]

A further link between OHS and industrial relations in the Australian context is a decision in 1986, in which a Federal tribunal was prepared to make extensive award provision relating to OHS in the vehicle building industry, and thereby effectively overriding the provisions of state preventative legislation.[55] However, the circumstances surrounding this award were exceptional and there seems little likelihood that they will be repeated.[56]

In contrast to these European and Australian models, the USA's approach to worker participation in OHS issues has been much more muted. An element of tripartism is evident in that the National Advisory Committee on Occupational Safety and Health did involve representatives of management and labour, as well as occupational safety and occupational health professionals, and the public, but there was very little effort to involve workers directly at workplace level.[57] At present:

... the great majority of USA's employees play no real role in the enforcement of health and safety regulations, and the USA still has by a considerable margin, the worst OHS rates of western industrialized nations.[58]

The limitations of the USA's approach, which relies instead on an adversarial strategy and an army of inspectors to identify breaches of the Act and issue citations, are now increasingly recognized.[59] Significantly, the Clinton Administration's proposed reforms to the OSH Act seek to involve the workforce directly in OHS issues in a number of important ways.[60]

First, as noted above, the Comprehensive Occupational Safety and Health Reform Bill, not yet passed by Congress, will require employers to establish and maintain safety and health programmes to reduce or eliminate hazards and prevent injuries and illness to employees. These programmes will include employee education and training. Second, the Bill requires employers with eleven or more full time employees to establish safety and health committees made up of employee representatives and up to an equal number of employer representatives. Third, the Bill allows affected employees to more actively participate in Commission proceedings by authorizing employee challenges to, and commission review of, penalties and violations. Finally, the Bill prohibits employers from penalizing an employee for taking ***bona fide*** health and safety action or for refusing to undertake unsafe work that would expose the employee to a danger of injury or serious impairment to health.

Even under the present law, there is nothing to prevent workers raising OHS issues in the broader context of collective bargaining and there may be significant benefits in them doing so. As Ashford[61] puts it:

Collective bargaining has the potential to go far beyond the mandates of the OSH Act by obliging employers to interact closely with workers rather than merely complying with loosely enforced and inadequate government standards. The negotiation process enables different local and industry wide needs to be met, particularly where hazards are extensive. Further, it may move the responsibility for occupational health and safety out of the sole hands of management and thus encourage the participation of workers in the process of controlling technology in the workplace.

Occasionally, workers in the USA have taken advantage of their rights to bargain over OHS, but the complex nature of the collective bargaining process, and the fear that jobs will be lost if strict OHS standards are introduced, have deterred workers from taking industrial action except in extreme circumstances.

In summary, provisions relating to broad worker participation in OHS have a long history, particularly in Scandinavia, and have been increasingly adopted in the UK and in Australia. The most effective vehicle for achieving direct worker participation in OHS is demonstrably that of worker elected safety representatives.[62] Despite initial employer concern that such provisions might be abused and used as an industrial relations weapon to extract concessions in other areas such as wages, there is little evidence that this has occurred in practice. On the contrary, there is considerable evidence that, at least in larger workplaces with safety conscious and effective worker organizations, worker participation can play a substantial role in improving OHS. It is significant that the

USA, the one country that has not attempted to involve the workforce directly in improving OHS, is now belatedly contemplating doing so.

Legal requirements to conduct regulatory impact statements and to review legislation periodically

Jurisdictions overseas are increasingly taking steps to assess the value of legislation in cost-benefit terms prior to its enactment, and some now review the impact of existing legislation periodically. However, much of this is done on an *ad hoc* basis, and, of the countries under review, it is only the USA (and Canada at Federal level[63]) that have sought systematically to establish cost-benefit analysis.

In 1981, President Reagan's Executive Order 12291 required agencies to conduct cost-benefit studies of all proposed regulations and legislation with major impacts on industry. However, in a landmark decision in the same year, it was held that the OSHA was not required to apply a cost- benefit analysis when promulgating cotton dust standards under the OSH Act 1970,[64] although it was required by the statute to produce a feasibility analysis. The latter is required by virtue of the OSH Act section 6, whereby the OSHA must consider the economic feasibility of compliance and the overall effect on the nation's economy in the promulgation of standards.[65]

This decision had important ramifications for other standards promulgated under the OSH Act, given the Supreme Court's indication that health and safety is of 'greater concern' than costs under the statute. Nevertheless, in subsequent years, both the Reagan and Bush administrations continued to place considerable emphasis on the costs of regulation, and showed a reluctance to implement new regulations unless benefits in excess of the projected costs could be clearly demonstrated. Indeed, the Reagan Administration institutionalized the concept of 'regulatory impact analysis' across broad areas of social policy through executive order and statutory requirements. The unsuccessful conservatively inspired Safety and Health Improvement and Regulatory Reform Bill of 1995 proposed that OSHA standards would have to satisfy additional risk assessment and cost-benefit analysis criteria, and peer reviews.[66]

In the UK, it is arguable that, by implication, the general duty provision of the HSW Act, with its standard of 'reasonable practicability' (which involves a balancing between risk, cost of prevention, and damage), implies some form of cost-benefit analysis by the courts in considering the expense and trouble that is reasonable in the avoidance of risk. As

Asquith LJ put it, in the landmark decision of *Edwards* v. *National Coal Board*,[67] an employer, prior to an accident, must consider the measures necessary and sufficient to prevent any breach, and then whether these were 'reasonably practicable':

> . . . in that a computation must be made by the owner, in which the quantum of risk is placed on the one scale and the sacrifice involved in the measures necessary for averting the risk (whether in money time or trouble) is placed on the other; and that if it be shown that there is a gross disproportion between them— the risk being insignificant in relation to the sacrifice—the defendants discharge the onus on them.

Although the present UK government has undertaken a recent and broad-ranging review to simplify legislation by removing unnecessary regulations,[68] the thrust of future legislative action is likely to be in response to EU Directives. Accordingly the government's position in terms of cost-benefit approaches is 'to seek to influence new EC legislation at the earliest possible stage, and in particular by ensuring that benefits are likely to outweigh costs, that proposals are properly evaluated in advance'.[69] Recently, however, the HSC has been taking steps to implement a recommendation that it should publish appraisals of the cost to industry and employers of the measures it proposes, together with estimates of the gains from such measures.[70]

In Sweden, the legislative history of the 1977 Act suggests that some weighing of costs and benefits is a material consideration in interpreting the law:

> Obviously we are not talking about a one step transformation of the working environment without any consideration of economic and technical resources . . . The interpretation of the law should develop taking into consideration the need to improve the working environment as well as the economic resources which are available at each point in time.[71]

Cost-benefit analysis also occurs *ad hoc* when the National Board 'has issued an internal regulation which mandates that an analysis of the costs of implementation, along with other factors, be conducted in connection with all directives issued by the Agency'.[72]

Finally, there is section 14 of the Swedish Code of Statutes (amending the Government Agencies Ordinance of 1987, and promulgated 16 June 1994), which requires an authority, before adopting a proposition, to consider whether it is the most appropriate measure, and to investigate the financial and other consequences of the provision. However, this investigation may be deferred until after a provision has been adopted if there is danger to the environment, health, or personal safety if the provision is not adopted. There is also a duty to continually review existing provisions to see if they are necessary and suitably framed.

In Denmark, the procedures for impact assessments were formalized in 1981. The Work Environment Council are required to provide a calculation of economic costs and benefits to society with proposals for regulations, when they are submitted to the Ministry of Labour for approval. From 1983, the Danish Labour Inspectorate have been required to make a preliminary estimate on costs and benefits, before the proposal is submitted to the Council for discussion. The obligation to conduct economic impact assessments covers all rules and standards but not administrative guidelines. There are no requirements concerning the evaluation of existing instruments.

All of the Australian OHS statutes qualify the general duty provisions with the expression 'reasonably practicable', or an equivalent,[73] which, as noted above, introduces a form of cost-benefit analysis by the courts in determining the appropriate measures required to minimize workplace risks. Most of the regulatory impact statements required in Australia to date have not involved comprehensive cost-benefit analyses. Some jurisdictions (for example, New South Wales, Victoria, Queensland, South Australia, and Tasmania) make provision, either in their OHS statutes or general procedures for making subordinate legislation, for some form of regulatory impact analysis.[74] In 1995, the Council of Australian Governments (COAG) agreed on 'Guidelines and Principles for National Standard Setting and Regulatory Action'. The guidelines require standard setting bodies, such as the NOHSC, to conduct a Regulatory Impact Statement for each new or amended standard.[75]

There are major differences of opinion about the value of impact statements and similar strategies. Those who utilize impact statements understandably tend to argue that they provide a rational basis for determining whether proposed (or existing) regulations are socially or economically justifiable. On the other hand, there are serious criticisms of the sort of cost-benefit or regulatory impact statements that have been attempted so far.[76]

Specifically, employers have little difficulty identifying the costs of new legislation to them, and also have an incentive to exaggerate those costs in an effort to block the proposed legislation. In contrast, the social benefits of new regulation are difficult to measure, since they are diffuse, and go unrecorded in the accounts of individual enterprises.[77] Moreover, it is almost always assumed that the costs of regulation are fixed. In practice, new legislation commonly promotes innovation, which has the effect not just of reducing those costs, but also saves substantial sums of money and thereby increases profits directly (for example, through new technology, substitute materials, and redesigned jobs).[78] Yet, this is not taken account of in the calculations. Finally, variables which are not easily quantified are commonly either ignored, undervalued, or given an arbitrary valuation in most impact statements (see the various attempts to value a life, where wildly different figures are produced by different organizations).

In essence, the direct costs are easily quantified and taken into account, whereas future benefits and soft variables are not, with the result that regulatory impact statements are commonly skewed in favour of the status quo and against new regulation. Against this background it has been argued that such statements tend to serve a political purpose, being used as a strategy to defeat the introduction of more stringent exposure standards to hazardous substances, under the guise of scientific neutrality.[79]

However, as McGarity[80] points out, 'if the public and regulatory beneficiaries are convinced that regulatory analysis is not being used cynically to reach particular substantive results, it can become an effective mechanism for achieving public accountability'. Specifically, he argues that it can help decision makers make rational and informed decisions (although it cannot fully inform or precisely point to rational conclusions), it can encourage the decision maker to articulate policy preferences and demonstrate to the public how those policy preferences were applied, and it can assist in assessing the advantages and disadvantages of particular regulatory options. Crucially, he concludes that the considerable disadvantages of such mechanisms can be avoided by recognizing that, alone, they cannot appropriately dictate regulatory results though they can achieve more modest ends such as setting agency priorities and structuring agency options.[81]

Relationship between regulation, complementary legislation, and common law provisions

This relationship has been examined exhaustively in the literature, and has been the subject of extensive study. With the exception of certain interest groups, most notably the legal profession, there is close to a consensus that there is no effective or positive contribution that the award of damages at common law is likely to make to accident prevention. Nor is there any positive relationship between provision for common law damages and either regulation or complementary legislation concerned to prevent occupational injury or disease.

Briefly, the award of common law damages was largely abolished in the USA as a trade-off for a guaranteed workers compensation system, and seems most unlikely ever to be revived. Similarly, in Denmark, the system of compensation is not fault-based but simply involves an evaluation by the National Board of Industrial Injuries as to the issue of compensation. In the UK, common law damages remain, but are not perceived to play any significant role in preventing accidents.[82] Nor is there any attempt to utilize common law in conjunction either with regulation or with complementary legislation.

In Australia, most of the workers compensation statutes restrict or prohibit workers' access to common law actions.[83] For example, common law actions against employers have been abolished in the Commonwealth (for economic loss only), Victoria (after 12 November 1977), South Australia, and in the Northern Territory. Queensland, Tasmania, and the Australian Capital Territory permit unrestricted access to the common law, while the remaining workers compensation statutes (and the Commonwealth in relation to non-economic loss) provide only limited common law rights to compensation. The New South Wales, Victorian (prior to 12 November 1977), and Western Australian statutes enact threshold requirements (such as death, 'serious injury', or 'serious disability') before a worker can take common law actions. The Commonwealth Act (for non-economic loss only) and the New South Wales Act require workers to elect whether to receive common law damages or statutory lump sum payment for permanent disability. With the exception of the South Australian and Western Australian Acts, the Australian workers compensation statutes make it compulsory for employers to insure against their common liability.

The reasons for this total lack of enthusiasm for common law provisions as a preventative tool are not hard to find. In particular, employers usually insure against the possibility of common law damages awards against them (and are often compelled by law to do so). This substantially reduces the financial incentive for employers to avoid work injuries. Any potential effect that civil actions for damages might have on accident prevention is further reduced by the fact that only a very small proportion of injury cases ever get to court. Ignorance, difficulties of proof, and the high cost of litigation largely account for this. Finally, fear of litigation sometimes results in cover ups and a reluctance to take remedial action after an accident in case this action is treated as tacit admission that previous measures were inadequate. For these, and other reasons,[84] common law damages are not an effective accident prevention measure, either alone or in conjunction with other measures.

Processes used to harmonize standards, implement nationally uniform standards, and to ensure mutual recognition of regulation

Of the countries under consideration, the USA and Australia have Federal systems and substantive difficulties in harmonizing standards, implementing nationally uniform standards, and ensuring mutual recognition of regulation. These countries have each adopted quite different approaches to resolve these problems.

In the USA, the approach taken under the OSH Act 1970 was to enact legislation at Federal level that would apply throughout the country (see above). Accordingly, the USA does not experience major problems of harmonization, national uniformity, or of mutual recognition within its territorial limits. As indicated above, the most significant difficulties are concerned with uniformity of enforcement and are caused by the delegation of responsibility for administration of the OSH Act to the individual states.

Since 1990 there has been a concerted attempt to harmonize OHS standards in Australia.[85] The Ministers of Labour Advisory Committee (MOLAC) comprising Commonwealth, state, and territory Ministers of Labour, in 1990 agreed that, 'as far as practicable, any standards endorsed by the NOHSC will be accepted as minimum standards and implemented in the state/territory jurisdiction as soon as possible after endorsement'.

A National Standards Summit of representatives of the NOHSC and state and territory tripartite OHS commissions and advisory councils was held in April 1991 to review the standard setting process. In November 1991, the premiers of the states and Chief Ministers of the territories reached an agreement that they would 'achieve nationally uniform OHS standards and uniform standards in relation to dangerous goods by the end of 1993'.[86] The Commonwealth subsequently supported the decision. The basic strategy was for MOLAC to work towards standardizing the parent Acts, while the NOHSC would develop new national standards and codes of practice, which state and territory governments would uniformly adopt as regulations and codes of practice, replacing their detailed and prescriptive regulations. All states and territories would be bound to implement OHS standards agreed to by a majority of six or more state and territory Ministers.

In December 1991 the NOHSC established a tripartite National Uniformity Taskforce to develop a strategy for harmonizing existing standards and to fast-track priority new standards by the end of 1993. The Taskforce comprised of representatives from Victoria, New South Wales, South Australia, The Australian Council of Trade Unions, the Australian Confederation of Industry (now Australian Chamber of Commerce and Industry), and the NOHSC. The Taskforce established a National Occupational Health and Safety Framework to guide national standards development activity. This framework identified hazard areas where national uniformity was to be achieved, and further areas where further investigation was required to determine whether it was necessary to develop national standards. The Taskforce declared seven key first-order priorities for achieving national uniformity: plant; certification of users and operators of industrial equipment; workplace hazardous substances; occupational noise; manual handling; major hazardous facilities; and storage and handling of dangerous goods. By the end of 1996 the NOHSC had declared six of these standards (all except dangerous goods) and, apart from major hazard facilities, most had been adopted by the States and Territories. In 1997 it became clear that, with the new Liberal-National Party government in power at Federal level, the NOHSC would not be leading the push for nationally uniform OHS standards. There is, however, still scope for the promotion of national uniformity of OHS standards through the Labour Ministers Council.

In the European context, the EU's 1989 'Framework' Directive for the Introduction of Measures to Encourage Improvements in Safety and Health of Workers should also be noted. This furthers the policy of the Single European Act of encouraging improvements in health and safety of workers through 'the harmonisation of conditions in this area, while maintaining the improvements made'. It lays down a series of principles that employers in each of the member countries should apply in developing protective and preventive measures. These include priority to the avoidance rather than the control of risk, and the importance of combating risk at source rather than through ameliorative measures.[87]

Five 'daughter' Directives cover workplace conditions, safe use of work equipment, manual handling, personal protective equipment, and display screen equipment. Since 1992 further Directives affecting health and safety at work have emanated from the EC, covering matters such as protection of workers from risks related to exposure to biological agents at work, minimum safety and health requirements at temporary or mobile construction sites, measures to encourage improvements in the safety and health at work of pregnant workers, and workers who have recently given birth, or who are breast feeding, the minimum requirements for the provision of safety signs at work, and other safety and health issues.

Each Directive creates a legal relationship between the EU and the member state that 'is binding as to the result achieved upon each member state to which it is addressed, but shall leave to the national authorities the choice of form and methods'.[88] It was originally believed that Directives did not provide any directly enforceable rights for individuals. However, a series of landmark decisions of the European Court of Justice have established that private individuals can enforce Directives in certain circumstances, which now arguably extend to employees in the private sector.[89]

Different member (or prospective member) countries are responding in different ways to EU Directives. For example, Sweden is in the process of amending about 55 of its 210 or so statutory instruments, and repealing more than 70. Denmark has also placed a high priority on achieving uniform work environment rules at a high level and regularly amends rules made under the Working Environment Act in order to comply with EU Directives. In contrast, the UK position is to avoid disrupting the basic framework of the 1974 legislation, and to minimize change to the most recent regulations, while continuing to modernize outdated regulations in a manner consistent with the EU Directive. In introducing regulations to meet the relevant Directive the intention is generally not to go beyond it, so as to minimize the impact of alterations in the law.

There remain doubts about both the efficiency and the adequacy of the pattern created by the single European Act, which gives primary responsibility to the home country regulator, subject to an agreed floor of minimum standards, while retaining also the residual right of the host country to regulate in the 'public interest'. Arguably this may leave workers inadequately protected, potentially lead to a deregulatory 'race to the bottom', and create uncertainty for producers due to continued regulatory diversity and lack of clarity as to the permissible scope of host country rules protecting the 'general good'.[90] There are fears that the gains made through the implementation at national level of EU Directives from 1989 to 1992 are currently under threat. Mounting pressure from employers and many member states for deregulation has meant that legislative activity in OHS is minimal. Little has been done to revise Directives that need updating. There are also suggestions that the new European Commission action programme on OHS focuses on non-legislative measures, none of which appear innovative, and which fall short of proposals put forward by employers and trade unions in 1992.[91]

Approach taken to ILO conventions

The ILO plays an important role in setting standards in respect of OHS. In respect of prevention, it has adopted a number of standards which address specific hazards (for example, the guarding of machinery, the working environment, benzene, and asbestos), and others which concern the problems of particular occupational groups (for example, seafarers, dock workers, and underground workers). Most important of all, in 1981 it adopted the *Occupational Safety and Health Convention* (No. 155). This is a very broad-based convention which specifies that each member shall, 'in consultation with the most representative organisations of employers and workers, formulate, implement and periodically review a coherent national policy on occupational safety, occupational health and the working environment'. The overall purpose is to prevent accidents and injuries to health by minimizing so far as reasonably practicable the causes of hazards inherent in the working environment. The Convention goes on to specify, in some detail, the areas that need to be addressed in order to achieve this outcome.[92]

However, ILO Convention No. 155 can only take effect when it is ratified by a member nation. Ratification is particularly important in a Federal system such as Australia where there may be constitutional constraints on the power of the Federal government to enact OHS legislation nationally.[93] Ratification of ILO Convention No. 155, in conjunction with the use of the external affairs power in section 51(29) of the Constitution, would enable the Commonwealth to regulate working conditions throughout Australia, and 'could provide a basis for preventative legislation [at the national level] at least as comprehensive as that envisaged by Robens, or as that adopted by any of the states'.[94] In its 1995 Report, the Australian Industry Commission recommended against uniform Australian Federal OHS legislation based upon a ratification of ILO Convention No. 155, preferring a 'template' approach for core elements of OHS legislation.

Unfortunately, ILO Convention No. 155 has only been ratified by a small minority of member states.[95] Only Sweden, of the countries currently being examined, has so far done so.

Types of instrument used to enact policy

The UK model under the HSW Act usefully illustrates the extent to which different instruments (for example, legislation, subordinate legislation, codes of practice/guidelines) can be used to achieve policy outcomes. The structure under the UK legislation (which is essentially reproduced in Australia) is for general duties, as described earlier, to be established by legislation. These lay down, in broad terms, the responsibilities of employers, employees, manufacturers, and others, in respect of health and safety at work. The structure and powers of the agencies responsible for administering and enforcing the law are also set out in the main enabling Act, as are the system of penalties, legal defences to liability, and procedural matters.

The bulk of highly detailed specific standards, which relate to particular types of workplaces or to particular processes or substances, are to be found in regulations. So also are the rights and responsibilities of safety representatives and safety committees. Since these detailed matters less commonly raise issues of principle they do not require the same degree of parliamentary scrutiny as do the general duties. The Act gives the Secretary of State very wide powers indeed to make regulations, including provision for the health and welfare of persons at work, and the protection of persons other than persons at work against risks to health and safety arising out of, or in connection with, the activities of persons at work. The technique of relying heavily on regulations to implement a parent Act is well established. The regulations are made by statutory instrument and Parliament has the power to annul them.[96] They are also subject to the supervisory jurisdiction of the courts. They are enforced in exactly the same way as the Act.[97] The general common law principle is that compliance with the specific requirements of a regulation is taken to be compliance with the more general correlative requirement expressed in the general duty.

The regulations are reinforced by codes of practice, which take on a far more important role than previously. These are approved by the HSC and are intended to provide practical guidance on how to comply with the general duties, with safety regulations, or with any of the pre-existing statutory provisions. The Commission may produce codes of its own initiative, adopt codes prepared by other bodies, or give a stamp of approval to codes which others have issued. Codes have been adopted on a wide range of matters, including giving guidance on safe levels of exposure in respect of health hazards, precautions to protect employees, safe methods of handling, and on health monitoring.

The legal status of the codes is important. While a failure of any person to observe a provision of an approved code of practice does not in itself render that person liable to criminal or civil proceedings, nevertheless where a person is alleged to have broken a general duty, a regulation, or any other relevant statutory provision, the fact that the accused had failed to observe a relevant code of practice may be taken as conclusive evidence of his failure to do all that is reasonably practicable, unless the court is satisfied that he has complied with his obligations in some other way. In effect, this provision creates a rebuttable presumption that breach of a code constitutes a breach of a general duty under the Act.

The different Australian jurisdictions have enacted provisions in relation to the standard setting in regulations and codes of practices which largely resemble the UK provisions. The Commonwealth, Victoria, South Australia, Tasmania, and the Northern Territory legislation follows the UK approach of a rebuttable presumption that breach of a code is a breach of a relevant duty or regulation. In the other jurisdictions, apart from Queensland, a code of practice provides evidence of what a reasonable employer would do to comply with a general duty or obligation in a regulation.

The Queensland OHS legislation takes a different approach to regulations and codes of practice (called advisory standards). When there is in force a regulation covering the risk, it must be followed to comply with the general duty. A person must follow relevant advisory standards; or adopt another method that identifies and manages exposure to risk. Where there is no guidance in the regulations or advisory standard, the person must take reasonable precautions and exercize proper diligence to ensure the obligation is discharged. It is a defence for the duty holder to show that he or she followed the relevant regulation or advisory standard, or, where there is no regulation or advisory standard about exposure to a risk, that she or he chose an appropriate way and took reasonable precautions and exercized proper diligence to prevent the contravention.

In Sweden, the principal mechanism for achieving policy outcomes (under the umbrella of the general 'framework' legislation described above) is regulations (ordinances),[98] coupled with sanctions. These can be promulgated very simply, without any of the sorts of notice, comments, or hearing requirements which characterize the USA's system of regulation. Swedish laws cannot be challenged in court as unconstitutional and the constitutionality of regulations issued by the National Board on Occupational Health and Safety may not be reviewed by a court. A further option is for the National Board to issue less stringent guidelines which require the Labour Inspectorate to issue a ruling prior to the imposition of sanctions. As noted earlier, the enabling Act and regulations are supplemented by agreements between employer and employee organizations.

In Denmark, various parts of the 1975 Act empower the Minister to make rules concerning OHS. These are formulated by way of administrative regulations issued by the Ministry of Labour. The process by which the rules are drafted is unusual, the employer and employee representative organizations drafting rules together with the Ministry of Labour. Prosecutions are usually based on the rules laid down in orders and regulations rather than on the general provisions of the Act.

The rules are supported by a system of non-statutory directions and announcements (guidelines, instructions, etc.) issued by the inspectorate. Directions enlarge upon the rules or offer guidance in complying with the rules. Announcements issued by the directorate of the inspectorate do not have force of law but assist inspectors by indicating to employers what measures need to be taken to comply with the law.[99]

In the USA, in contrast to the UK and Scandinavia, there exists a legalistic 'command and control' regulatory culture, with a consequent heavy emphasis on highly detailed specific regulations, capable of enforcement by inspectors with only very modest training and skills.[100] In particular, each employer has a duty to comply with specific standards promulgated under the OSH Act, and, in cases not covered by specific standards, the employer owes a general duty in terms described earlier.

There are three types of standards mandated by the OSH Act. First, there are interim standards consisting of Federal standards from other Acts and national consensus standards existing at the time of implementation of the OSH Act. Second are permanent standards to replace or augment the interim standards. Third are temporary emergency standards which may be issued immediately upon the finding of serious danger to employees.

As to the model of standards adopted, the OSHA has summed up its current approach as follows: 'in promulgating standards, OSHA seeks the optimum blend of performance language, which offers flexibility, and specification language, which provides exact guidance. To the extent that the agency achieves this objective, standards are easy to understand and follow and workers are effectively protected.'[101] Under the 'New OSHA', the OSHA:

> . . . is instilling common sense in its regulatory processes by involving employers, workers and safety and health professionals in regulatory partnerships, by basing protective standards on consensus wherever possible, and by rewriting old standards (as well as writing new ones) in plain language. As part of this effort OSHA is listening more closely than ever to the concerns of small business.[102]

There is no precise equivalent to the codes of practice approach of the UK and elsewhere, and the regulations, on which the USA system so heavily relies, are (apart from the first category above) required to go through a tortuous process of scrutiny and challenge before they are finally promulgated.[103] In the case of permanent standards this frequently involves a challenge to the validity of the standard in the USA Court of Appeals, which can be both very expensive and mean protracted delays. It is now common for standards to take more than five years from the OSHA's announced intent to regulate to the final rule.[104]

This process is widely believed to be one of the least successful aspects of the USA system. It has resulted in the OSHA producing less than thirty new health standards over more than twenty years, despite compelling scientific evidence of the need for greater worker protection from health hazards. The proposed reforms announced by the Clinton Administration seek to mitigate these problems by requiring the OSHA to respond to petitions for health and safety standards within ninety days of receipt, and, if the agency finds that a standard is warranted, to issue a proposed rule within twelve months of the petition and a final rule eighteen months later.

Enforcement powers

As we argue in the body of this book, legislation that is not enforced seldom fulfils its social objectives, and effective enforcement is vital to the successful implementation of OHS legislation. This section examines the enforcement powers of agencies, their representatives, employees, employee representatives, and individuals, comparing and contrasting the UK, USA, Scandinavian, and Australian experience.

In the UK, the main burden of enforcing the 1974 Act falls on the HSE, with day-to-day enforcement being the responsibility of the Health and Safety Inspectorate (with 1,500 inspectors and 600 specialist staff) which has brought together a number of previously separate inspectorates (with some non-industrial sector responsibilities being devolved onto local authorities). In common with most inspectorates, the HSE's budget has been reduced in recent years (see further our discussion of the ratio of inspectors to workplaces in Chapter 4). The inspectors have very extensive powers to enter premises, to examine equipment or materials, to conduct investigations, to require information and disclosure of documents.[105]

The HSE has a programme of preventive inspection in which it seeks to secure compliance with the OHS legislation across industry, assess the adequacy of the control of work risks by duty holders, encourage businesses to actively manage OHS as an integral part of their businesses, to provide information and advice on the precautions necessary to reduce workplace hazards, and to gather information about risks and management competence. It strives to achieve four criteria in its enforcement programmes: proportionality (of enforcement responses to the risk to health and safety posed by the breach); consistency (cases involving similar circumstances receive a similar response); transparency (communicating the framework of duties and enforcement approach to duty holders); and targeting (so that the most risk-generating activities and hazards are the primary focus of inspection and enforcement). We discuss the HSE's approach to inspection further in Chapter 4.

In order to enforce the Act, the inspectors have two major powers. First, they may institute a prosecution, alleging a specific breach of a relevant statutory provision. Where the court finds the breach proven, it may impose criminal sanctions on the wrongdoer (who may be an individual or a corporation). The main sanctions stipulated in the legislation are fines (which in the Crown Court may be any amount the court thinks just) or, in respect of a limited class of serious breaches, imprisonment for up to two years either in addition to or instead of a fine. The Act also empowers a court to order a person convicted of an offence to take specified steps to remedy the contravention. Proceedings may be brought against any director, manager, or similar officer of a body corporate where an offence committed by that body was either committed with their consent or connivance or was attributable to their neglect.[106]

The second major enforcement power involves the service by an inspector of a form of administrative notice requiring an unsatisfactory situation to be remedied. There are two forms of notice: the improvement notice and the prohibition notice. An improvement notice may be issued if there is a contravention of any of the relevant statutory provisions, directing a person to remedy the fault within a specified time. This notice would be served on the person who is deemed to be contravening the law, or it can be served on any person on whom responsibilities are placed, whether that person is an employer, an employed person, or a supplier of equipment or materials. A prohibition notice may be issued where an activity or state of affairs involves a risk of serious personal injury. The notice will direct the recipient to cease the activity either immediately or within a specified period. The notice can be served on the person undertaking that activity, or on the person in control of it at the time the notice is served.[107]

There are some 3,000 prosecutions each year, with prosecutions reserved for serious breaches, usually of a flagrant, wilful, or reckless nature. The inspectorate relies much more heavily on prohibition and improvement notices or on informal counselling in the large majority of circumstances. Prohibition and improvement notices have proved to be perhaps Robens's most successful innovation. They are preventative in nature, and allow action to be taken swiftly without the necessity of going to court. Between 5,000 and 6,500 improvement notices are issued each year, and around 4,000 prohibition notices, although fewer notices have been issued since 1995. Appeals are rare and infrequently successful. In contrast to the cumbersome and time-consuming nature of the traditional prosecution process, these orders offer a quick and simple mechanism capable of being used on the spot to deal with serious hazards immediately they are detected. Moreover, such orders are particularly flexible in that they do not necessarily specify how an employer may come into compliance, thereby leaving her or him free to choose the least-cost method and avoid unnecessary expense.[108]

Employees, employee representatives, and individuals do not have any enforcement powers under the Act. Specifically, section 38 makes it clear that no employee, trade union official, or interested private person may directly set the law in motion to secure compliance with the provisions of the Act, health and safety regulations, or pre-existing legislation. That is (apart from civil action brought by an injured employee for breach of existing statutory provisions or the regulations), employees and their representatives cannot take part directly in the enforcement of the Act.

One further power under UK legislation deserves mention, namely the power of the courts to make a compensation order requiring a convicted person to pay compensation for any personal loss, injury, or damage resulting from the offence (Powers of the Criminal Courts Act 1973).

Finally, the HSE in recent years has shown an increasing willingness to refer OHS contraventions resulting in workplace fatalities to the Crown Prosecution Service. Some of these cases are discussed further in Chapter 6 of this book.

The Australian jurisdictions have largely followed the UK approach to enforcement. In each jurisdiction enforcement is by a state OHS enforcement agency, which has broad powers to enter and inspect premises and to investigate contraventions of the OHS statues. The ratio of inspectors to workplaces varies between the different states and territories. For example, in 1993–4 it was 1:1,715 in New South Wales, 1:1,165 in Victoria, 1:848 in Queensland, and 1:391 in the Northern Territory.[109] Inspectors may enforce the OHS statutes using informal methods (advice, requirements, education), or administrative sanctions (improvement and prohibition notices, and in New South Wales, the Northern Territory, Queensland (and proposed in Victoria and Tasmania) on-the-spot fines). Formal prosecutions are initiated as a last resort, and in a minority of instances, usually where there has been a serious injury or a fatality.

Most prosecutions are taken against corporate employers, and, in a small number of cases, individual managers or directors are also prosecuted. In most jurisdictions, where a prosecution is successful, the courts can impose monetary fines for offences. The maximum fines vary in magnitude—from A\$40,000 for corporations in Victoria (increased, from 1998, to A\$250,000), to A\$550,000 for certain offences by corporations in New South Wales. In some jurisdictions, additional fines can be imposed for repeat offenders, and individual managers or directors can be imprisoned. In New South Wales, a sentencing court may order a convicted offender to remedy the cause of the offence. In most Australian jurisdictions courts can adjourn proceedings short of conviction, and can place offenders on a 'good behaviour bond.' In some jurisdictions (for example, Victoria) courts can impose fines without recording a conviction against the offender. In New South Wales there is explicit provision enabling a secretary of a trade union to bring a prosecution for contravention of the OHS statute, with appropriate powers for trade union representatives to investigate OHS contraventions. In all Australian jurisdictions there is also the possibility of a prosecution for manslaughter and related offences (such as negligently inflicting injury), although the rules governing the attribution of guilty intention to a corporate employer are, as in the UK, extremely restrictive. To date, there has only been one successful prosecution of a corporate employer for manslaughter.[110]

In Sweden, the Labour Inspectorate exercizes its broad powers to inspect premises (400 inspectors for roughly 38,000 workplaces, and a ratio of 1 inspector for every 10,500 employees),[111] to provide counselling and information to duty holders, to investigate work injuries, and to asses, in advance, plans of working premises, personnel facilities, work processes, and working methods. The inspectorate focuses its attention in areas of high priority, or by focusing on high-risk industries, and sets its priorities from knowledge of the hazards in different sectors and occupations, and of the conditions of particular workplaces.[112] Since the early 1990s, the National Board of Occupational Safety and Health and the Labour Inspectorate have been developing inspection methods known as 'systems supervision' as a special method for checking whether the systems created for IC (see above) within large and medium-sized companies are efficient. Inspection is targeted at the high-level management, and focuses on the system of organization of work, as well as random checks on actual working conditions to make certain that OHS management systems are being implemented.[113] We discuss this approach further in Chapter 5.

There is provision for mandatory sanctions, principally in the form of fines. However, until recently, Swedish inspectors perceived their role to be almost exclusively that of giving practical advice and of encouraging unions and management to co-operate on OHS issues. As a result, these sanctions were only issued after unreasonable and persistent delay by the employer, or the refusal to implement a change that had been ordered by the Labour Inspectorate.[114] Inspectors also have the power to issue a written order to correct a violation (work environment improvement notices) but, again, such orders in the past were issued only rarely. In practice, Swedish inspectors usually gave verbal instructions at the end of the conference with the employer that concludes their inspection, without resorting to any legally binding formal enforcement mechanism.[115]

However, a shift is taking place in Sweden, with the inspectorate undertaking a greater number of workplace inspections, and becoming increasingly willing to use coercive measures when it identifies troublesome workplaces.[116] These measures include the use of injunctions or prohibitions in order to achieve necessary modifications to the work environment.[117] At the time of issuing an injunction or prohibition, the inspector can set a contingent fine, which can be imposed by a court if the injunction or prohibition is not adhered to. Where a prohibition is violated, a safety delegate can suspend work. Where there is no contingent fine, the employer can be prosecuted and fined. Significantly, the percentage of inspections leading to the issuing of improvement orders has increased appreciably, due partly to improved procedures for the prioritization and selection of inspection projects.[118] More than 45 per cent of visits by inspectors result in written orders for improvements to the working environment. Of these, 2 per cent require further enforcement through injunctions or prohibitions proceedings.[119]

The Act includes provisions breach of which will result in the imposition of penalties, without the need for a prior injunction or prohibition. These include contraventions of requirements specified by the National Board of Occupational Safety and Health, furnishing inaccurate information to an inspector, and tampering with a safety device. From 1994, the National Board has been empowered to prescribe payment of a special charge (sanction penalties) for infringements of specified provisions, which eliminates penalty provisions for the contraventions concerned. The Board is required to indicate exactly how the charge will be calculated. This procedure obviates the need for formal prosecutions of these offences, which had been subject to backlogs in the courts.[120] There had also been suggestions that the police and prosecutors had not accorded high priority to work environment prosecutions.[121] In 1991 the Swedish Criminal Code was amended to include a new section on work environment offences. A person deliberately or negligently defaulting on their obligations under the Work Environment Act can be convicted of a work environment offence if, as a result, a person loses their life, is injured, or is exposed to danger. All in all, about twenty cases per year result in court proceedings.[122]

In Denmark, the Labour Inspectorate (the Danish WES), like inspectorates elsewhere, has very broad powers of entry into premises and of inspection. Rather than simply react to injuries and OHS problems, it aims to induce employers to take preventive measures, by requiring companies to have an objective, an ongoing documentation process, a cause and effect analysis, and problem-solving approaches. Until recently, this approach has not met with much success, although, since the implementation of the EC Framework Directive, the Working Environment Act has required all companies with particularly hazardous jobs to affiliate to the Occupational Health Services system, and to have a written workplace assessment (see above). This has enabled the inspectorate to require companies to enter agreements with the Occupational

Health Services (see above) setting out their future initiatives. The Occupational Health Services have been receiving greater resources, and providing organizations with OHS information and counselling, leaving the inspectorate with a supervisory and enforcement function.[123] The inspectorate conducts special preventive and goal-oriented campaigns, and aims to promote co-operation between management and workers in the enterprise in establishing OHS systems. Where enterprises work systematically to manage workplace risks, the inspectorate's role changes to one of guidance and spot checks.[124]

In 1998 the inspectorate announced that it would implement a new inspection strategy of differentiated inspection according to the ability of firms to conduct workplace assessments and other working environment activities. The new approach to inspection applies to general inspections only. Procedures for investigating injuries, fatalities, and complaints remain unchanged. The nature of the general inspection will depend on the extent to which the firm is implementing workplace assessment procedures. Where there are no workplace assessment procedures in place, the inspectorate will conduct a traditional, thorough inspection of the firm, and will take appropriate enforcement measure (see below), including guidance, instructions, prohibition, and prosecution where there are severe contraventions of well-known rules. Where the firm has begun to implement a workplace assessment system, the inspectorate will conduct a traditional inspection with traditional enforcement measures for acute problems, but will require the most important working environment problems to be incorporated into the workplace assessment processes. Where the firm has systematically implemented a working assessment process, and has integrated it with environment or quality control systems, the inspectorate will conduct a random inspection on one or more of the firm's departments, to ensure that the system of control is being implemented. The inspectorate will report on the important working environment problems not covered by the firm's workplace assessment. The firm will then present the inspectorate with an action plan. The inspectorate will monitor activities with random inspections. Where acute problems are found, or there are severe contraventions of well-known rules, the inspectorate will take the appropriate enforcement action.[125]

The ratio of inspectors to employees in Denmark is 1:9,000.[126] Each year about 55,000 enterprises are visited. On average 17,500 improvement notices are issued annually, and in 8,200 cases recommendations are issued.[127] As is also common elsewhere, the Danish inspectorate operates to a large extent informally, offering verbal instruction and advice to employers and using its considerable powers of discretion. Where it does decide to take formal action then three options are open. First, the inspector can issue a written notice (an improvement notice) requiring that matters contravening the law be remedied within a specified period of time. Second, an inspector may issue a prohibition notice requiring that immediate steps be taken to avoid imminent and serious risk to the health and safety of employees. Finally, an inspector may choose to prosecute, particularly where there has been a failure to comply with an inspector's decision. Since 1991 criminal prosecutions have increased substantially.[128] For example, in 1995, 530 cases were prosecuted, 13 of them individual cases against employees for infringing the Working Environment Act. The average fine in 1994 was DKK14,000. The highest ever fine imposed was DKK300,000.[129] Sanctions also include imprisonment, although this would be most unlikely in practice. Parliament has been considering a proposal to increase penalties for serious contraventions of OHS regulations which lead to fatalities.[130] The Ministry of Labour is also moving towards more frequent prosecutions of employees, to encourage individual employees to assume responsibility for safety without altering the fundamental responsibility of employers.[131]

In the USA, the OSH Act, like the UK legislation, gives broad powers to inspectors to enter premises, inspect, and question any person concerning possible violations. However, in most other respects, the USA approach to enforcement differs substantially from that of the UK, the Scandinavian countries, and Australia.

The OSHA's resources do not permit it to conduct frequent inspections of the 5 million workplaces for which it has responsibility. Commentators have noted that at present resource levels the OSHA will be able to inspect every workplace once every 109 years.[132] It focuses inspections on the bigger employers, and the most dangerous hazards, and limits inspections of employers who voluntarily adopt health and safety programmes.[133] When, however, contraventions are detected, OSHA inspectors (called 'compliance officers') have very little discretion, and in most circumstances are required to take some formal action ('first-instance citations') once they have identified a violation, thereby reducing the possibility of an inspector being 'captured' or corrupted by employers. In contrast, under the UK, Scandinavian, and Australian approach, inspectors rely very heavily on informal interaction with employers, seeking to function primarily through education and persuasion rather than prosecution, which is often treated as a last resort, and almost as an admission of failure.[134]

In the USA, after completing an inspection, an inspector must report any violation to the Area Director, who decides whether to issue a citation. In some cases the inspector may issue it directly. Notices may be issued in lieu of citations for *de minimus* violations. A citation must describe the specific nature of the violation and establish a reasonable time for the abatement of the condition in violation. Within a reasonable time after issuing the citation, the Secretary of Labor must notify the employer of any proposed penalties. The employer then has a limited period within which to contest the proposed penalties or citation. If the employer fails to do so, the violation and the assessed penalty are deemed final and are not subject to review. If the employer contests the citation or penalty, the case goes to an administrative law judge. The judge's decision is final unless one of the three members of the Review Commission exercizes a statutory right of directing review.[135]

The OSH Act provides a range of monetary penalties for violation of the Act or rules promulgated under it. Any employer who fails to correct a violation for which a citation has been issued, within the period permitted for abatement, may be assessed for a civil penalty for each day the violation continues. Wilful or repeated violations merit the highest penalties. Greater penalties are provided if an employer receives a citation for a serious violation (i.e. where there is a serious probability that death or serious bodily injury could result). Even if a violation is specifically determined not to be of a serious nature, an employer will still receive a civil penalty. Criminal prosecutions are relatively rare. An employer who commits a violation that results in the death of an employee is liable to be prosecuted under the OSH Act, resulting in a fine of up to US$500,000 and/or up to six months' imprisonment.[136] The average number of criminal cases referred to the Department of Justice annually was 7 in the period 1978–80, and 2.8 in the period 1982–7. In 1991 the number of prosecutions was 16.[137]

In 1995 the average penalty imposed by the OSHA for a 'serious violation' was US$763 (11 per cent of the maximum). The average penalty for 'other than serious violations' was US$52 (7 per cent of the maximum).[138]

The Clinton Administration's proposed reforms would authorize the OSHA to require an employer to take immediate action where a hazard poses an immediate danger of death or serious injury, with fines of up to US$50,000 per day for failure to take corrective action. Criminal penalties would be increased to a maximum of ten years in gaol for wilful violations that cause death, and a maximum of five years for those that cause serious bodily injury. The Bill also establishes a new minimum penalty of US$1,000 for each serious violation and directs that this money be used to increase funding for the OSHA programme.

The OSH Act does allow employees to become involved in the administrative and enforcement activities under the Act. In particular it gives employees the right to request an OSHA inspection, to accompany an OSHA inspector, to obtain a review if an inspector fails to issue a citation after employees have formally alleged violations, to appeal if the abatement period appears unreasonably long, and a number of ancillary rights.

It will be apparent from the above description, that the USA's approach to enforcement is considerably more punitive, deterrence-oriented, litigious, and confrontationist, than either the UK, Australia, or the Scandinavian countries examined. In part, the USA's approach to enforcement is a product of that country's history and in particular of hostility that business commonly shows towards any state intrusion into its affairs. In this environment of conflict between business and government, a legalistic deterrence-oriented approach is at least understandable, even if its efficacy may be questioned. In countries and cultures where such antagonism does not exist to anything like the same degree, a confrontational strategy of enforcement makes little sense.[139]

Nevertheless, in recent years, as we describe in Chapters 3 and 4 of this book, the OSHA has increased its co-operative approaches, as part of 'the New OSHA' promoted by the Clinton Administration. The OSHA offers employers a choice between traditional enforcement and new approaches, such as partnership, compliance assistance, and training. It offers special assistance to small business. It has a new targeting system based on worksite specific data. It is focusing less on individual technical violations, and instead examining employers' OHS programmes.[140] In the past few years further pressure has come from the Republican majority in Congress which has proposed numerous amendments to the OSH Act to promote, even require, increased co-operation by the OSHA, and to reduce the punitive aspects of OSHA enforcement. These have included requiring 50 per cent of OSHA resources to be devoted to co-operative programmes, exemptions of some employers from routine inspections, restrictions on the OSHA's authority to conduct inspections in response to employee complaints, the elimination of first instance citations, and limitations on the size of penalties that the OSHA can assess.[141]

NOTES

1 *Report of the Committee on Health and Safety at Work 1970–1972* (HMSO, London, 1972).

2 N. A. Ashford, *Crisis in the Workplace: Occupational Disease and Injury* (MIT Press, Cambridge, Mass., 1976), p. 15.

3 See further D. A. Ballam, 'The OSHA's preemptive effect on state criminal prosecutions of employers for workplace deaths and injuries' (1988) 26 *American Business Law Journal*; and R. H. Sand, 'OSHA takes a step forward on preemption—and two steps back on generic regulation' (1992/3) 18 *Employee Relations Law Journal*.

4 J. L. Hirsch, *Occupational Safety and Health Handbook; an employer guide to OSHA laws, regulations and practices* (Butterworths Legal Publications, Austin, USA, 1993).

5 See R. Johnstone, *Occupational Health and Safety Law and Policy* (LBC Information Services, Sydney, 1997), pp. 80–4. See also A. Brooks, *Occupational Health and Safety Law in Australia*, 4th edn. (CCH, Sydney, 1993).

6 For a comprehensive analysis of the employer's duties under the HSW Act, see F. Wright, *Law of Health and Safety at Work* (Sweet & Maxwell, London, 1997), Ch. 4.

7 See further A. Brooks, *Occupational Health and Safety Law in Australia*, 4th edn. (CCH, Sydney, 1993); and R. Johnstone, *Occupational Health and Safety Law and Policy* (LBC Information Services, Sydney, 1997), Chs. 5 and 6.

8 See R. Johnstone, *Occupational Health and Safety Law and Policy* (LBC Information Services, Sydney, 1997), Table 5.1; and W. B. Creighton, W. J. Ford, & R. J. Mitchell, *Labour Law: Text and Materials*, 2nd edn. (Law Book Company, Sydney, 1993), Table 40.1.

9 See N. Gunningham, *Safeguarding the Worker* (Law Book Company, Sydney, 1984), pp. 340–3.

10 Most notably in the Comprehensive Occupational Health and Safety Reform Bill of 1993.

11 N. A. Ashford, *Crisis in the Workplace: Occupational Disease and Injury* (MIT Press, Cambridge, Mass., 1976), p. 161.

12 An Ordinance on Internal Control of the Working Environment (AFS 1992: 6).

13 National Board of Occupational Safety and Health, Job Modification and Rehabilitation Ordinance (AFS 1994:1).

14 D. Walters, 'Preventive Services in Occupational Health and Safety in Europe: Developments and Trends in the 1990s' (1997) 27 *International Journal of Health Services* 247 at 252.

15 Swedish Working Environment Association, *17 Questions & Answers Based on Questions Frequently Asked by Foreign Visitors* (Swedish Working Environment Association, Stockholm, 1996).

16 Working Environment Act 1975, Part 4.

17 We are grateful to Per Langaa Jensen for notifying us of this development. Workplace risk assessment requirements were originally introduced by an executive order of the Danish Ministry of Labour in 1993.

18 European Foundation for the Improvement of Living and Working Conditions, 'Problems of implementing the newly amended Work Environment Act' (1997) Aug. *Euroline.*

19 European Foundation for the Improvement of Living and Working Conditions, Walters, D. (ed.), *The Identification and Assessment of Occupational Health and Safety Strategies in Europe*, Vol. 1 (Office for the Official Publications of the European Communities, Luxembourg, 1996), pp. 35–7.

20 For a discussion of the UK provisions, see F. Wright, *Law of Health and Safety at Work* (Sweet & Maxwell, London, 1997), Ch. 4. For the Australian provisions, see R. Johnstone, *Occupational Health and Safety Law and Policy* (LBC Information Services, Sydney, 1997), Ch. 6.

21 S. Dawson, P. Willman, A. Clinton, & M. Bamford, *Safety at Work: The Limits of Self-Regulation* (Cambridge University Press, Cambridge, 1988).

22 See P. James & D. Walters, 'Non-union Rights of Involvement: The Case of Health and Safety at Work' (1997) 26 *Industrial Law Journal* 35.

23 P. James, 'Reforming British Health and Safety Law: a framework for discussion' (1992) 21 *Industrial Law Journal* 83 at 91.

24 S. Dawson, P. Willman, A. Clinton, & M. Bamford, *Safety at Work: The Limits of Self-Regulation* (Cambridge University Press, Cambridge, 1988), p. 85.

25 Swedish Working Environment Association, *17 Questions & Answers Based on Questions Frequently Asked by Foreign Visitors* (Swedish Working Environment Association, Stockholm, 1996).

26 E. Tucker, 'Worker participation in health and safety regulation: lessons from Sweden' (1992) 37 *Studies in political economy* 95.

27 Swedish Working Environment Association, *17 Questions & Answers Based on Questions Frequently Asked by Foreign Visitors* (Swedish Working Environment Association, Stockholm, 1996).

28 E. Tucker, 'Worker participation in health and safety regulation: lessons from Sweden' (1992) 37 *Studies in political economy* 95; and J. E. Korostoff, L. M. Zimmerman, & C. E. Ryan, 'Rethinking the OSHA Approach to Workplace Safety: A Look at Worker Participation in the Enforcement of Safety Regulations in Sweden, France and Great Britain' (1991) *13 Comparative Labor Law Journal* 45.

29 E. Tucker, 'Worker participation in health and safety regulation: lessons from Sweden' (1992) 37 *Studies in political economy* 108.

30 European Agency for Safety and Health at Work, *Priorities and Strategies in Occupational Safety and Health Policy in the Member States of the European Union* (European Agency for Safety and Health at Work, Bilbao, Spain, 1997), p. 176.

31 J. E. Korostoff, L. M. Zimmerman, & C. E. Ryan, 'Rethinking the OSHA Approach to Workplace Safety: A Look at Worker Participation in the

Enforcement of Safety Regulations in Sweden, France and Great Britain' (1991) *13 Comparative Labor Law Journal* 45.

32 The Swedish Institute, *Fact Sheets on Sweden: Occupational Safety and Health* (Swedish Institute, Stockholm, Aug. 1996).

33 E. Tucker, 'Worker participation in health and safety regulation: lessons from Sweden' (1992) 37 *Studies in political economy* 95; and K. Frick, 'Can Management Control Health and Safety at Work?' (1990) 11 (3 Aug.) *Economic and Industrial Democracy*.

34 J. E. Korostoff, L. M. Zimmerman, & C. E. Ryan, 'Rethinking the OSHA Approach to Workplace Safety: A Look at Worker Participation in the Enforcement of Safety Regulations in Sweden, France and Great Britain' (1991) *13 Comparative Labor Law Journal* 45.

35 E. Tucker, 'Worker participation in health and safety regulation: lessons from Sweden' (1992) 37 *Studies in Political Economy* 95 at 111.

36 Swedish Occupational Safety and Health Administration (The National Board of Occupational Safety and Health and the Labour Inspectorate), *A Report on activities in 1991/1992* (1992).

37 We are grateful to Kaj Frick for this information.

38 See A. B. Martin, A. J. Linehan, & I. Whitehouse, *The Regulation of Health and Safety in Five European Countries: Denmark, France, Germany, Spain and Italy* (HSE, London, 1996), Ch. 1.

39 European Foundation for the Improvement of Living and Working Conditions, 'Problems of implementing the newly amended Work Environment Act' (1997) Aug. *Euroline*.

40 European Foundation for the Improvement of Living and Working Conditions, 'Employers rejoin Danish health and safety system' (1998) June *Euroline*.

41 Environmental Resources Ltd. for the Commission of the European Communities, *The law and practice concerning occupational health in the member States of the European Community*, Vol. 2, Denmark, Federal Republic of Germany (Graham and Trotman Ltd., 1985).

42 European Foundation for the Improvement of Living and Working Conditions, Walters, D. (ed.), *The Identification and Assessment of Occupational Health and Safety Strategies in Europe*, Vol. 1 (Office for the Official Publications of the European Communities, Luxembourg, 1996), p. 32.

43 A. B. Martin, A. J. Linehan, & I. Whitehouse, *The Regulation of Health and Safety in Five European Countries: Denmark, France, Germany, Spain and Italy* (HSE, London, 1996), Ch. 1, para. 4.1.2.

44 Ibid. paras 1.1 and 4.1.2.

45 Arbejdsministeriet, *Working Environment Regulation in Denmark* (Danish Ministry of Labour, 1994).

46 We are grateful to Per Langaa Jensen for notifying us of these changes.

47 European Foundation for the Improvement of Living and Working Conditions, 'Employers opt out of the Danish health and safety system' (1997) June *Euroline*.

48 European Foundation for the Improvement of Living and Working Conditions, 'Employers rejoin Danish health and safety system' (1998) June *Euroline*.

49 A. B. Martin, A. J. Linehan, & I. Whitehouse, *The Regulation of Health and Safety in Five European Countries: Denmark, France, Germany, Spain and Italy* (HSE, London, 1996), Ch. 1, para. 4.1.3.

50 European Foundation for the Improvement of Living and Working Conditions, Walters, D. (ed.), *The Identification and Assessment of Occupational Health and Safety Strategies in Europe*, Vol. 1 (Office for the Official Publications of the European Communities, Luxembourg, 1996), pp. 32–3.

51 See W. B. Creighton, W. J. Ford, & R. J. Mitchell, *Labour Law: Text and Materials*, 2nd edn. (Law Book Company, Sydney, 1993), Table 40.2; and R. Johnstone, *Occupational Health and Safety Law and Policy* (LBC Information Services, Sydney, 1997), Ch. 4.

52 See W. B. Creighton, W. J. Ford, & R. J. Mitchell, *Labour Law: Text and Materials*, 2nd edn. (Law Book Company, Sydney, 1993), Table 40.2; R. Johnstone, *Occupational Health and Safety Law and Policy* (LBC Information Services, Sydney, 1997), pp. 483–95.

53 See W. B. Creighton, W. J. Ford, & R. J. Mitchell, *Labour Law: Text and Materials*, 2nd edn. (Law Book Company, Sydney, 1993); R. Johnstone, *Occupational Health and Safety Law and Policy* (LBC Information Services, Sydney, 1997), Ch. 9; and W. B. Creighton, & P. Rozen, *Occupational Health and Safety Law in Victoria* (Federation Press, Sydney, 1997), Ch. 8.

54 W. Pearse & C. Refshauge, 'Workers Health and Safety in Australia: An Overview' (1987) 17 *International Journal of Health Services* 635.

55 *A.M.I. Toyota v. A.D.S.T.E.* (1986) 17 IR 1. See also N. Gunningham, 'The Role of Federal Awards in Regulating Occupational Health and Safety Standards' (1990) 3 *Australian Journal of Labour Law* 54; and W. B. Creighton, W. J. Ford, & R. J. Mitchell, *Labour Law: Text and Materials*, 2nd edn. (Law Book Company, Sydney, 1993).

56 W. B. Creighton & P. Rozen, *Occupational Health and Safety Law in Victoria* (Federation Press, Sydney, 1997), p. 17.

57 J. L. Hirsch, *Occupational Safety and Health Handbook; an employer guide to OSHA laws, regulations and practices* (Butterworths Legal Publications, Austin, USA, 1993); and T. O. McGarity, & S. A. Shapiro, *Workers at Risk: The failed promise of occupational safety and health administration* (Praeger, Westport, 1993).

58 J. E. Korostoff, L. M. Zimmerman, & C. E. Ryan, 'Rethinking the OSHA Approach to Workplace Safety: A Look at Worker Participation in the Enforcement of Safety Regulations in Sweden, France and Great Britain' (1991) *13 Comparative Labor Law Journal* 45 at 47.

59 D. Vogel, *National Styles of Regulation* (Cornell University Press, Ithaca, NY, 1986); and E. Bardach & R. Kagan, *Going by the Book: The problems of regulatory unreasonableness* (Temple University Press, Philadelphia, 1982).

60 J. E. Korostoff, L. M. Zimmerman, & C. E. Ryan, 'Rethinking the OSHA Approach to Workplace Safety: A Look at Worker Participation in the Enforcement of Safety Regulations in Sweden, France and Great Britain' (1991) *13 Comparative Labor Law Journal* 45.

61 N. A. Ashford, *Crisis in the Workplace: Occupational Disease and Injury* (MIT Press, Cambridge, Mass., 1976), p. 31.

62 D. Dawson, P. Willman, A. Clinton, & M. Bamford, *Safety at Work: The Limits of Self-Regulation* (Cambridge University Press, Cambridge, 1988).

63   In Canada, regulatory impact statements or similar mechanisms are mandatory at Federal government level, and are published with the new regulation in the government gazette. Similar mechanisms are not prevalent in
the provinces and are not required in the major jurisdiction of Ontario.

64   T. Fisher, 'A "Quality" Approach to Occupational Health, Safety and
Rehabilitation' (1991) 7(1) *Journal of Occupational Health and Safety—Australia
and New Zealand* 23.

65   See further J. Berger, 'The Hoechst Dispute: A Paradigm Shift in Occupational Health and Safety' in M. Quinlan (ed.), *Work and Health: The origins,
management and regulation of occupational illness* (Macmillan, Melbourne, 1993);
T. A. McGarity & S. A. Shopiro (1993).

66   For a general discussion of OHS law-making in the US, see K. Williams,
'Making Rules for Occupational Health in Britain, the USA and the EC'
(1997) *Anglo-American Review* 82.

67   [1949] 1 KB 704 at 712.

68   Health and Safety Commission, *Review of Health and Safety Legislation,* Main
Report (Health and Safety Executive, London, 1994).

69   Health and Safety Commission, *Plan of Work, 1994/1995* (HSC, London,
1994).

70   F. Wright, *Law of Health and Safety at Work* (Sweet and Maxwell, London,
1997). See also K. Williams, 'Making Rules for Occupational Health in
Britain, the USA and the EC' (1997) *Anglo-American Review* 82.

71   Proposition 1976/1977:149 AML[1977]301.

72   B. J. Fleischauer, 'Occupational Safety and Health Law in Sweden and the
United States: Are There Lessons to be Learned by Both Countries?' (1983)
6 Hastings International and Comparative Law Review 283 at 330.

73   R. Johnstone, *Occupational Health and Safety Law and Policy* (LBC Information
Services, Sydney, 1997), pp. 202–20.

74   Ibid. pp. 281–9.

75   See the Industry Commission, Australia, *Work, Health and Safety,* Vol. II
(AGPS Canberra, 1995), Appendix E; and R. Johnstone, *Occupational
Health and Safety Law and Policy* (LBC Information Services, Sydney, 1997),
pp. 160–1.

76   On the advantages and disadvantages of cost-benefit analysis as applied to
OHS and environment see T. O. McGarity, 'Regulatory analysis and
regulatory reform' (1987) 65 *Texas Law Review* 7; and R. Hahn & J. Hurd.
The Costs & Benefits of Regulation—Review & Synthesis' (1991) 8 *Yale
Journal of Regulation* 233–78 and Industry Commission, Australia, *Work,
Health and Safety,* Vol. II (AGPS, Canberra, 1995), Appendix D.

77   T. O. McGarity, 'Regulatory analysis and regulatory reform' (1987) 65
Texas Law Review 7, 1243 at 1279 and 1281.

78   M. Porter, *The competitive advantage of nations* (Free Press, 1990).

79   T. Moss, 'Cost-benefit analysis and the cotton dust standard' (1982) 35
Rutgers Law Review.

80   T. O. McGarity, 'Regulatory analysis and regulatory reform' (1987) 65
Texas Law Review 7, 1243 at 1279 and 1330.

81   See also T. O. McGarity and S. A. Shopiro (1993) pt. 5.

82   P. Cain, *Atiyah's accident compensation and the law* (Martin and Weidenfeld,

London, 1987); and J. Stapleton, *The Disease and Injury Debate* (Clarendon Press, Oxford, 1986).

83 R. Johnstone, *Occupational Health and Safety Law and Policy* (LBC Information Services, Sydney, 1997), Table 11.6.

84 See *Report of the Committee on Health and Safety at Work 1970–1972* (HMSO, London, 1972) (Robens Report); E. Pearson, *Report of the Royal Commission on Civil Liability and Compensation* (HMSO, London, 1978).

85 See R. Johnstone, *Occupational Health and Safety Law and Policy* (LBC Information Services, Sydney, 1997), pp. 305–13; and A. Brooks, *Occupational Health and Safety Law in Australia*, 4th edn. (CCH, Sydney, 1993), pp. 940–51.

86 Premiers and Chief Ministers, *Communique*, Adelaide, 21–2 Nov. 1991.

87 See further, R. F. Eberlie, 'The new health and safety legislation of the European Community' (1990) 19 *Industrial Law Journal* 81.

88 Treaty of Rome, Article 189.

89 *Francovich v. Italian Republic* [1992] IRLR 84; and F. Wright, *Law of Health and Safety at Work* (Sweet & Maxwell, London, 1997), pp. 40–1.

90 S. E. Katz, 'The general good and the Second Banking Directive: a major loophole?' (1993) 166 *Butterworths Journal of International Banking and Finance Law*.

91 European Foundation for the Improvement of Living and Working Conditions, 'Conference highlights differences in national implementation of health and safety Directives' (1997) Dec. *Euroline*. For a discussion of the impact of the 1997 Treaty of Amsterdam on the 'Social Chapter', see C. Barnard, 'The United Kingdom, the 'Social Chapter' and the Amsterdam Treaty' (1997) 26 *Industrial Law Journal* 275.

92 W. B. Creighton, 'Occupational Health and Safety Regulation: The Role of ILO Standards', in M. Quinlan (ed.), *Work and Health: The Origins, Management and Regulation of Occupational Illness* (Macmillan, Melbourne, 1993). See also the accompanying *Occupational Safety and Health Recommendation* 1981, No. 164.

93 N. Gunningham, *Safeguarding the Worker* (Law Book Company, Sydney, 1984); and R. Johnstone, *Occupational Health and Safety Law and Policy* (LBC Information Services, Sydney, 1997), pp. 88–100.

94 W. B. Creighton, W. J. Ford, & R. J. Mitchell, *Labour Law: Text and Materials*, 2nd edn. (Law Book Company, Sydney, 1993).

95 W. B. Creighton, 'Occupational Health and Safety Regulation: The Role of ILO Standards', in M. Quinlan (ed.), *Work and Health: The Origins, Management and Regulation of Occupational Illness* (MacMillan, Melbourne, 1993).

96 See K. Williams, 'Making Rules for Occupational Health in Britain, the USA and the EC' (1997) *Anglo-American Review* 82.

97 See F. Wright, *Law of Health and Safety at Work* (Sweet & Maxwell, London, 1997), Ch. 5.

98 Ordinances are published in the Code of Statutes of the National Board of Occupational Safety and Health. The Code also includes recommendations concerning the implementation of ordinances.

99 Environmental Resources Ltd. for the Commission of the European Communities, *The law and practice concerning occupational health in the member States of*

the European Community, Vol. 2, Denmark, Federal Republic of Germany (Graham and Trotman Ltd., 1985).

100   D. Vogel, *National Styles of Regulation* (Cornell University Press, Ithaca, NY, 1986); and S. Kelman, S, *Regulating America, Regulating Sweden* (MIT Press, Cambridge, Mass. 1982).

101   L. Davey, 'Specification versus performance language in safety standards' (1990) Fall *Job Safety and Health Quarterly*.

102   G. R. Watchman, *Statement to Subcommittee on Public Health and Safety of the Senate Committee, Labor and Human Resources*, 10 July 1997.

103   See K. Williams, 'Making Rules for Occupational Health in Britain, the USA and the EC' (1997) *Anglo-American Review* 82.

104   See also D. R. Cherrington, 'The race to the courthouse: conflicting views toward the judicial review of OSHA standards' (1994) 95 *Brigham Young University Law Review*.

105   F. Wright, *Law of Health and Safety at Work* (Sweet & Maxwell, London, 1997), pp. 114–18.

106   Ibid. pp. 120–52.

107   Ibid. pp. 153–68.

108   See further S. Dawson, P. Willman, A. Clinton, & M. Bamford, *Safety at Work: The Limits of Self-Regulation* (Cambridge University Press, Cambridge, 1988); and F. Wright, *Law of Health and Safety at Work* (Sweet & Maxwell, London, 1997), Ch. 6.

109   Industry Commission, Australia *Work, Health and Safety* (AGPS, Canberra, 1995), Vol. II, p. 423.

110   For further detail of the Australian provisions, see R. Johnstone, *Occupational Health and Safety Law and Policy* (LBC Information Services, Sydney, 1997), Ch. 8.

111   European Foundation for the Improvement of Living and Working Conditions, 'Conference highlights differences in national implementation of health and safety Directives' (1997) Dec. *Euroline*.

112   European Foundation for the Improvement of Living and Working Conditions, Walters, D. (ed.), *The Identification and Assessment of Occupational Health and Safety Strategies in Europe*, Vol. 1 (Office for the Official Publications of the European Communities, Luxembourg, 1996), pp. 178–9.

113   Swedish National Board of Occupational Health and Safety, *Newsletter*, 4/ 1989, 2/1994, and 2/1995.

114   B. J. Fleischauer, 'Occupational Safety and Health Law in Sweden and the United States: Are There Lessons to be Learned by Both Countries?' (1983) 6 *Hastings International and Comparative Law Review* 283 at 298.

115   Ibid. 283.

116   Swedish National Board of Occupational Health and Safety, *Newsletter*, 4/ 1992, 2/1993, 6/1993, and 3/1995.

117   Swedish Occupational Safety and Health Administration (The National Board of Occupational Safety and Health and the Labour Inspectorate), *A Report on activities in 1991/1992* (1992), p. 9.

118   Ibid. p. 8.

119   The Swedish Institute, *Fact Sheets on Sweden: Occupational Safety and Health* (Swedish Institute Stockholm, Aug. 1996).

120 Swedish National Board of Occupational Health and Safety, *Newsletter*, 2/ 1995 and 1/1997.

121 Ibid. 2/1995.

122 The Swedish Institute, *Fact Sheets on Sweden: Occupational Safety and Health* (Swedish Institute, Stockholm, Aug. 1996).

123 European Foundation for the Improvement of Living and Working Conditions, Walters, D. (ed.), *The Identification and Assessment of Occupational Health and Safety Strategies in Europe*, Vol. 1 (Office for the Official Publications of the European Communities, Luxembourg, 1996), pp. 33–4.

124 European Agency for Safety and Health at Work, *Priorities and Strategies in Occupational Safety and Health Policy in the Member States of the European Union* (European Agency for Safety and Health at Work, Bilbao, Spain, 1997), p. 19.

125 P. L. Jensen, 'Workplace assessment in Denmark—a genuine success for risk assessment?' Paper delivered to *Policies for Occupational Health and Safety Management Systems and Workplace Change Conference*, Amsterdam 21–4 Sept. 1998.

126 European Foundation for the Improvement of Living and Working Conditions, 'Conference highlights differences in national implementation of health and safety Directives' (1997) Dec. *Euroline*.

127 European Agency for Safety and Health at Work, *Priorities and Strategies in Occupational Safety and Health Policy in the Member States of the European Union* (European Agency for Safety and Health at Work, Bilbao, Spain, 1997), p. 17.

128 A. B. Martin, A. J. Linehan, & I. Whitehouse, *The Regulation of Health and Safety in Five European Countries: Denmark, France, Germany, Spain and Italy* (HSE, London, 1996), Ch. 1, para. 6.5.3.

129 European Foundation for the Improvement of Living and Working Conditions, 'Danish building and construction sites are hazardous workplaces' (1997) Apr. *Euroline*.

130 European Agency for Safety and Health at Work, *Priorities and Strategies in Occupational Safety and Health Policy in the Member States of the European Union* (European Agency for Safety and Health at Work, Bilbao, Spain, 1997), p. 19.

131 European Foundation for the Improvement of Living and Working Conditions, 'Danish building and construction sites are hazardous workplaces' (1997) Apr. *Euroline*.

132 Bureau of National Affairs, Inc., 'Current OSHA Inspection Rate Once Every 109 Years, AFL–CIO Says' (1998) 27(7) *Occupational Safety and Health News* 1663–4.

133 S. Shapiro & R. S. Rabinowitz, 'Punishment versus Cooperation in Regulatory Enforcement: A Case Study of OSHA' (1997) 49 *Administrative Law Review* 713 at 745.

134 S. Dawson, P. Willman, A. Clinton, & M. Bamford, *Safety at Work: The Limits of Self-Regulation* (Cambridge University Press, Cambridge, 1988); and K. Hawkins, *Environment and Enforcement: Regulation and the Social Definition of Pollution* (Oxford University Press, Oxford, 1984).

135 For a comprehensive analysis of the OSHA's enforcement powers and

strategies, see S. Shapiro & R. S. Rabinowitz, 'Punishment versus Cooperation in Regulatory Enforcement: A Case Study of OSHA' (1997) 49 *Administrative Law Review* 713.

136 See, generally, M. T. Cimino, 'Criminal Prosecution of Workplace Safety
Violations' (1992) 94 *West Virginia Law Review* 1007 at 1992; and J. L. Hirsch,
*Occupational Safety and Health Handbook; an employer guide to OSHA laws, regula-
tions and practices* (Butterworths Legal Publications, Austin, USA, 1993).

137 R. Kagan, Criminal Prosecution for Regulatory Offenses, Workshop on
Regulatory Law Enforcement, Socio-legal Center, Ohio State University 20–1 May 1993.

138 T. H. McQuiston, R. C. Zakocs, & D. Loomis, 'The Case for Stronger
OSHA Enforcement—Evidence from Evaluation Research' (1998) 88
American Journal of Public Health 1022 at 1022.

139 D. Vogel, *National Styles of Regulation* (Cornell University Press, Ithaca, NY,
1986); and S. Kelman, *Regulating America, Regulating Sweden* (MIT Press,
Cambridge, Mass. 1982).

140 G. R. Watchman, *Statement to Subcommittee on Public Health and Safety of the
Senate Committee, Labor and Human Resources*, 10 July 1997.

141 S. Shapiro, & R. S. Rabinowitz, 'Punishment versus Cooperation in Regulatory Enforcement: A Case Study of OSHA' (1997) 49 *Administrative Law
Review* 713 at 755–61.

Selected Bibliography

Aalders, M., 'Regulation and In-Company Environmental Management in the Netherlands' (1993) 15(2) *Law and Policy* 75.

Aalders, M., & Wilthagen, T., 'Moving Beyond Command and Control: Reflexivity in the Regulation of Occupational Safety, Health and the Environment' (1997) 19(4) *Law and Policy* 415.

American Bar Association, *3 Standards for Criminal Justice* (Little, Brown, Boston, 1980).

Arbejdsministeriet, *Working Environment Regulation in Denmark* (Danish Ministry of Labour, 1994).

Arlen, J., 'The Potentially Perverse Effects of Corporate Criminal Liability' (1994) 23 *Journal of Legal Studies* 833.

Ashford, N. A., *Crisis in the Workplace: Occupational Disease and Injury* (MIT Press, Cambridge, Mass., 1976).

Ashford, N. A., et al., 'The Encouragement of Technological Change for Preventing Chemical Accidents: Moving firms from secondary prevention and mitigation to primary prevention'. A Report to the US Environmental Protection Agency, Centre for Technology, Policy and Industrial Development at MIT, Cambridge, Mass., July 1993.

Australian Law Reform Commission, *Sentencing of Federal Offenders*, Report No. 15 Interim (AGPS, Canberra, 1980).

—— *Sentencing Penalties*, Discussion Paper No. 30 (AGPS, Canberra, 1987).

—— *Compliance with the Trade Practices Act 1994*, Report No. 68 (AGPS, Canberra, 1994).

Australian Manufacturing Council/Manufacturing Advisory Group (NZ), *Leading the Way: A study of best manufacturing practice in Australia and New Zealand* (1994).

Ayres, I., & Braithwaite, J., *Responsive Regulation: Transcending the Deregulation Debate* (Oxford University Press, Oxford, 1992).

Baldwin, R., *Rules and Government* (Oxford University Press, Oxford, 1995).

Baldwin, R., & Daintith, T., *Harmonisation and Hazard: Regulating Health and Safety in the European Workplace* (Graham and Trotman, London, 1992).

Ballam, D. A., 'The OSHA's preemptive effect on state criminal prosecutions of employers for workplace deaths and injuries' (1988) 26 *American Business Law Journal*.

Bardach, E., & Kagan, R., *Going by the Book: The problem of regulatory unreasonableness* (Temple University Press, Philadelphia, 1982).

Barnard, C., 'The United Kingdom, the 'Social Chapter' and the Amsterdam Treaty' (1997) 26 *Industrial Law Journal* 275.

Barrett, P., & James, P., 'Do Safety Policies Assist Workplace Safety?' (1982) July *Health and Safety at Work* 24.

Bartal, A., & Thomas, L., 'Direct and Indirect Effects of Regulation' (1985) 28 *Journal of Law and Economics* 1.

Baysinger, B., 'Organizational Theory and the Criminal Liability of Organizations' (1992) 71 *Boston University Law Review* 341.

Beardsley, D., *Incentives for Environmental Improvement: An Assessment of Selected Innovative Programs in the States and Europe* (Global Environmental Management Initiative, Washington DC, 1997).

Berger, J., 'The Hoechst Dispute: A Paradigm Shift in Occupational Health and Safety', in M. Quinlan (ed.), *Work and Health: The origins, management and regulation of occupational illness* (Macmillan, Melbourne, 1993).

Bergman, D., *Deaths at Work: Accidents or Corporate Crime. The Failure of Inquests and the Criminal Justice System* (Workers' Educational Association, London, 1991).

——'Corporate sanctions and corporate probation' (1992) 25 *New Law Journal* 1312.

——*The Perfect Crime? How Companies Get Away with Manslaughter in the Workplace* (HASAC, Birmingham, West Midlands, 1994).

——'Government is Weak on Corporate Crime Says Law Campaigner' (1997) December *Health and Safety* 34.

Biagi, M., 'From Conflict to Participation in Safety: Industrial Relations and the Working Environment in Europe 1992' (1990) 6 *The International Journal of Comparative Labour Law and Industrial Relations* 67.

Bixby, M., 'Workplace Homicide: Trends, Issues and Policy' (1991) 70 *Oregon Law Review* 333.

Blake, G., *TQM and Strategic Environmental Management in Executive Enterprises* (Publications Co. Inc. 1992).

Blewett, V., 'OHS Best Practice Column: Benchmarking OHS: A tool for best practice' (1995) 11(3) *Journal of Occupational Health Safety—Australia and New Zealand* 237.

Blewett, V., & Shaw, A., 'Quality Occupational Health and Safety?' (1996) 12(4) *Journal of Occupational Health & Safety—Australia and New Zealand* 481.

Block, M. K., 'Optimal Penalties, Criminal Law and the Control of Corporate Behaviour' (1992) 71 *Boston University Law Review* 395.

Bonner, J. P., & Forman, B. N., 'Bridging the Deterrence Gap: Imposing Criminal Penalties on Corporations and their Executives for Producing Hazardous Products' (1993) 1 *San Diego Justice Journal* 1.

Bottomley, B. A., 'Systems Approach to Prevention', Paper presented at *Future Safe Conference*, Sydney, May 1994.

Box, S., *Power, Crime and Mystification* (Tavistock, London, 1983).

Braithwaite, J., *To Punish or Persuade: Enforcement of coal mine safety* (State University of New York Press, Albany, United States, 1985).

——*Crime, Shame and Reintegration* (Cambridge University Press, Cambridge, 1989).

——'Foreword', in J. A. Sigler & J. E. Murphy (eds.), *Corporate Lawbreaking &
Interactive Compliance: resolving the regulation-deregulation dichotomy* (Quorum Books,
US, 1991).

——*Improving Regulatory Compliance: Strategies and Practical Applications in OECD
Countries*, Regulatory Management and Reform Series No. 3 (OECD, Paris,
1993).

——'Responsive Business Regulatory Institutions', in C. A. J. Coady and C. J.
C. Sampford (eds.), *Business Ethics and the Law* (The Federation Press, Sydney,
1993).

Braithwaite, J., & Fisse, B., 'Varieties of Responsibility and Organizational
Crime' (1985) 7 *Law and Policy* 315.

Braithwaite, J., & Geis, G. (1982) 28 *Crime and Delinquency* 282.

Braithwaite, J., & Grabosky, P., *Occupational Health and Safety Enforcement in Australia*
(Australian Institute of Criminology, Canberra, 1985).

——*Of Manners Gentle: Enforcement Strategies of Australian Business Regulatory Agencies*
(Oxford University Press and the Australian Institute of Criminology,
Melbourne, 1986).

Braithwaite, J., & Makkai, T., 'Testing an Expected Utility Model of Corporate
Deviance' (1991) 25 *Law and Society Review* 7.

——'Trust and Compliance' (1994) 4 *Policing & Society* 1.

Brooks, A., 'Rethinking Occupational Health and Safety Legislation' (1988)
September *The Journal of Industrial Relations* 348.

——*Occupational Health and Safety Law in Australia*, 4th edn. (CCH, Sydney, 1993).

Brown, R. M., 'Administrative and Criminal Penalties in the Enforcement of
Occupational Health and Safety Legislation' (1992) 30(3) *Osgoode Hall Law
Journal* 691.

Brunton, N., 'Directors, Companies and Pollution in Western Australia' (1995) 3
Environmental and Planning Law Journal 159.

Bucy, P. H., 'Organizational Sentencing Guidelines: The Cart Before the Horse'
(1993) 71 *Washington University Law Quarterly* 329.

Burdeu, M., 'Policy Into Practice: Integrating health and safety management
into management systems', Paper presented at the *OH&S in the Public Sector
Conference*, Sydney, 31 August and 1 September 1995.

Bureau of National Affairs, 'Current OSHA Inspection Rate Once Every 109
Years, AFL–CIO Says' (1998) 27(7) *Occupational Safety and Health News* 1663.

Burgess, W., *'Best Practice' and OH&S, Victorian Institute of Occupational Health and
Safety* (Ballarat, Victoria, 1997).

Burke, N., 'Gaining Organisational Commitments to OH&S by Integrating
Safety Onto Your Business Plans', Paper presented at *Proactive OH&S
Management Conference*, Sydney, 9 and 10 March 1994.

Busey, J. B., 'US Federal Administration', an announcement before the
Aero Club of Washington DC (27 March 1990) (1991) 12 *Cardozo Law Review*
1327.

Cain, P., *Atiyah's accident compensation and the law* (Martin and Weidenfeld, London, 1987).

California Law Review, 'Criminal Sentences for Corporations: Alternative Fining Mechanisms' (1985) 73 *California Law Review* 443.

Canadian Law Reform Commission, *Criminal Responsibility for Group Action*, Working Paper 16 (Information Canada, Ottawa, 1976).

Carson, W. G., 'White Collar Crime and the Enforcement of Factory Legislation' (1970) 10 *British Journal of Criminology* 383.

—— 'Some Sociological Aspects of Strict Liability and the Enforcement of Factory Legislation' (1970) 33 *Modern Law Review* 396.

—— 'Symbolic and Instrumental Dimensions of Early Factory Legislation', in R. G. Hood (ed.), *Crime, Criminology and Public Policy* (Heinemann, London, 1974).

—— 'The Conventionalisation of Early Factory Crime' (1979) 7 *International Journal of the Sociology of Law* 37.

—— 'The Institutionalization of Ambiguity: Early British Factory Acts', in G. Geis and E. Stotland (eds.), *White Collar Crime: Theory and Research* (Sage, London, 1980).

—— *The Other Price of Britain's Oil* (Martin Robertson, UK, 1982).

—— 'Hostages to History: Some Aspects of the Occupational Health and Safety Debate in Historical Perspective', in W. B. Creighton and N. Gunningham (eds.), *The Industrial Relations of Occupational Health and Safety* (Croom Helm, Sydney, 1985).

Carson W. G., & Johnstone, R., 'The Dupes of Hazard: Occupational Health and Safety and the Victorian Sanctions Debate' (1990) 26 *Australian and New Zealand Journal of Sociology* 126.

Cebon, P., *The Missing Link: Organisational behaviour as a key element in energy/environment regulation and university energy management*, unpublished thesis, MA, Massachusetts Institute of Technology.

Chambliss, W. J., 'Types of Deviance and the Effectiveness of Legal Sanctions' (1967) *Wisconsin Law Review* 703.

Chang, R., *TQM Fever*, an interview presented by Business Report on ABC National Radio, July 1995.

Chappell, D., & Norberry, J., 'Deterring Polluters: The Search for Effective Strategies' (1990) 13 *University of New South Wales Law Journal* 97 at 108.

Cherrington, D. R., 'The race to the courthouse: conflicting views toward the judicial review of OSHA standards' (1994) 95 *Brigham Young University Law Review*.

Cialdini, R. B., *Influence: Science & Practice*, 2nd edn. (Scott Foresman & Co., Glenview, Ill., 1988).

Cimino, M. T., ' Criminal Prosecution of Workplace Safety Violations' (1992) 94 *West Virginia Law Review*, 1007.

Clinard, M. B., & Yeager, P. C., *Corporate Crime* (1980).

Clinton, B. (President), & Gore, A. (Vice-President), *The New OSHA: Reinventing Worker Safety and Health*, National Performance Review, White House, Washington DC, May 1995.

Coffee, J. C., '"No Soul to Damn: No Body to Kick": An Unscandalized Inquiry into the Problem of Corporate Punishment' (1981) 79 *Michigan Law Review* 386.

—— 'Environmental Crime and Punishment', Thursday February 1994, *New York Law Journal* 5, 10, and 29.

Coffee, J. C., Gruner, R., & Stone, C., 'Standards for Organizational Probation: A Proposal to the United States Sentencing Commission' (1988) 10 *Whittier Law Review* 77.

Cole, C. (ed.), *The Death and Life of the American Quality Movement* (Oxford University Press, New York, 1995).

Collins, M., 'OSHA's Safety and Health Program Standard' (1998) April *The Synergist 24.*

Colvin, E., 'Corporate Personality and Criminal Liability' (1995) 6 *Criminal Law Forum* 1.

Comcare Australia, *Analysis of Data from the 1994–95 Planned Investigation Program* (AGPS, Canberra, 1996).

Committee on Safety and Health, *Report of the Committee on Safety and Health at Work 1970–72* (HMSO, London, 1972).

Cowan, A., 'Note: Scarlett Letters for Corporations? Punishment by Publicity Under the New Sentencing Guidelines' (1992) 65 *University of Southern California Law Review* 2387.

Creighton, W. B., 'Occupational Health and Safety Regulation: The Role of ILO Standards', in M. Quinlan (ed.), *Work and Health: The Origins, Management and Regulation of Occupational Illness* (Macmillan, Melbourne, 1993).

Creighton, W. B., Ford, W. J., & Mitchell, R. J., *Labour Law: Text and Materials*, 2nd edn. (Law Book Company, Sydney, 1993).

Creighton, W. B., & Rozen, P., *Occupational Health and Safety Law in Victoria* (Federation Press, Sydney, 1997).

Croall, H., 'Mistakes, Accidents, and Someone Else's Fault: The Trading Offender in Court' (1988) 15 *Journal of Law and Society* 293.

—— *White-Collar Crime* (Open University Press, Buckingham, 1992).

Cullen, Lord (Chairman), *Piper Alpha Inquiry* (HMSO, London, 1990).

Curcio, A. A., 'Painful Publicity: An Alternative Punitive Damages Sanction' (1996) 45 *De Paul Law Review* 341.

Damaska, M., *The Faces of Justice and State Authority: A Comparative Approach to the Legal Process* (Yale University Press, New Haven and London, 1986).

Davey, L., 'Specification versus performance language in safety standards' (1990) Fall *Job Safety and Health Quarterly.*

Davies, T., & Mazurek, J., *Industry Incentives for Environmental Improvement* (Global Environmental Management Initiative, Washington DC, 1997).

Dawson, D., Willman, P., Clinton, A., & Bamford, R., *Safety at Work: The Limits of Self-Regulation* (Cambridge University Press, Cambridge, 1988).

Dawson, P., 'Managing Quality in the Multi-Cultural Workforce', in W. Wilkinson & H. Willmott (eds.), *Making Quality Critical: New perspectives on organizational change* (Routledge, London, 1995), p. 190.

Deming, W. E., *Quality, Productivity and Competitive Position* (MIT Press, Cambridge, Mass., 1982).

——— *Out of the Crisis* (Cambridge University Press, Cambridge, 1986).

Deturbide, M. E., 'Corporate Protector or Environmental Safeguard? The emerging role of the environmental audit' (1995) 5 *Journal of Environmental Law and Policy* 1.

Devine, P. J., 'The Draft Organization Sentencing Guidelines for Environmental Crimes' (1995) 20 *Columbia Journal of Environmental Law* 249.

Diver, C., 'A Theory of Regulatory Enforcement' (1980) 28(3) *Public Policy* 257.

Dobbs, D. B., 'Ending Punishment in "Punitive" Damages: Deterrence Measured Remedies' (1989) 40 *Alabama Law Review* 831.

Douglas, R., & Laster, K., *Victim Information and the Criminal Justice System: Adversarial or Technocratic Reform* (La Trobe University, Melbourne, 1994).

Eagle, A., 'Raising the Profile of Health and Safety at Work' (1997) December *Health and Safety* 8.

Eberlie, R. F., 'The new health and safety legislation of the European Community' (1990) 19 *Industrial Law Journal* 81.

Edelman, P. T., 'Corporate Criminal Liability for Homicide: The Need to Punish Both the Corporate Entity and its Officers' (1987) 92 *Dickinson Law Review* 193.

Edgar, A., 'Directors' Liability' (1997) 141 *Solicitors' Journal* 328.

Ellickson, R., 'Bringing Culture and Human Frailty to Rational Actors: A Critique of Classical Law and Economics' (1989) 65 *Chicago-Kent Law Review* 23.

Elliott, E. D., 'Environmental TQM: Anatomy of a pollution control program that works!' (1994) 92 *Michigan Law Review* 1840.

Else, D., *Enhanced Cohesion and Co-ordination of Occupational Health and Safety Training in Australia*, Report to the Minister for Industrial Relations (March 1992).

Emery, F. E., & Thorsud, E., *Democracy at Work* (Nijhof, Leiden, 1976).

Emmett, E. A., 'New Directions for Occupational Health and Safety in Australia' (1992) 8(4) *Journal of Occupational Health & Safety—Australia and New Zealand* 293.

——— *National Uniformity/Regulatory Reform: A more effective approach to occupational health and safety*, Occasional Paper No. 4, Worksafe, Sydney, May 1993.

——— 'Regulatory Reform in Occupational Health and Safety in Australia', a paper presented at *The Institute of Public Affairs Conference on Risk, Regulation and Responsibility*, Sydney, July 1995.

Emmett, E., & Hickling, C., 'Integrating Management, Systems & Risk Management Approaches' (1995) 11(6) *Journal of Occupational Health & Safety—Australia and New Zealand* 617.

Ennis, P. I., 'Environmental Audits: Protective shield, or smoking guns?' (1992) 42 *Washington University Journal of Urban and Contemporary Law*, 389.

Environment Institute of Australia, (1995) *Newsletter*, No. 31, December, 6.

Environmental Resources Ltd. for the Commission of the European Communities, *The law and practice concerning occupational health in the member States of the European Community*, Vol. 2, Denmark, Federal Republic of Germany (Graham and Trotman Ltd., 1985).

European Agency for Safety and Health at Work, *Priorities and Strategies in Occupational Safety and Health Policy in the Member States of the European Union* (European Agency for Safety and Health at Work, Bilbao, Spain, 1997).

European Foundation for the Improvement of Living and Working Conditions, *Second European Survey on Working Conditions* (Dublin, 1996).

—— -*Policies on Health and Safety in Thirteen Countries of the European Union: The European Situation*, Vol. II (Office of Official Publications of the European Communities, Luxembourg, 1996).

Euroline 'Danish building and construction sites are hazardous workplaces' (1997) April.

——-'Employers opt out of the Danish health and safety system' (1997) June.

——'Problems of implementing the newly amended Work Environment Act' (1997) August.

——'Conference highlights differences in national implementation of health and safety Directives' (1997) December.

——'Union proposals to enforce health and safety legislation at the workplace' (1998) January.

——'Public employers ordered to act against difficult working conditions' (1998) February.

——'Penalties for labour law infringements under review' (1998) February.

——'Employers rejoin Danish health and safety system' (1998) June.

European Trade Union Technical Bureau for Health and Safety, July (1996) 3 *TUTB Newsletter*.

Evans, J. R., & Lindsay, W. M., *The Management and Control of Quality* (West Publishing Company, US, 1989).

Field, S., & Jorg, N., 'Liability and Manslaughter: Should We Be Going Dutch?' [1991] *Criminal Law Review* 156.

Fiorelli, P. E., 'Fine Reductions Through Effective Ethics Programs' (1992) 56 *Albany Law Review* 403.

Fiorelli, P. E., & Rooney, C. J., 'The Environmental Sentencing Guidelines for Business Organizations: Are There Murky Waters in Their Future?' (1995) 22 *Environmental Affairs* 481.

Fiorino, D. J., 'Towards a New System of Environmental Regulation: The case for an industry sector approach' (1996) 26 *Environmental Law* 457.

Fisher, T., 'A "Quality" Approach to Occupational Health, Safety and Rehabilitation' (1991) 7(1) *Journal of Occupational Health & Safety—Australia and New Zealand* 23.

Fisse, B., 'The Use of Publicity as a Criminal Sanction against Business Corporations' (1971) 8 *Melbourne University Law Review* 107.

—— 'Community Service as a Sanction Against Corporations' [1981] *Wisconsin Law Review* 970.

—— 'Reconstructing Corporate Criminal Law: Deterrence, Retribution, Fault and Sanctions' (1983) 56 *Southern California Law Review* 1141.

—— *Howard's Criminal Law*, 5th edn. (Law Book Company, Sydney, 1990).

—— 'Sentencing Options against Corporations' (1990) 1 *Criminal Law Forum* 211.

—— 'Corporate Criminal Responsibility' (1991) 15 *Criminal Law Journal* 166.

—— 'Recent Developments in Corporate Criminal Law and Corporate Liability to Monetary Penalties' (1991) 13 *University of New South Wales Law Review* 1.

—— 'Individual and Corporate Criminal Responsibility and Sanctions Against Corporations', in R. Johnstone (ed.), *Occupational Health and Safety Prosecutions in Australia: Overview and Issues* (Centre for Employment and Labour Relations Law, The University of Melbourne, 1994).

—— 'Corporations, Crime and Accountability' (1995) 6 *Current Issues in Criminal Justice* 378.

Fisse, B., & Braithwaite, J., *The Impact of Publicity on Corporate Offenders* (State University of New York Press, Albany, New York, 1983).

—— 'The Allocation of Responsibility for Corporate Crime: Individualism, Collectivism, and Accountability' (1988) 11 *Sydney Law Review* 469.

— — *Corporations, Crime and Accountability* (Cambridge University Press, Sydney, 1993).

Flagstad, K., *The Functioning of the Internal Control Reform*, Doctoral Thesis, NTH, Trondheim, 1995.

Fleischauer, B. J., 'Occupational Safety and Health Law in Sweden and the United States: Are There Lessons to be Learned by Both Countries?' (1983) 6 *Hastings International and Comparative Law Review* 283.

Foster, A., 'Safety—Voluntary Program Exposes Industry Divide' (1998) 160(17) *Chemical Week* 34.

Fox, F., 'Corporate Sanctions: Scope for a New Eclecticism' (1982) 24 *Malaya Law Review* 26.

Fox, R., & Freiberg, A., 'Silence is not Golden: The functions of prosecutors in sentencing in Victoria' (1987) 61 *Law Institute Journal* 554.

French, P., 'Publicity and the Control of Corporate Conduct', in B. Fisse & P. French (eds.), *Corrigible Corporations and Unruly Law* (1985).

Freyer, D. H., 'Corporate Compliance Programs for FDA-Regulated Companies: Incentives for Their Development and the Impact of the Federal Sentencing Guidelines for Organizations' (1996) 51 *Food and Drug Law Journal* 225.

Frick, K., 'Can Management Control Health and Safety at Work?' (1990) 11 (3 August) *Economic and Industrial Democracy*.

—— Enforced Voluntarism—purpose, means and goals of systems control, National Institute for Working Life, Solna, Workshop on *Integrated Control/Systems Control*, Dublin, 29–30 August 1996.

Frostberg, C., *Internal Control of the Working Environment*, Analysis and Planning Division, National Board of Occupational Health and Safety, Sweden, December, 1994.

Gallagher, C., 'Occupational Health and Safety Management' in *Belts to Bytes*, Conference Proceedings, Work Cover, Adelaide, Australia, 1994.

Geis, G. (ed.), *On White Collar Crime: Offences in business, politics, and the professions* (Free Press, New York, 1982).

Genn, H., 'Business Responses to the Regulation of Health and Safety in England' (1993) 15 *Law & Policy* 219.

George, J., 'How Can OH&S Contribute To Overall Efficiency At The Enterprise Level?', Paper presented at the *Strategic OH&S Management Conference*, Sydney, 13 and 14 September 1993.

Glasbeek, H., *The Maiming and Killing of Workers: The one-sided nature of risk taking in capitalism*, Jurisprudence Centre Working Papers, Department of Law, Carleton University, Ottawa, 1986.

——'A Role for Criminal Sanctions in Occupational Health and Safety', in *New Developments in Labour Law, Meredith Memorial Lectures*, 1988 (Yvon Blais, Montreal, 1989).

—— 'Occupational Health and Safety Law: Criminal Law as a Political Tool' (1998) 11 *Australian Journal of Labour Law* 95.

Glendon, I., 'Risk Management in the 1990s: Safety Auditing' (1995) 11(6) *Journal of Occupational Health & Safety —Australia and New Zealand* 569.

Glendon, I., & Booth, R., 'Risk Management in the 1990s: Measuring management performance in occupational health and safety' (1995) 11(6) *Journal of Occupational Health & Safety—Australia and New Zealand* 559.

Global Environmental Management Initiative (GEMI), *Total Quality Environmental Management* (Washington DC, 1992).

Gorbatoff, T. G., 'OSHA Criminal Penalty Reform Act: Workplace Safety May Finally Become a Reality' (1991) 39 *Cleveland State Law Review* 551.

Gordon, B. R., 'Employee Involvement in the Enforcement of the Occupational Health and Safety Laws of Canada and the United States' (1994) 15(4) *Comparative Labor Law Journal* 527.

Grafe-Buckens, A., 'Old and New EMAS: Challenges for the European Eco-Management and Audit Scheme' (1997) 6(11) *European Environmental Law Review* 300.

Gray, W, B., & Scholz, J. T., 'OSHA Enforcement and Workplace Injuries: A behavioural approach to risk assessment' (1990) *Journal of Risk & Uncertainty* 283.

—— 'Analysing the equity and efficiency of OSHA enforcement' (1991) 3(3) *Law and Policy* 185.

—— 'Does Regulatory Enforcement Work? A Panel Analysis of OHSA Enforcement' (1993) 27 *Law and Society Review* 177.

Groenewegen, P., & Vergagt, P., 'Environmental issues as threats and opportunities for technological innovation' (1991) 3(1) *Technological Analysis and Strategic Management* 43.

Gross, J. A., 'The Broken Promises of the National Labor Relations Act and the Occupational Safety and Health Act: Conflicting values and conceptions of rights and justice' (1998) 73(1) *Chicago-Kent Law Review* 351.

Gruner, R. S., 'To Let the Punishment Fit the Organization: Sanctioning Corporate Offenders through Corporate Probation' (1988) 16 *American Journal of Criminal Law* 1.

—— 'Just Punishment and Adequate Deterrence for Organizational Misconduct: Scaling Economic Penalties Under the New Corporate Sentencing Guidelines' (1992) 66 *Southern California Law Review* 225.

—— 'Towards an Organizational Jurisprudence: Transforming Corporate Criminal Law Through Federal Sentencing Reform' (1994) 36 *Arizona Law Review* 407.

Gruner, R. S., & Brown, L. M., 'Organizational Justice: Recognizing and Rewarding the Good Corporate Citizen' (1996) *The Journal of Corporation Law* 731.

Gunningham, N., *Safeguarding the Worker* (Law Book Company, Sydney, 1984).

—— 'Negotiated Compliance—A Case of Regulatory Failure' (1987) 9 *Law and Policy* 69.

—— 'The Role of Federal Awards in Regulating Occupational Health and Safety Standards' (1990) 3 *Australian Journal of Labour Law* 54.

—— 'Environmental Auditing: Who audits the auditors?' (1993) 10(4) *Environmental and Planning Law Journal*, 229.

—— 'Beyond Compliance: Management of environmental risk', in Boer, Fowler & Gunningham (eds.), *Environmental Outlook: Law and policy* (ACEL, Federation Press, 1994).

—— 'Environment, Self-Regulation, and the Chemical Industry: Assessing Responsible Care' (1995) 17(1) *Law and Policy* 57.

—— 'From Compliance to Best Practice in OHS: The roles of specification, performance and systems-based standards' (1996) 9(3) *Australian Journal of Labour Law* 221.

—— 'Towards Innovative Occupational Health and Safety Regulation' (1998) 40(2) *The Journal of Industrial Relations* 204.

—— 'Towards Effective and Efficient Enforcement of Occupational Health and Safety Regulations: Two Paths to Enlightenment' (1998) 19(4) *Comparative Labor Law & Policy Journal*.

—— 'Environmental Management Systems and Community Participation: Rethinking Chemical Industry Regulation', forthcoming (1998) 16(2) *UCLA Journal of Environmental Law and Policy*.

——'Integrating Management Systems and Occupational Health and Safety Regulation', forthcoming (1999) *Journal of Law and Society*.

Gunningham, N., & Grabosky, P., *Smart Regulation: Designing Environmental Policy* (Clarendon Press, Oxford, 1998).

Gunningham, N., Johnstone, R., & Rozen, P., *Enforcement Measures for Occupational Health and Safety in New South Wales: Issues and Options* (WorkCover Authority of New South Wales, Sydney, 1996).

Gunningham, N., & Prest, J., 'Environmental audit as a regulatory strategy' (1993) 15(4) *Sydney Law Review* 492.

Gustavsen, B., & Hunnius, G., *New Patterns of Work Reform: The Case of Norway* (The University Press, Oslo, 1981).

Hadfield, G., 'Problematic Relations: Franchising and the Law of Incomplete Contracts' (1990) 42 *Stanford Law Review* 927.

Hahn, R., & Hurd, J., 'The Costs & Benefits of Regulation—Review & Synthesis' (1991) 8 *Yale Journal of Regulation* 233.

Haines, F., *Corporate Regulation: Beyond 'Punish or Pursuade'* (Clarendon Press, Oxford, 1997).

Hall, G., 'Victim Impact Statements: Sentencing on Thin Ice' (1992) 15 *New Zealand Universities Law Review* 143.

Harper, P., *Turning Rhetoric into Practice: Integrating TQM into plant operations* (Executive Enterprises Publications, 1992).

Harrell, M., 'Organizational Environmental Crime and the Sentencing Reform Act of 1984: Combining Fines with Restitution, Remedial Orders, Community Service and Probation to Benefit the Environment While Punishing the Guilty' (1995) 6 *Villanova Environmental Law Journal* 243.

Harvard Law Review, Note, 'Growing the Carrot: Encouraging Effective Corporate Compliance' (1996) 109 *Harvard Law Review* 1783.

Hawkins, K., *Environment and Enforcement: Regulation and the Social Definition of Pollution* (Oxford University Press, Oxford, 1984).

Health and Safety Commission (UK), *Health and Safety Commission Annual Report 1988/89* (HMSO, London, 1990).

——*Plan of Work for 1994/95*, Volume 1 Main Report and Volume 2 Summary (HMSO, Sheffield, UK, 1994).

Health and Safety Commission (UK), *Review of Health and Safety Legislation*, Main Report (Volume 1) and Summary of findings and of the Commission's response (Volume 2) (HMSO, UK, 1994).

Health & Safety Executive (UK), *Successful Health and Safety Management* HS(G)65 (HMSO, London, 1991).

——*The Costs of Accidents at Work* (HMSO, London, 1993).

——*Review of Health and Safety Legislation* (HMSO, London, 1994).

——*Health and Safety in Small Firms* (HMSO, London, 1995).

—— *The Regulation of Health and Safety in Five European Countries*, Contract Research Report 84 (HMSO, London, 1996).

Hemmings, N., 'The New South Wales Experiment: The Relative Merits of Seeking to protect the Environment through the Criminal Law by Alternative Means' (1993) October *Commonwealth Law Bulletin* 1987.

Hickey, J., & Spangaro, C., *Judicial Views About Pre-Sentence Reports* (Judicial Commission of New South Wales and the New South Wales Probation Service, Sydney, 1995).

Hill, S., 'From Quality Circles to Total Quality Management', in A. Wilkinson & H. Willmott (eds.), *Making Quality Critical: New perspectives on organizational change* (Routedge, London, 1995).

Hillary, R., 'Environmental Reporting Requirements Under the EU: Eco-management and audit scheme (EMAS)' (1995) 15(4) *The Environmentalist* 293.

Hilmer, F., et al., *Working Relations: A fresh start for Australian enterprises* (Business Council of Australia, 1992).

Hirsch, J. L., *Occupational Safety and Health Handbook: An employer guide to OSHA laws, regulations and practices* (Butterworths Legal Publications, Austin, 1993).

Hodges, J., 'Eight Years of Robens Style Legislation in Queensland—What have we learnt?', Paper presented to *Productivity, Ergonomics and Safety Conference*, Ergonomics Society of Australia, 24–27 November 1996, Canberra.

Home Office, Research and Planning Unit, *Unit Fines: Experiments in Four Courts*, Paper 59 (Home Office, London, 1990).

Hopkins, A., 'Patterns of Prosecution', in R. Johnstone (ed.), *Occupational Health and Safety Prosecutions in Australia: Overview and Issues* (Centre for Employment and Labour Relations Law, The University of Melbourne, 1994).

—— —'Compliance With What?: The fundamental regulatory question' (1994) *34 British Journal of Criminology* 431.

——— *Making Safety Work: Getting Management Commitment to Occupational Health and Safety* (Allen & Unwin, Sydney, 1995).

Hopkins, A., & Hogan, L., 'Influencing Small Business to Attend to Occupational Health and Safety' (1998) 14(3) *Journal of Occupational Health & Safety—Australia and New Zealand* 237.

Huff, K. B., 'The Role of Corporate Compliance Programs in Determining Corporate Criminal Liability: A Suggested Approach' (1996) *Columbia Law Review* 1252.

Humphrey, H. III, 'Public/Private Environmental Auditing Agreements: Finding better ways to promote voluntary compliance' (1994) 3 *Corporate Conduct Quarterly* 1.

Hutter, B. M., 'Regulating Employers and Employees: Health and Safety in the Workplace' (1993) 20(4) *Journal of Law and Society* 452.

——— *Compliance: Regulation and Environment* (Clarendon Press, Oxford, 1997).

Industry Commission, Australia, *Work, Health & Safety: Inquiry into Occupational*

Health & Safety Volume I Report and Volume II Appendices, Report No. 47 (AGPS, Canberra, 1 September 1995).

Jacobs, S., 'The Future of Regulatory Reform', Paper presented at *From Red Tape to Results: International perspectives on regulatory reform*, Conference organized by the New South Wales Government, Sydney, June 1995.

James, P., 'Reforming British Health and Safety Law: A framework for discussion' (1992) 21 *Industrial Law Journal* 83.

James, P., & Walters, D., 'Non-union Rights of Involvement: The Case of Health and Safety at Work' (1997) 26 *Industrial Law Journal* 35.

Jeffress, C., Assistant Secretary, Occupational Safety and Health Administration, *Statement Before the Subcommittee on Oversight and Investigations*, Committee on Education and the Workforce, US House of Representatives, 8 May 1998.

Jensen, P. L., *Internal Control in Denmark*, Technical University of Denmark, Lyngby, Workshop on Integrated Control/Systems Control, Dublin, 29–30 August 1996.

—— 'Workplace assessment in Denmark—a genuine success for risk assessment?', Paper delivered to *Policies for Occupational Health and Safety Management Systems and Workplace Change Conference*, Amsterdam, 21–24 September 1998.

Jensen, P. L., Stranddorf, J., & Moller, N., *Developing Safety Management in Practice*, Department of Working Environment, Technical University of Denmark, Lyngby, Denmark, undated.

Johansson, B., *The motivation for working environment activities among heads of small firms* (TULEA, Lulea, 1994).

Johnston, J., 'Plying Industry's Green Standard' (1993) 21(2) *New Scientist* 10.

Johnstone, R. (ed.), *Occupational Health and Safety Prosecutions in Australia* (Centre for Employment and Labour Relations Law, The University of Melbourne, Melbourne, 1994).

—— *The Court and the Factory: The Legal Construction of Occupational Health and Safety offences in Victoria*, unpublished Ph. D. thesis, the University of Melbourne, 1994.

—— 'The Legal Construction of Occupational Health and Safety Offences in Victoria: 1983–1991', in R. Johnstone (ed.), *Occupational Health and Safety Prosecutions in Australia* (Centre for Employment and Labour Relations Law, The University of Melbourne, Melbourne, 1994).

—— (ed.), *New Directions in Occupational Health and Safety Prosecutions: The Individual Liability of Corporate Officers, and Prosecutions for Manslaughter and Related Offences* (Centre for Employment and Labour Relations Law, Working Paper No. 9, The University of Melbourne, 1996).

Johnstone, R. (ed.), 'Prosecutions in the Light of the Industry Commission's Report' (1996) October *Law Institute Journal* 54.

—— *Occupational Health and Safety Law and Policy: Text and materials* (LBC Information Services, Sydney, 1997).

—— 'Occupational Health and Safety Prosecutions in Australia: Rethinking

State Enforcement of Occupational Health ad Safety Statutes', in R. Mitchell and J. W. (eds.), *Facing the Challenge in the Asia Politic Region: Contemporary Themes and Issues in Labour Law* (Centre for Employment and Labour Relations Law, the University of Melbourne, Occasional Monograph No. 5, 1997), pp. 193–214.

——'Does the Crime Match the Punishment?' (1997) June *Complete Safety* 21-3.

Junger-Tas, J., *Alternatives to Prison Sentences: Experiences and Developments*, (RDC—Dutch Ministry of Justice/Krugler Publications, Amsterdam, 1994).

Juran, J., *Juran on Planning for Quality* (Free Press, New York, 1988).

Justice, *Sentencing: A Way Ahead* (Justice, London, 1989).

Kaasen, K., 'Post Piper Alpha: Some reflections on offshore safety regimes from a Norwegian perspective' (1991) 9(4) *Journal of Energy and Natural Resource Law*, 281.

Kagan, R., Criminal Prosecution for Regulatory Offenses, Workshop on **Regulators Law Enforcement**, Socio-Legal Center, Ohio State University, 20–21 May 1993.

——'Regulatory enforcement' in D. Rosenbloom & R. Schwartz (eds.), *Handbook of Regulation and Administrative Law*, (Dekker, New York, 1994).

Kamp, A., & Le Blansch, K., 'Integrating Management of OHS and the Environment: Participation, Prevention and Control', Paper delivered to *Policies for Occupational Health and Safety Management Systems and Workplace Change Conference*, Amsterdam, 21–24 September 1998.

Kast, F. E., & Rosenzweig, J. E., *Organisation and Management: A systems approach*, 2nd edn., (McGraw-Hill, Tokyo, 1974).

Katz, S. E., 'The general good and the Second Banking Directive: A major loophole?' (1993) 166 *Journal of International Banking and Finance Law*.

Kean Report (Committeee of Inquiry into Australia's Standards and Conformance Infrastructure), *Linking Australia Globally: Overview* (AGPS, Canberra, 1995).

Kelman, S., *Regulating America, Regulating Sweden* (MIT Press, Cambridge, Mass., 1982).

Kennedy, K. C., 'A Critical Appraisal of Criminal Deterrence Theory' (1983) *88 Dickinson Law Review* 1.

Koch, C., & Nielsen, K., *Danish Working Environmental Regulation: How reflexive—How political?* A *Scandinavian case*, Working Paper, Technical University of Denmark, Lyngby, Denmark, June 1996.

Korostoff, J. E., Zimmerman, L. M., & Ryan, C, E., 'Rethinking the OSHA Approach to Workplace Safety: A Look at Worker Participation in the Enforcement of Safety Regulations in Sweden, France and Great Britain' (1991) *13 Comparative Labor Law Journal* 45.

Kovach, K. A., Hamilton, N. G., Alston, T. M., & Sullivan, J. A., 'OSHA and the Politics of Reform: An analysis of OSHA reform initiatives before the 104th Congress' (1997) 34 *Harvard Journal of Legislation* 169.

Kracht, R. L., 'Comment: A Critical Analysis of the Proposed Sentencing Guidelines for Organisations Convicted of Environmental Crimes' (1995) 40 *Villanova Law Review* 513.

Kriesberg, S. M., 'Decisionmaking Models and the Control of Corporate Crime' (1976) 85 *Yale Law Journal* 1091.

Laing, R., *Report of the Inquiry into Operations of the Occupational Health and Safety and Welfare Act 1984* (Western Australian Government, Perth, 1992).

Larsson, T., Systems Control Development in Sweden, SAMU, Institute for Social Science, Uppsala, Workshop on **Integrated Control/Systems Control**, Dublin, 29–30 August 1996.

La Trobe/Melbourne Occupational Health and Safety Project (W. G. Carson, W. B. Creighton, C. Henenberg & R. Johnstone), *Victorian Occupational Health and Safety: An assessment of law in transition* (Department of Legal Studies, La Trobe University, Victoria, 1989).

Laufer, W., 'Culpability and the Sentencing of Corporations' (1992) 71 *Nebraska Law Review* 1049.

Lauredsen, P., Working Paper, Stavanger College, Rogaland Research, Stavanger, 1995.

Law Commission, *Codification of the Criminal Law: Strict Liability and the Enforcement of the Factories Act 1961*, Published Working Paper No. 30 (HMSO, London, 1970).

———*Legislating the Criminal Code: Involuntary Manslaughter* (HMSO, London, 1997).

Lederman, F., 'Criminal Law, Perpetrator and Corporation: Rethinking a Complex Triangle' (1985) 76 *Journal of Law and Criminology* 285.

Leigh, L. H., *The Criminality of Corporations in English Law* (Weidenfeld & Nicolson, 1969).

Lemkin, J. M., 'Deterring Environmental Crime Through Flexible Sentencing: A Proposal for the New Organisational Environmental Sentencing Guidelines' (1996) 84 *California Law Review* 307.

Levine, S. P., & Dyjack, D. T., 'Critical Features of an Auditable Management System for an ISO 9000-Compatible Occupational Health and Safety Standard' (1997) 58(4) *American Industrial Hygiene Association Journal* 291.

Lewis-Beck, M-S, & Alford, J. R., 'Can Government Regulate Safety: The Coal Mine Example' (1980) 76 *American Political Science Review* 745.

Lindoe, P., *Internal Control—Conflicting Interests between Bureaucratic Control and Participatory Co-operation*, Doctoral Thesis, University of Trondheim, Norwegian Institute of Technology, 1992.

———*Internal Control* (Stavanger College, Rogaland Research, Stavanger, 1995).

———Quality Management Systems: Between regulation and self-regulation, Workshop on **Quality Management**, Kristiansand, April 1997.

—— 'Self-regulation of Occupational Health and Safety in Norway: A revitalisation of the Working Environment Act', Stavanger College, Rogaland Resarch, Stavanger, undated.

Lindsay, F. D., 'Successful Health and Safety Management: The contribution of management audit' (1992) 15 *Safety Science* 387.

Lopez, M. S., 'Application of the Audit Privilege to Occupational Safety and Health Audits: Lessons learned from environmental audits' (1996) 12(2) *Journal of Natural Resources and Environmental Law* 21.

Martin, A., Linehan, A., & Whitehouse, I., *The Regulation of Health and Safety in Five European Countries: Denmark, France, Germany, Spain, and Italy with a supplement on recent developments in the Netherlands*, HSE Contract Research Report No. 84/1996 (UK Health and Safety Executive, London, 1996).

Mayhew, C., Ferris, R., & Harnett, C., *An Evaluation of the Impact of Targeted Interventions On the OHS Behaviours of Small Business Building Industry Owners/Managers/Contractors* (National Health and Safety Commission, Australia, 1997).

Mayhew, C., Quinlan, M., & Bennett, L., *The Effects of Subcontracting/Outsourcing on Occupational Health and Safety*, Executive Summary, Industrial Relations Research Centre, University of New South Wales, Sydney, Australia, 1996.

McAvoy, P. W., *OSHA Safety Regulation*, Report of the Presidential Task Force (American Enterprise Institute, Washington DC, 1977).

McBarnet, D. J., *Conviction: Law, the State and the Construction of Justice* (Macmillan, London, 1981).

McClean, J., Parks, J., & Kidder, D. L., '"Till Death Do Us Part": Changing work relationships in the 1990s' (1994) 1 *Trends in Organisational Behaviour* 111.

McGarity, T. O., 'Regulatory analysis and regulatory reform' (1987) 65 *Texas Law Review* 7.

McGarity, T. O., & Shapiro, S. A., *Workers at Risk: The failed promise of occupational safety and health administration* (Praeger, Westport, 1993).

McQuiston, T. H., Zakocs, R. C., & Loomis, D., 'The Case for Stronger OSHA Enforcement—Evidence from Evaluation Research' (1998) 88 *American Journal of Public Health* 1022.

Mendeloff, J., 'The Role of OSHA Violations' (1984) 26 *Journal of Occupational Medicine* 353.

—— *Regulating Safety* (MIT Press, Cambridge, Mass., 1979). 'The Role of OSHA Violations' (1984) 26 *Journal of Occupational Medicine* 353.

——*A Preliminary Evaluation of the 'Top-200' Program in Maine*, Report to the Office of Statistics, Occupational Safety and Health Administration, US Department of Labor, Washington DC, March 1996.

Mentzger, M. B., & Schwenk, C. R., 'Decision Making Models, Devil's Advocacy, and the Control of Corporate Crime' (1990) 28 *American Business Law Journal* 323.

Merritt, A., *Guidebook to Australian Occupational Health and Safety Laws*, 2nd edn (CCH Australia Ltd., Sydney, 1986).

Merrylees, P., 'Identifying and Implementing the Key Elements in an Effective Compliance System', Paper presented at the *Environmental Management Systems, Programs and Due Diligence Conference*, Sydney, 27 and 28 March 1995.

Metzger, M., 'Corporate Criminal Liability for Defective Products: Policies, Problems, and Prospects' (1984) 73 *Geo. L J* 1.

—— 'Orgnisations and the Law' (1987) 25 *American Business Law Journal* 257.

Mialon, M., 'Safety at Work in French Firms and the Effect of the European Directives of 1989' (1990) 6 *The International Journal of Comparative Labour Law and Industrial Relations* 129.

Miester, D. J., 'Criminal Liability for Corporations That Kill' (1990) 64 *Tulane Law Review* 919.

Miller, J. L., & Anderson, A. B., 'Updating the Deterrence Doctrine' (1986) 77 *The Journal of Criminal Law and Criminology* 418.

Minister, M., 'Federal Facilities and the Deterrence Failure of Environmental Laws: The Case for Criminal Prosecution of Federal Employees' (1994) 18 *Harvard Environmental Law Review* 137.

Mintzberg, H., *The Structuring of Organisations* (Prentice-Hall, Englewood Cliffs, NJ, 1979).

Mitchell, D., 'Victim Impact Statements: A brief examination of their implementation in Victoria' (1996) *Current Issues in Criminal Justice* 163.

Moore, J., 'Corporate Culpability Under the Federal Sentencing Guidelines' (1992) 34 *Arizona Law Review* 743.

Morris, J. M., The Structure of Criminal Law and Deterrence' (1986) *Criminal Law Review* 524.

Moss, T., 'Cost-benefit analysis and the cotton dust standard' (1982) 35 *Rutgers Law Review*.

Mukatis, W. A., & Brinkman, P. G., 'Managerial Liability for Health, Safety and Environmental Crime: A Review and Suggested Approach to the Problem' (1987) 25 *American Business Law Journal* 323.

Nadarajah, R., 'Ontario's Policy on Access to Environmental Evaluations: The creation of audit privilege' (1998) 7(3) *Journal of Environmental Law and Practice* 311.

Nagel, I. H., & Swenson, W. M., 'The Federal Sentencing Guidelines for Corporations: Their Development, Theoretical Underpinnings, and Some Thoughts on their Future' (1993) 71 *Washington University Law Quarterly* 205.

National Board of Occupational Health and Safety (Sweden), 'The Costs of Ill-health' (1992)(2) *Newsletter* 1.

—— *Internal Control of the Working Environment in Small Enterprises: A Monitoring Project of the Occupational Health and Safety Administration, 1995–1997* (Stockholm, March, 1997).

National Commission on Reform of Criminal Laws, *Study Draft of a New Criminal Code* (Washington, 1970).

Needleman, C., OHSA at the Crossroads: Conflicting Frameworks for Regulating Occupational Health and Safety in the United States', Paper delivered to *Policies for Occupational Health and Safety Management Systems and Workplace Change Conference*, Amsterdam 21–24 September 1998.

New South Wales Law Reform Commission, *Sentencing*, Discussion Paper 33 (Sydney, 1996).

Nichols, T., & Armstrong, P., *Safety or Profit: Industrial accidents and the conventional wisdom* (Falling Water Press, Bristol, 1973), reproduced in T. Nichols (ed.), *The Sociology of Industrial Injury* (Mansell, London, 1997).

Nonet, P., & Selznick, P., *Law and Society in Transition: Towards responsive law* (Harper and Row, New York, 1978).

Norrie, A., *Crime, Reason and History: A Critical Introduction to Criminal Law* (Weidenfeld & Nicolson, London, 1993).

Novak, E. F., 'Sentencing the Corporation' (1995) 21 (Summer) *Litigation* 31.

Nunes, J. W., 'Comment: Organizational Sentencing Guidelines: The Conundrum of Compliance Programs and Self-reporting (1995) 27 *Arizona State Law Journal* 1039.

Occupational Health and Safety Small Business Forum, *Occupational Health and Safety Small Business Forum: A forum to assist small business occupational health and safety performance improve* (Australian Chamber of Commerce and Industry, Unley, SA, 1996).

Orts, W. E., 'Reflexive Environmental Law' (1995) 89(4) *Northwestern University Law Review* 1227.

Osborne, J., & Zairi, M., *Total Quality Management and the Management of Occupational Health and Safety*, Health and Safety Executive, Research Report 153 (London, UK, 1997).

Owens, L., & Lantis, K., 'Safety Management Systems and Incentives' (1996) 12(5) *Journal of Occupational Health & Safety—Australia and New Zealand* 597.

Packer, H. L. *The Limits of the Criminal Process* (Stanford University Press, Stanford, 1968).

Palomares-Soler, M., & Thimme, P. M., 'Environmental Standards: EMAS and ISO 14001 Compared' (1996) 5(8/9) *European Environmental Law Review* 24.

Parliament of New South Wales, Legislative Council, Standing Committee on Law and Justice, *Report of the Inquiry into Workplace Safety*, Interim Report, Report No. 8 (New South Wales Government, December 1997).

Pearce, F., 'Corporations and Accountability', in P. Stenning (ed.), *Accountability for Corporate Crime*, (University of Toronto Press, Toronto, 1995).

Pearce, F., & Tombs, S., 'Hazards, Law and Class: Contextualizing the Regulation of Corporate Crime' (1997) 6 *Social and Legal Studies* 79.

Pearse, W., & Refshauge, C., 'Workers Health and Safety in Australia: An Overview' (1987) 17 *International Journal of Health Services* 635.

Pearson, E., *Report of the Royal Commission on Civil Liability and Compensation* (HMSO, London, 1978).

Pendergrass, J., & Pendergrass, J. jun., 'Beyond Compliance: A call for EPA regulation of voluntary efforts to reduce pollution' (1991) 21 *Environmental Law Reporter*.

Perrow, C., *Normal Accidents: Living with high risk technologies* (Basic Books, New York, 1984). .

Perry, C. C., 'Government Regulation of Coal Mine Safety: Effects of Spending under Strong and Weak Law' (1982) 10 *American Politics Quarterly* 303.

Plastics Processing Industry Training Board, *Evaluation of Written Safety Policies in Polymer Process Companies* (1983).

Polinsky, A. M., & Shavell, S., 'Should Employees be Subject to Fines and Imprisonment Given the Existence of Corporate Liability?' (1993) 13 *International Review of Law and Economics* 239.

Polk, K., Haines, F., & Perrone, S., 'Homicide, Negligence and Work Death: The Need for Legal Change' in M. Quinlan (ed.), *Work and Health: The Origins, Management and Regulation of Occupational Illness* (Macmillan, Melbourne, 1993).

Porter, M., *The competitive advantage of nations* (Free Press, 1990).

Pragnall, B., *Occupational Health Committees in NSW: An analysis of the AWIRS data*, ACIRRT, Working Paper No. 31, Sydney, March 1994.

Premiers and Chief Ministers, *Communique*, Adelaide, 21–22 November, 1991.

President's Council on Environmental Quality (PCEQ) (Quality Environmental Sub-committee), *Total Quality Management: A framework for pollution prevention* (Washington DC, 1993).

Quinlan, M., *The Development of Occupational Health and Safety Control Systems in a Changing Environment*, Paper presented to Workshop on Integrated Control/ Systems Control, Foundation for the Improvement of Living and Working Conditions, Dublin, 1996.

——*Reshaping Regulation and OHS Management Systems: Lessons from Europe*, (forthcoming).

Quinlan, M., & Bohle, P., *Managing Occupational Health and Safety in Australia* (Macmillan, Melbourne, 1991).

Rappaport, A., & Flaherty, M., *Multilateral Corporation and the Environment* (Centre for Environmental Management, Tufts University, 1991).

Rees, J., *Reforming the Workplace: A study of self-regulation in occupational health and safety* (University of Pennsylvania Press, 1988).

Reiley, R. A., 'The New Paradigm: ISO 14000 and its regulatory reform' (1997) 22 *The Journal of Corporation Law* 535.

Reiner, I., & Chatten-Brown, J., 'When is it not an Accident but a Crime?' (1989) 17 *Northern Kentucky Law Review* 84.

—— 'Deterring Death in the Workplace: The Prosecutor's Perspective' (1989) 17 *Law Medicine and Health Care* 23.

Rest, K., & Ashford, N., *Occupational Health and Safety in British Columbia: An Administrative Inventory of the Prevention Activities of the Workers' Compensation Board* (Ashford Associates, Cambridge, Mass., 1997).

Richard, Justice K. Peter, *The Westray Story: A Predictable Path to Disaster*, Report of the Westray Public Inquiry, Province of Nova Scotia, 1997.

Ridley, A., & Dunford, L., 'Corporate Killing—Legislating for Unlawful Death' (1997) 26 *Industrial Law Journal* 99.

Robens Committee (Committee on Safety and Health at Work), *Report of the Committee on Health and Safety at Work 1970–1972* (HMSO, London, 1972).

Robinson, B., 'Victorian Legislation', Paper presented at the *Environmental Management Systems, Programs and Due Diligence Conference*, Sydney, 27 and 28 March 1995.

Saksvik, O., *The Norwegian Internal Control Regulation*, SINTEF IFIM, Institute for Social Research in Industry, Trondheim, Workshop on Integrated Control/Systems Control, Dublin, 29–30 August 1996.

Saksvik, O., & Nytro, K., 'Implementation of Internal Control of Health, Environment and Safety in Norwegian Enterprises', Paper presented to *Seventh European Congress on Work and Organisational Psychology*, Gyor, Hungary, April 1995.

——*Implementation of Internal Control of Health, Environment and Safety: An Evaluation and a Model for Implementation* (SINTEF IFIM, Institute for Social Research in Industry, Trondheim, Norway, 1995).

Sand, R. H., 'OSHA takes a step forward on preemption—and two steps back on generic regulation' (1992/3) 18 *Employee Relations Law Journal.*

Scholz, J. T., 'Cooperation, Deterrence and the Ecology of Regulatory Enforcement' (1984) 18 *Law and Society Review* 179.

——'Cooperative Regulatory Enforcement and the Politics of Administrative Effectiveness' (1991) 85(1) *American Political Science Review* 115.

Schwenk, C. R., 'Decision Making Models, Devil's Advocacy, and the Control of Corporate Crime' (1990) *28 American Business Law Journal* 323.

Scot, A., 'Europe Weighs Its Standards Options—ISO Versus EMAS' (1997) 159(13) *Chemical Week* 33.

Secretary of State for Employment (UK), *Building Business . . . Not Barriers,* a White Paper presented to Parliament (HMSO, London, 1986).

Shannon, H., et al., 'Overview of the Relationship Between Organisational and Workplace Factors and Injury Rates' (1997) 26(3) *Safety Science* 201.

Shapiro, S. S., & Rabinowitz, R. S., 'Punishment Versus Cooperation in Regulatory Enforcement: A case study of OSHA' (1997) 49(4) *Administrative Law Review* 713.

Shavell, S., 'Criminal Law and the Optimal Use of Nonmonetary Sanctions as a Deterrent' (1985) 85 *Columbia Law Review* 1232.

Shaw, A., 'What Works? The strategies which help to integrate OHS management within business development and the role of the outsider', Paper delivered to *Policies for Occupational Health and Safety Management Systems and Workplace Change Conference*, Amsterdam, 21–24 September 1998.

Shikawa, K., *What Is Total Quality Control? The Japanese Way* (Prentice-Hall, Englewood Cliffs, NJ, 1985).

Sigler, J. A., & Murphy., J. E. (eds.), *Corporate Lawbreaking & Interactive Compliance: Resolving the regulation-deregulation dichotomy* (Quorum Books, US, 1991).

Skaar, S., Lindoe, P., Claussen, T., Jersin, E., & Tinmannsvik, R., *Internkontroll— orkenvandring eller veien til det forjettede land? [Internal Control—wandering in the desert or the road to the promised land?]* (SINTEF IFIM, Trondheim, 1994).

Slapper, G., 'Crime without conviction' (1992) February 14 *New Law Journal* 192.

——'Corporate Manslaughter: An Examination of the Determinants of Prosecutorial Policy' (1993) 2 *Social and Legal Studies* 423.

——'A Corporate Killing' (1994) 144 *New Law Journal* 1735.

Smith, M., 'The Company Behind Bars' (1995) February *Health and Safety at Work* 8.

Smith, S., 'An Iron Fist in a Velvet Glove: Redefining the Role of Criminal Prosecution in Creating an Effective Environmental Enforcement System' (1995) 19 *Criminal Law Journal* 12.

——'Doing Time for Environmental Crimes: The United States Approach to Criminal Enforcement of Environmental Laws (1995) *Environmental and Planning Law Journal* 168.

South Australia, Criminal Law and Penal Methods Reform Committee, *The Substantive Criminal Law*, Fourth Report (South Australian Government Printer, Adelaide, 1977).

Spector, B., & Beer, M., 'Beyond TQM Programmes' (1994) 7 *Journal of Organisational Change* 63.

Spiegelhoff, T. L., 'Limits on Individual Accountability for Corporate Crimes' (1984) 67 *Marquette Law Review* 604.

Stace, R., 'TQM and the Role of Internal Audit' (1994) 64(6) *Australian Accountant* 26.

Standards Association of Australia, Joint Technical Committee SF/1, Occupational Health and Safety Management. *Occupational Health and Safety Management Systems: General guidelines on principles, systems and supporting techniques* (Standards Australia, Homebush, NSW, 1997).

Stapleton, J., *The Disease and Injury Debate* (Clarendon Press, Oxford, 1986).

Steinzor, R., 'Reinventing Environmental Regulation: The Dangerous Journey from Command to Self-Control' (1998) 22 *Harvard Environmental Law Review* 103.

Stessens, G., 'Corporate Criminal liability: A Comparative Perspective' (1994) 43 *International and Comparative Law Quarterly* 493.

Stewart, R. S., 'Regulation, Innovation, and Administrative Law: A conceptual framework' (1981) 69(5) *California Law Review* 1256.

Stone, C., *Where the Law Ends: The Social Control of Corporate Behaviour* (Harper and Row, New York, 1975).

—— 'A Slap on the Wrist for the Kepone Mob' (1977) 22 *Business and Society Review* 4.

—— 'Corporate Regulation: The Place of Social Responsibility'. in B. Fisse & P. French (eds.), *Corrigible Corporations and Unruly Law* (Trinity University Press, San Antonio, 1987).

Stuart, D., 'Punishing Corporate Criminals with Constraint' (1995) *Criminal Law Forum* 219.

Sutherland, E., *e-Collar Crime: The Uncut Version* (Yale University Press, New Haven, 1983).

Swedish Institute, *Fact Sheets on Sweden: Occupational Safety and Health* (Swedish Institute, Stockholm, August 1996).

Swedish National Board of Occupational Health and Safety, *Newsletter*, 4/1989, 4/1992, 2/1993, 6/1993, 2/1994, 2/1995, 3/1995, 1/1997.

—— 'The Costs of Ill-health' (1992) 2 *Newsletter* 1.

Swedish Occupational Safety and Health Administration (The National Board of Occupational Safety and Health and the Labour Inspectorate), *A Report on activities in 1991/1992* (1992).

Swedish Working Environment Association, *17 Questions & Answers Based on Questions Frequently Asked by Foreign Visitors* (Swedish Working Environment Association, Stockholm, 1996).

Temby, I., 'The Role of the Prosecutor in the Sentencing Process' (1986) 10 *Criminal Law Journal* 199.

Teubner, G., 'Substantive and Reflexive Elements in Modern Law' (1983) 17 *Law & Society Review* 239.

Teubner, G., Farmer, L., & Murphy, D. (eds.) *Environmental Law and Ecological Responsibility: The concept and practice of ecological self-regulation* (Wiley, UK, 1994).

Tibor, T., & Feldman, I., *Implementing ISO 14000: A practical, comprehensive guide to the ISO 14000 environmental management standards* (McGraw-Hill, New York, 1997).

Tombs, S., 'Law, Resistance and Reform: "Regulating" Safety Crimes in the UK' (1995) 4 *Social and Legal Studies* 343.

Tonry, M., *Sentencing Matters* (Oxford University Press, New York, 1996).

Ts'ai, Lim Wen, 'Corporations and the Devil's Dictionary: The problem of individual responsibility for corporate crimes' (1990) 12 *Sydney Law Review* 311.

Tucker, E., 'Worker participation in health and safety regulation: lessons from Sweden' (1992) 37 *Studies in Political Economy* 95.

—— *Worker Health and Safety Struggles: Democratic possibilities and constraints*, Paper presented at the International Symposium on the Social and Economic Aspects of Democratisation of Contemporary Society, Moscow, 12–17 October 1992.

—— 'The Westray Mine Disaster and its Aftermath: The Politics of Causation' (1995) 10 *Canadian Journal of Law and Society* 91.

——'Defeat Goes On: An Assessment of Third Wave Health and Safety Legislation' in P. Stenning (ed.), *Accountability for Corporate Crime* (University of Toronto Press, Toronto, 1995).

Tuckman, A., 'The Yellow Brick Road: Total Quality Management and the Restructuring of Organizational Culture' (1994)15 (3) *Organizational Studies* 727.

United States, Congress, House of Representatives, Committee on Small Business, *The SAFE Act: How third party consultations have worked where OSHA has failed: Hearing before the Committee on Small Business* (US GPO, Washington DC, 1998).

United States Department of Labor, Occupational Safety and Health Administration, *Field Inspection Reference Manual*, 26 September 1994.

United States Environment Protection Agency (USEPA), *Environmental Auditing Policy Statement*, 51 Fed. Reg. 25004, 9 July 1986.

United States General Accounting Office, *Occupational Safety and Health Options* (Washington DC, 1990).

——*Regulatory Reform: Implementation of the small business advocacy review panel requirements*, Report to Congressional Requestors, GGD-98-36, The Office, Washington DC, 1998.

United States, Sentencing Commission, *Sentencing Guidelines for Organizational Defendants*, Preliminary Draft (United States Government Printing Office, Washington DC, 1991).

Van Someren, T. C. R. van der Kolk, J. ten Have, K., & Calkoen, P. T. (KPMG Environment/IVA), *Company Environmental Management Systems, Interim Evaluation 1992*, Summary on behalf of Ministerie van Volkshuisvesting Ruimtelijk Ordening en Milieubeuheer, The Hague/Tilburg, January 1993.

Van Waarden, F., den Hertog, J., Vinke, H., & Wilthagen, T., *Prospects for Safe and Sound Jobs: The impact of future trends on costs and benefits of occupational safety and health* (Dutch Ministry of Social Affairs and Employment, Gravenhage, 1997).

Verma, D. K., 'Occupational Health and Safety Trends in Canada, in Particular in Ontario' (1996) 40(4) *Annals of Occupational Hygiene* 477.

Vogel, D., *National Styles of Regulation* (Cornell University Press, Ithaca, NY, 1986).

Vogel, L., *Prevention at the Workplace* (Trade Union Technical Bureau, Brussels, 1998).

Walters, D. (ed.), *The Identification and Assessment of Occupational Health and Safety Strategies in Europe*, Volume 1: The National Situations (European Foundation for the Improvement of Living Conditions, Dublin, 1996).

——'Preventive Services in Occupational Health and Safety in Europe: Developments and Trends in the 1990s' (1997) 27 *International Journal of Health Services* 247.

Walton, M., *The Deming Management Method* (Dodd Mead & Co., New York, 1986).

Watchman, G. R., *Statement to Subcommittee on Public Health and Safety of the Senate Committee, Labor and Human Resources*, 10 July 1997.

Watson, J. L., 'The "New" OSHA: Reinventing Worker Safety and Health' (1998) 12(3) *Natural Resources and Environment* 183.

Webb, J., 'Quality Management and the Management of Quality', in A. Wilkinson & H. Willmott (eds.), *Making Quality Critical: New perspectives on organizational change* (Routledge, London, 1995).

Wells, C., 'Manslaughter and Corporate Crime' (1989) 139 *New Law Journal* 931.

—— *Corporations and Criminal Responsibility* (Clarendon Press, Oxford, 1993).

—— 'Corporate Manslaughter: A Cultural and Legal Form' (1995) 6 *Criminal Law Forum* 45.

—— 'The Corporate Manslaughter Proposals: Pragmatism, Paradox and Peninsularity' [1996] *Criminal Law Review* 545.

Whiting, J. F., & Horrigan, H. R., 'Validating OHS Performance', Paper presented at the *Proactive OH&S Management Conference*, Sydney, 9 and 10 March 1994.

Wilkinson, A., & Willmott, H. (eds.), *Making Quality Critical: New perspectives on organizational change* (Routledge, London, 1995).

Williams, B. L., & Kavanaugh, K., 'Compliance programs and Federal Organizational Sentencing guidelines' (1993) June *Res Gestae* 558.

Williams, K, 'Making Rules for Occupational Health in Britain, the USA and the EC' [1997] *Anglo-American Review* 82.

Wilthagen, T., 'Reflexive Rationality in the Regulation of Occupational Health and Safety', in R. Rogowski and T. Wilthagen, *Reflexive Labour Law* (Kluwer, Deventer, 1994).

—— 'External Regulation of Internal Control: Outside looking in, but how and by whom?', Hugo Sinzheimer Institute, University of Amsterdam, Workshop on **Integrated Control/Systems Control**, Dublin, 29–30 August 1996.

Woodrow, K., 'The Proposed Federal Environmental Sentencing Guidelines: A Model for Corporate Environmental Compliance Programs' (1994) 25 *Environmental Reporter* 325.

Woolf, T., 'The Criminal Code Act 1995 (Cth)—Towards a Realist Vision of Corporate Criminal Liability' (1997) 21 *Criminal Law Journal* 257.

Woolfson, C., Foster, J., & Beck, H., '"*Paying for the Piper": Capital & Labour in Britains' Offshore Oil Industry*' (Mansell Publishing Ltd., UK 1997).

WorkCover Authority (NSW) (Working Party of Representatives of the NSW Self Insurers and the WorkCover Authority), *A Quality Approach to Occupational Health and Safety and Rehabilitation for Self Insurer: Quality OH&S system model and system audit guidelines*, Internal WorkCover Report (undated).

Worksafe Australia, *OHS Good for Business* (AGPS, 1995).

Workytch, R., & Van Sandt, C., 'Occupational Health and Safety Management in the United States and Japan: The Du Pont Model and the Toyota Model', Paper delivered to *Policies for Occupational Health and Safety Management Systems and Workplace Change Conference*, Amsterdam, 21–24 September 1998.

Wright, C., 'A Fallible Safety System: Institutionalised Irrationality in the Offshore Oil and Gas Industry' (1994) February *The Sociological Review* 79.

Wright, F., *Law of Health and Safety at Work* (Sweet & Maxwell, London, 1997).

Yellen, D., & Mayer, C. J., 'Coordinating Sanctions for Corporate Misconduct: Civil or Criminal Punishment?' (1992) 29 *American Criminal Law Review* 961.

Yohay, S. C., 'OSHA Compels Disclosure of Safety and Health Audits: Smart enforcement or misguided policy?' (1993) 18 *Employee Relations Law Journal* 663.

Zagrocki, E., 'Federal Sentencing Guidelines: The Key to Corporate Integrity or Death Blow to Any Corporation Guilty of Misconduct' (1992) 30 *Duquesne Law Review* 331.

Zimolong, G., 'What Successful Companies Do', Paper presented at *Safety In Action Conference*, Melbourne, Australia, February 1998.

Zimring, F., & Hawkins, G., *Deterrence: The Legal Threat of Crime Control* (The University of Chicago Press, Chicago, 1973).

Index

Printed in the United Kingdom
by Lightning Source UK Ltd.
102625UKS00002B/23